Algal Systems for Resource Recovery from Waste and Wastewater

Algal Systems for Resource Recovery from Waste and Wastewater

Edited by
Piet N. L. Lens and Amitap Khandelwal

IWA PUBLISHING

Published by **IWA Publishing**
Unit 104–105, Export Building
1 Clove Crescent
London E14 2BA, UK
Telephone: +44 (0)20 7654 5500
Fax: +44 (0)20 7654 5555
Email: publications@iwap.co.uk
Web: www.iwapublishing.com

First published 2023

British Library Cataloguing in Publication Data
A CIP catalogue record for this book is available from the British Library

ISBN: 9781789063530 (paperback)
ISBN: 9781789063547 (eBook)
ISBN: 9781789063554 (ePub)

Doi: 10.2166/9781789063547

This eBook was made Open Access in December 2023.

Contents

doi: 10.2166/9781789063547_xiii

Preface

In the past few decades, our planet has witnessed an unprecedented surge in population, carbon emissions, and the demand for essential resources, particularly energy and water. This exponential growth has come at a cost: a staggering increase in waste and wastewater, presenting formidable challenges to our environment and sustainability. Faced with this urgent dilemma, the imperative to develop innovative technologies for resource recovery has never been more critical. Amidst this challenge, algae, tiny yet extraordinary photosynthetic organisms, have emerged as potent microbiota in the quest for environmental solutions. In this context, this book *Algal Systems for Resource Recovery from Waste and Wastewater* testifies to the pivotal role algae can play in addressing some of the world's most pressing issues.

In recent years, algae-based wastewater treatment has made significant strides. Rigorous research validated the integration of specific algae strains into existing treatment plants, elevating their efficiency. Cutting-edge technologies, such as advanced photobioreactors and real-time monitoring systems, empowered precise control and seamless automation. Symbiotic systems and the dual-purpose utilization of harvested algae for biofuel production bolstered economic viability. Scalable implementations and widespread commercialization swiftly followed successful pilot programs. Ongoing cutting-edge research continues to sharpen the focus on efficiency enhancements, new strain exploration, and integration of other modern technologies such as anaerobic digestion and bioelectrochemical systems, promising an unwavering and sustainable technical solution to the pressing issue of wastewater pollution.

Within this book, we embark on a profound exploration of various algae-based systems, unveiling their transformative potential and transition from laboratory trials to real world solutions. Wastewaters, rich in resources like phosphorus, demand efficient nutrient removal for the development of a circular bioeconomy. Algae-based treatment systems achieve both wastewater clean-up and valuable biomass production. Algae have a unique ability to absorb pollutants or transform them into sustainable bioproducts. Their capacity to convert wastewater into valuable biomass and value-added commodities opens doors to a multitude of applications, ranging from the production of sustainable biofuels to the creation of nutrient-rich animal feed and fertilizers. This book chronicles the remarkable journeys of scientists and researchers from around the globe to unlock the potential of these tiny organisms. It presents the current status, major challenges and recent scientific innovations in algae-based technologies for waste remediation and nutrient recovery.

Authored by experts and researchers at the forefront of algal biology, bioprocess engineering, and environmental science, this comprehensive volume aims to provide an authoritative resource for academics, researchers, industry professionals, and policymakers. Its pages will empower readers with knowledge about the latest advancements, challenges, and breakthroughs in the use of algae for wastewater treatment and energy recovery. Each contributed chapter is presented on a stand-alone basis, so that the reader will find it helpful to consider only the theme of each chapter. There are nevertheless many connections between what may at first seem to be quite different topics. As in all the books of the *Integrated Environmental Technology* series, one of our purposes was to draw out and emphasize these interdisciplinary links. For this reason, a comprehensive index is included to facilitate cross-referencing. We hope that the work described in this book will inspire those working in the field and will encourage those who are beginning to investigate it.

We wish to thank all contributors to this book for their valuable contributions by sharing their expertise in the various chapters. We also thank all past and present co-workers as well as all collaborators who joined in unravelling different areas of the application of algae in environmental technology as described in this book, especially those at National University Ireland Galway and UNESCO-IHE. We would also like to thank all the reviewers who put a lot of effort into improving the quality of this book. In addition, the national and international granting agencies who supported our work on various aspects of algal based pollutant removal and resource recovery over the years are gratefully acknowledged, in particular the Science Foundation Ireland (SFI), who financially supported the open access publication of this book through the SFI Research Professorship Programme *Innovative Energy Technologies for Biofuels, Bioenergy and a Sustainable Irish Bioeconomy* (IETSBIO[3]; grant number 15/RP/2763) and the Department of Foreign Affairs (DFA) under the SDG Challenge project *Floating Treatment Wetland* (grant number SFI/21/FIP/SDG/9933). We are also grateful to the editorial team of IWA Publishing, in particular Mark Hammond, Andrew Peart and Katharine Allenby for their help and editorial support in realizing this book.

Amitap Khandelwal
Piet N. L. Lens
University of Galway, Ireland

List of Contributors

Acien, F. G.
Department of Chemical Engineering - CIESOL, University of Almeria, Almeria, Spain.

Ali, Sameh Samir
Biofuels Institute, School of the Environment and Safety Engineering, Jiangsu University, Zhenjiang, China.
and
Botany Department, Faculty of Science, Tanta University, Egypt.

Assemany, Paula
Federal University of Lavras (Universidade Federal de Lavras/UFLA), Department of Environmental Engineering, Lavras, MG, Brazil.

Braga, Matheus Q.
Federal University of Viçosa (Universidade Federal de Viçosa/UFV), Department of Civil Engineering, Campus Universitário, Viçosa, MG, Brazil.

Calijuri, Maria Lúcia
Federal University of Viçosa (Universidade Federal de Viçosa/UFV), Department of Civil Engineering, Campus Universitário, Viçosa, MG, Brazil.

Castro, Jackeline S.
Federal University of Viçosa (Universidade Federal de Viçosa/UFV), Department of Civil Engineering, Campus Universitário, Viçosa, MG, Brazil.

Chandra, Rashmi
Department of Civil and Environmental Engineering, University of Waterloo, Waterloo, Canada.

Chhabra, Meenu
Department of Biosciences and Bioengineering, Indian Institute of Technology, Jodhpur, India.

Fernández-Sevilla, J. M.
Department of Chemical Engineering - CIESOL, University of Almeria, Almeria, Spain.

Ferreira, Jéssica
Federal University of Viçosa (Universidade Federal de Viçosa/UFV), Department of Civil Engineering, Campus Universitário, Viçosa, MG, Brazil.

Gama, Rafael C. N.
Federal University of Viçosa (Universidade Federal de Viçosa/UFV), Department of Civil Engineering, Campus Universitário, Viçosa, MG, Brazil.

Gómez, C.
Department of Chemical Engineering - CIESOL, University of Almeria, Almeria, Spain.

Guzman, J. L.
Department of Informatics - CIESOL, University of Almeria, Almeria, Spain.

Harish
Department of Botany, Mohanlal Sukhadia University, Udaipur, Rajasthan, India-313001

He, Q. Peter
Department of Chemical Engineering, Auburn University, Alabama, United States.

Hooshmand, Masoomeh
Department of Civil and Environmental Engineering, University of Waterloo, Waterloo, Canada.

Khandelwal, Amitap
School of Biological and Chemical Sciences, University of Galway, Ireland.

Kornaros, Michael
Laboratory of Biochemical Engineering & Environmental Technology, Department of Chemical Engineering, University of Patras, Greece.

Lawrence, James
School of Biological and Chemical Sciences, University of Galway, Ireland.

Lens, Piet N. L.
School of Biological and Chemical Sciences, University of Galway, Ireland.

Lorentz, Juliana F.
Federal University of Viçosa (Universidade Federal de Viçosa/UFV), Department of Civil Engineering, Campus Universitário, Viçosa, MG, Brazil.

Majhi, Pritikrishna
DBT-BIRAC E-YUVA Centre, GIETU, Gunupur, Odisha, India-765022

Magalhães, Iara B.
Federal University of Viçosa (Universidade Federal de Viçosa/UFV), Department of Civil Engineering, Campus Universitário, Viçosa, MG, Brazil.

Marangon, Bianca B.
Federal University of Viçosa (Universidade Federal de Viçosa/UFV), Department of Civil Engineering, Campus Universitário, Viçosa, MG, Brazil.

Minkina, Tatiana
Academy of Biology and Biotechnology, Southern Federal University, Rostov-on-Don, Russia.

Mishra, Akanksha
Department of Biosciences and Bioengineering, Indian Institute of Technology, Jodhpur, India.

Morillas, A.
Department of Chemical Engineering - CIESOL, University of Almeria, Almeria, Spain.

Murphy. Jerry D.
Civil, Structural and Environmental Engineering, School of Engineering and Architecture, University College Cork, Cork, Ireland.

Nordio, R.
Department of Chemical Engineering - CIESOL, University of Almeria, Almeria, Spain.

Oliva, Armando
Department of Civil, Architectural and Environmental Engineering, University of Naples Federico II, Via Claudio 21, 80125, Naples, Italy.

Pereira, Alexia S. A. P.
Federal University of Viçosa (Universidade Federal de Viçosa/UFV), Department of Civil Engineering, Campus Universitário, Viçosa, MG, Brazil.

Pessi, Bruno Assis
Department of Civil and Environmental Engineering, São Paulo State University (UNESP), Bauru, São Paulo, Brazil.

Pispas, Konstantinos
Laboratory of Biochemical Engineering & Environmental Technology, Department of Chemical Engineering, University of Patras, Greece.

Pithania, Ruchi
Division of biological sciences, University of California, San Diego, 92093-USA.

Rada-Ariza, Angelica M.
IHE Institute for Water Education, Delft, The Netherlands.

Rodriguez, E.
Department of Informatics - CIESOL, University of Almeria, Almeria, Spain.

Sakarika, Myrsini
Center for Microbial Ecology and Technology, Ghent University, Belgium.

Sanchez, A.
Department of Chemical Engineering - CIESOL, University of Almeria, Almeria, Spain.

Saxena, Pallavi
Academy of Biology and Biotechnology, Southern Federal University, Rostov-on-Don, Russia.
and
Division of Research and Development, Lovely Professional University, Punjab, India.

Silva, Thiago A.
Federal University of Viçosa (Universidade Federal de Viçosa/UFV), Department of Civil Engineering, Campus Universitário, Viçosa, MG, Brazil.

Sventzouri, Eirini
Laboratory of Biochemical Engineering & Environmental Technology, Department of Chemical Engineering, University of Patras, Greece.

Tarigholizadeh, Sarieh
Academy of Biology and Biotechnology, Southern Federal University, Rostov-on-Don, Russia.

van der Steen, Peter
IHE Institute for Water Education, Delft, The Netherlands.

Vazquez, Carlos Lopez
IHE Institute for Water Education, Delft, The Netherlands.

Velasquez Orta, Sharon
Department of Chemical Engineering, Newcastle University, England, United Kingdom.

Wang, Jin
Department of Chemical Engineering, Auburn University, Alabama, United States.

Zouhayr, A.
Aqualia, Madrid, Spain.

Part 1

Process Fundamentals

doi: 10.2166/9781789063547_0003

Chapter 1

Algal systems for resource recovery from waste and wastewater

*A. Khandelwal and P. N. L. Lens**

University of Galway, University Road, H91 TK33, Galway, Ireland
**Corresponding author: piet.lens@universityofgalway.ie*

ABSTRACT

This chapter provides an overview of the book. The introduction highlights the need for algal-based technologies in waste management and resource recovery in order to boost the circular bioeconomy globally. The book is divided into four parts, consisting of twelve chapters in total, which provide a detailed description of topics ranging from process fundamentals to up-to-date information on various modern algal-based technologies for waste remediation, nutrient recovery, and simultaneous energy generation. The book is suitable for students, research professionals and policymakers who are working in the domain of environmental engineering/sciences, wastewater treatment and renewable energy.

As a consequence of the swift proliferation of the global economy and population, the availability of water resources for direct human consumption has become insufficient. Forecasts indicate a projected 40% global water deficit by 2030, which gives rise to critical challenges for both society and economic advancement (Kandasamy *et al.*, 2023). This scarcity is primarily attributed to escalating water demands, the contamination of existing water supplies, and a lack of efficient technologies for water recycling. As a result, the imperative of water remediation is bound to assume a central role on the international stage, demanding urgent attention and action.

Historically, wastewater treatment arising from diverse industries has predominantly relied on the implementation of chemical processes such as flocculation, disinfection, oxidation, and neutralization and physical techniques, including grit chamber, floatation, and screening (Chojnacka *et al.*, 2020; Kurniawan *et al.*, 2022). Despite their widespread use, these chemical and physical treatment methodologies remain financially burdensome and generate substantial volumes of slurry or sludge, thereby requiring supplementary treatment steps. Moreover, the wastewater treatment processes are energetically expensive and demand trained staff for the operation of treatment facilities, which are associated with considerable capital costs for infrastructure development (Kandasamy *et al.*, 2023).

Consequently, scientists and researchers are currently exploring alternative approaches for wastewater treatment and nutrient recovery, centering on the utilization of microalgae. These innovative methods hold the promise of providing an environmentally friendly and sustainable means of treating wastewater, potentially enabling the recovery of nutrients as high as 95% (Moradi & Saidi, 2022). Microalgae growing in wastewater can facilitate the production of biomass, which contains valuable components such as carbohydrates, proteins, lipids, and other valuable biomolecules that can be utilized in the production of third-generation biofuels (Shearian Sattari *et al.*, 2022). Several modes such as open ponds, photobioreactors, and advanced culture systems, are being considered to foster the cultivation of microalgae, offering diverse and promising pathways for their effective implementation (Kandasamy *et al.*, 2023; Khandelwal *et al.*, 2023).

The successful cultivation of microalgae in diverse industrial wastewaters, along with the efficacy of the effluent treatment processes, is contingent upon achieving an optimal nutrient load and composition within the wastewater. In instances where the nitrogen:phosphorus (N:P) ratios in the water are reduced, certain strains of *Cladophora* have demonstrated enhanced efficiency in removing nutrients from the environment (Sandani *et al.*, 2020). On the other hand, algal families characterized by higher N:P ratios, such as *Pseudanabaena*, exhibit more effective nutrient removal capabilities (Kandasamy *et al.*, 2023). Nonetheless, a comprehensive study involving filamentous benthic algae has indicated that for the specific context of municipal wastewater nutrient removal, the optimal N:P ratios should fall within the range of 5:1 to 15:1, 7:1 to 10:1, and 7:1 to 20:1 for *Cladophora*, *Klebsormidium*, and *Pseudanabaena*, respectively (Valchev & Ribarova, 2022). Generally, the various strains of algae do not respond similarly to different N:P ratios, leading to varying impact on their nutrient removal capabilities.

The shift toward a circular bioeconomy, which emphasizes resource diversification, has provided the impetus for transforming conventional wastewater treatment processes capable of handling various waste streams. The transition has gained momentum, and the increasing enthusiasm can be credited to the dynamic and evolving nature of microalgal-based wastewater treatment solutions. Overcoming critical barriers related to nutrient assimilation and achieving increased microalgae growth rates have rendered microalgae-based wastewater treatment a compelling and powerful alternative to traditional methods (Khan *et al.*, 2022). In this context, this book explores the potential applications of algal biomass in wastewater remediation and bioenergy production. The book is divided into the following four parts.

1.1 PROCESS FUNDAMENTALS

This chapter discusses the cultivation of microalgae in wastewater, their metabolic modelling to analyze the growth rate (Chapter 2) and their interaction with bacteria (Chapter 3). To advance sustainable wastewater treatment technology, a comprehensive investigation is proposed, focusing on the symbiotic bacterio-algal relationship (Chapter 3) and the role of quorum sensing signal molecules in shaping the integrated wastewater treatment solution involving both algae and bacterial processes. This segment aims to lay the groundwork for refining algae–bacteria based wastewater treatment methods through various approaches.

These findings are expected to offer valuable insights for promoting sustainable economic and environmental development. Additionally, the utilization of synergistic bacterial–algal wastewater treatment technologies has the potential to contribute toward lowering the carbon emissions (Hena *et al.*, 2021). By combining these approaches, the research endeavors to pave the way for more effective and environment-friendly wastewater treatment practices, with the ultimate goal of fostering sustainable development and mitigating environmental impacts.

Furthermore, the basics of macroalgae-based biorefinery are also discussed in detail (Chapter 4), which makes the book suitable for every phycologist. This part majorly focuses on the following three aspects: (1) metabolic modelling of algal growth using waste as substrate, (2) synergistic approach of algae–bacteria for efficient wastewater treatment and selection of key microalgae and bacterial species in wastewater treatment systems, and (3) use of of macroalgae to produce fertilizers, feed (additives), and other value-added products.

1.2 ALGAL-BASED WASTEWATER TREATMENT

Although the use of microalgae for wastewater treatment was proposed in the last century, the technology was not sufficiently efficient and robust to be applied at a commercial scale. Only recent advances in the knowledge of biological systems, the engineering of the reactors and the harvesting and processing of the produced biomass allow the development of the first industrial demonstrations (Acién *et al.*, 2016). Facilities of several hectares are already in operation demonstrating the feasibility of this technology (Nguyen *et al.*, 2022). However, challenges remain for the further improvement and enlargement of these systems. They are related to (a) the improvement of knowledge and management of the biological system, (b) the development of adequate strategies for the allocation and implementation of large-scale facilities, (c) the definition of optimal operational conditions, including the development of non-assisted systems capable to operate under variable environmental conditions, and (d) the development of adequate routes for biomass valorization (Acién *et al.*, 2016).

Large efforts which are being devoted to solving these challenges and thus to making this technology reliable for industrial applications, are detailed in Chapter 5. Furthermore, possibilities and challenges in coculturing methanotrophs with microalgae for wastewater treatment are discussed in Chapter 6. Part 2 summarizes the status, major challenges, and potential contribution of microalgae-related wastewater treatment processes.

1.3 VALORIZATION OF ALGAL BIOMASS BY INTEGRATING WITH DIFFERENT TECHNOLOGIES

Microalgal systems play a crucial role in shifting the perspective of wastewater from being seen as disposable waste to being recognized as a valuable resource capable of yielding new value-added products. This shift toward a more sustainable approach brings together significant environmental and economic potential, endorsing the principles of a circular economy (Amaro *et al.*, 2023). Through the production of bioenergy and bioproducts, these systems contribute to the energy–environment nexus, paving the way for a sustainable closed-loop economy (Bele *et al.*, 2023).

Given the pressing challenges of global water scarcity and the escalating costs associated with wastewater treatment, numerous research works and government projects have emerged, exploring the application of microalgal systems for wastewater treatment while concurrently extracting valuable biomass resources. Specifically, managing manure poses significant difficulties and expenses for livestock and poultry operations, particularly in cold climate regions (Bele *et al.*, 2023). Addressing these challenges necessitates adopting sustainable approaches to nutrient management, reuse, and recycling, which can not only generate additional income for farmers but also enhance agricultural environmental sustainability.

The integration of green innovations, such as algae cultivation, bioelectrochemical systems (BES), and anaerobic digestion emerges as a key strategy to recover nutrients, complete utilization of manure, and make the overall process more sustainable. Part 3 aims to comprehensively explore the potential of integrating microalgae into the growing biogas and wastewater industry along with the potential of BES for simultaneous waste remediation, algae cultivation, and power generation. It seeks to identify opportunities and challenges inherent in this approach and reviews the prospective bioproducts, such as bioelectricity arising from BES (Chapter 7), biogas (Chapter 8), and bioethanol (Chapter 9). Such integration represents a transformative approach that harnesses the vast untapped potential of waste, aligning with the principles of the circular economy and advancing the sustainable development goals.

1.4 ALGAL BIOTECHNOLOGY

The microalgae biorefinery presents a promising and sustainable solution for producing biofuels and a diverse array of bulk chemical products. Extensive efforts have been dedicated to utilizing microalgae biomass in biorefineries to advance sustainable development, primarily due to their abundant bioactive constituents (Okeke *et al.*, 2022). Part 4 provides a comprehensive review of potential strategies aimed at enhancing microalgae biorefinery obtaining high-value-added renewable products (Chapters 10 and 11) and optimizing the transformation of microalgae-based technologies into economically viable products (Chapter 12). The focus is on ensuring the long-term viability of the processes, taking into account both economic feasibility and environmental considerations.

Moreover, ongoing research explores the integration of microalgae biorefineries with other eco-friendly alternatives, such as microalgae-based bioplastics, which opens up new possibilities for synergistic applications (Okeke *et al.*, 2022). The microalgae biorefining process holds the promise of becoming a key element of green technology, facilitating the biosynthesis of a broad range of valuable biofuels and biochemical products, further reinforcing the outlook for sustainable and environmentally friendly solutions to recover resources from waste and wastewater.

REFERENCES

Acién F. G., Gómez-Serrano C., Morales-Amaral M. M., Fernández-Sevilla J. M. and Molina-Grima E. (2016). Wastewater treatment using microalgae: how realistic a contribution might it be to significant urban wastewater treatment? *Applied Microbiology and Biotechnology*, **100**(21), 9013–9022, https://doi.org/10.1007/s00253-016-7835-7

Amaro H. M., Salgado E. M., Nunes O. C., Pires J. C. M. and Esteves A. F. (2023). Microalgae systems – environmental agents for wastewater treatment and further potential biomass valorisation. *Journal of Environmental Management*, **337**, 117678, https://doi.org/10.1016/j.jenvman.2023.117678

Bele V., Rajagopal R. and Goyette B. (2023). Closed loop bioeconomy opportunities through the integration of microalgae cultivation with anaerobic digestion: a critical review. *Bioresource Technology Reports*, **21**, 101336, https://doi.org/10.1016/j.biteb.2023.101336

Chojnacka K., Witek-Krowiak A., Moustakas K., Skrzypczak D., Mikula K. and Loizidou M. (2020). A transition from conventional irrigation to fertigation with reclaimed wastewater: prospects and challenges. *Renewable and Sustainable Energy Reviews*, **130**, 109959, https://doi.org/10.1016/j.rser.2020.109959

Hena S., Gutierrez L. and Croué J. P. (2021). Removal of pharmaceutical and personal care products (PPCPs) from wastewater using microalgae: a review. *Journal of Hazardous Materials*, **403**(June 2020), 124041, https://doi.org/10.1016/j.jhazmat.2020.124041

Kandasamy S., Narayanan M., Raja R., Devarayan K. and Kavitha R. (2023). The current state of algae in wastewater treatment and energy conversion: a critical review. *Current Opinion in Environmental Science and Health*, **33**, 100469, https://doi.org/10.1016/j.coesh.2023.100469

Khan A. A., Gul J., Naqvi S. R., Ali I., Farooq W., Liaqat R., AlMohamadi H., Štěpanec L. and Juchelková D. (2022). Recent progress in microalgae-derived biochar for the treatment of textile industry wastewater. *Chemosphere*, **306**, 135565, https://doi.org/10.1016/j.chemosphere.2022.135565

Khandelwal A., Chhabra M. and Lens P. N. L. (2023). Integration of third generation biofuels with bio-electrochemical systems: current status and future perspective. *Frontiers in Plant Science*, **14**(February), 1–14, https://doi.org/10.3389/fpls.2023.1081108

Kurniawan S. B., Ahmad A., Imron M. F., Abdullah S. R. S., Othman A. R. and Hasan H. A. (2022). Potential of microalgae cultivation using nutrient-rich wastewater and harvesting performance by biocoagulants/bioflocculants: mechanism, multi-conversion of biomass into valuable products, and future challenges. *Journal of Cleaner Production*, **365**(June), 132806, https://doi.org/10.1016/j.jclepro.2022.132806

Moradi P. and Saidi M. (2022). Biodiesel production from *Chlorella vulgaris* microalgal-derived oil via electrochemical and thermal processes. *Fuel Processing Technology*, **228**, 107158, https://doi.org/10.1016/j.fuproc.2021.107158

Nguyen L. N., Aditya L., Vu H. P., Johir A. H., Bennar L., Ralph P., Hoang N. B., Zdarta J. and Nghiem L. D. (2022). Nutrient removal by algae-based wastewater treatment. *Current Pollution Reports*, **8**(4), 369–383, https://doi.org/10.1007/s40726-022-00230-x

Okeke E. S., Ejeromedoghene O., Okoye C. O., Ezeorba T. P. C., Nyaruaba R., Ikechukwu C. K., Oladipo A. and Orege J. I. (2022). Microalgae biorefinery: an integrated route for the sustainable production of high-value-added products. *Energy Conversion and Management*, **16**, 100323, https://doi.org/10.1016/j.ecmx.2022.100323

Sandani W. P., Nishshanka G. K. S. H., Premaratne R. G. M. M., Nanayakkara Wijayasekera S. C., Ariyadasa T. U. and Premachandra J. K. (2020). Comparative assessment of pretreatment strategies for production of microalgae-based biodiesel from locally isolated *Chlorella homosphaera*. *Journal of Bioscience and Bioengineering*, **130**(3), 295–305, https://doi.org/10.1016/j.jbiosc.2020.03.004

Shearian Sattari M., Ghobadian B. and Gorjian S. (2022). A critical review on life-cycle assessment and exergy analysis of Enomoto bio-gasoline production. *Journal of Cleaner Production*, **379**(P1), 134387, https://doi.org/10.1016/j.jclepro.2022.134387

Valchev D. and Ribarova I. (2022). A review on the reliability and the readiness level of microalgae-based nutrient recovery technologies for secondary treated effluent in municipal wastewater treatment plants. *Processes*, **10**(2), 399, https://doi.org/10.3390/pr10020399

doi: 10.2166/9781789063547_0009

Chapter 2

Metabolic modelling of microalgae for wastewater treatment

B. Assis Pessi

São Paulo State University (UNESP), School of Engineering, Bauru, Brazil

ABSTRACT

Wastewater treatment using microalgae presents a promising approach for sustainable and efficient removal of pollutants. However, the complexity of metabolic networks involved in microalgae metabolism poses challenges for computational analysis. This chapter explores network reduction methods, specifically focusing on the application of the DRUM (dynamic reduction of unbalanced metabolism) framework, to streamline the modelling of microalgae-based wastewater treatment systems. This chapter describes the general core metabolism of microalgae, reviews methods of metabolic network reduction, and ends it with the application of a case study. The DRUM framework divides the complete metabolic network into subnetworks where the quasi-steady-state assumption (QSSA) holds, reducing the number of state variables and simplifying the kinetic modelling. By calculating the elementary flux modes (EFMs) for each subnetwork, macroscopic reactions are derived, representing the collective behaviour of internal reactions within the subnetworks. To demonstrate the effectiveness of the DRUM framework, a case study based on *Chlorella* sp. microalgae is presented. The study focuses on treating volatile fatty acid waste, a common byproduct of dark fermentation. The reduced metabolic model, obtained using the DRUM framework, accurately captures the dynamics of microalgae growth and medium concentration. This chapter underscores the significance of network reduction methods in optimizing microalgae-based wastewater treatment systems. These reduction methods pave the way for further advancements in the development and optimization of microalgae-based wastewater treatment technologies.

Keywords: microalgae, wastewater, metabolism, modelling, model reduction.

2.1 INTRODUCTION

Microalgae are unicellular organisms capable of growing autotrophically with solar energy through photosynthesis. Some species can also grow heterotrophically by absorbing a source of organic carbon compounds, such as glucose and acetate. They have emerged as a promising solution for wastewater treatment due to their ability to remove pollutants and nutrients while simultaneously producing valuable biomass. Microalgae are also very important organisms in the carbon cycle of the planet, being responsible for 40% of global fixation of carbon. They can be used to produce a variety of products such as proteins, vitamins, cosmetics, feedstock, and food (Barsanti & Gualtieri 2018).

Despite advances made in microalgae research during the past decades, the production of microalgae at industrial scale is still limited. Different bottlenecks explain the limited use of these processes in comparison with their potential, in particular, the economic and environmental profitability must still be improved. In this context, metabolic modeling plays a crucial role in understanding and improving the efficiency of microalgae-based wastewater treatment systems.

The use of mixotrophic growth, provided that the substrate is cheap and with a low environmental impact, is a promising way to increase productivity. This is of key interest if the substrate is a waste, and the process is used at the same time to produce biomass and to treat wastewater. By analyzing the metabolic pathways involved in the utilization of waste substrates, metabolic models can guide the optimization of cultivation conditions and nutrient supplementation strategies to enhance microalgae productivity and pollutant removal efficiency. This approach not only addresses the economic viability of microalgae-based wastewater treatment, but also offers a sustainable solution for waste management (Castillo *et al.*, 2021).

The use of metabolic networks provides a solid foundation to model microalgae growth in complex environments, such as wastewater. However, because of the complexity of these metabolic networks, in particular genome-scale ones, the utilization of methods to reduce the size of metabolic models proves to be essential for optimizing the efficiency of microalgae-based wastewater treatment. By employing metabolic models, it is possible to gain insights into the utilization of waste substrates, guide cultivation conditions, and design nutrient supplementation strategies to enhance microalgae productivity and pollutant removal.

Consideration of cultivation methods and bioreactor design is also essential, even in the case of metabolic models, especially regarding the modeling of the effects of light on the growth. The two most widespread processes for producing microalgae are closed photobioreactors and open raceways (Schade and Meier 2019). Photobioreactors can lead to a higher production output with better resistance to biological contaminants, but the energy input necessary for mixing and cooling strongly penalizes the economic and environmental balances (Tan *et al.*, 2018). The more rustic raceways are a simpler and cheaper way for producing microalgae outdoors. They need less energy input and the functional design is simpler. The drawback is the higher contamination in the culture by grazers, bacteria, viruses, or even other competitive microalgae species (Williams *et al.*, 2010, Mata *et al.*, 2010). Both of these cultivation methods may operate in batch, continuous, or even fed-batch conditions.

In this chapter, we focus on the theoretical approach, using modeling in order to optimize the system's efficiency. Such approaches have proven to be efficient in many different biotechnological applications. In the microalgae field, they are probably even more important to rationally manage the complexity of these nonlinear systems, which are exposed to weather fluctuations, affecting light and temperature. The development of numerical models is thus a prerequisite for understanding and managing these dynamical systems, involving several time scales, and permanently submitted to different perturbations. There is a need to bridge the gap between the detailed metabolic knowledge in the cell, and the necessity for control to keep a limited model complexity. Reducing metabolic models is difficult in a framework of permanent environmental fluctuations, maintaining the cell far from the balanced growth conditions which are generally the rule in metabolic modeling. Going from a metabolic model to a mechanistic model that can support process control is, therefore, a challenging objective. This chapter mentions different approaches to meet this goal.

2.2 MAIN METABOLIC PATHWAYS

In this section, we aim to elucidate the key metabolic pathways associated with microalgae, particularly focusing on relevant literature. The fundamental significance of microalgae lies in their ability to undergo growth through the utilization of photosynthetic energy. This process involves the absorption of carbon dioxide (CO_2) via the photosynthetic pathway, whereby energy derived from photons, whether from artificial or solar light sources, is captured.

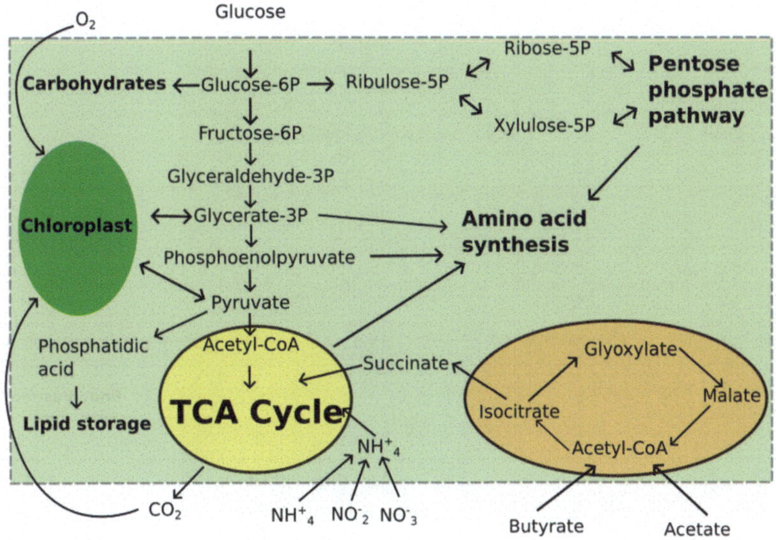

Figure 2.1 Main metabolic pathways of microalgae, demonstrating growth using as carbon substrates: butyrate, acetate, glucose, and CO_2.

A brief summary of the metabolism is important to understand model results and the soundness of these numerical results. Furthermore, a better comprehension also enables modelers to grasp what can be possibly achieved with the model and formulate subsequent steps accordingly. Figures 2.1 and 2.2 provide a simplified representation of the principal metabolic reactions. The eukaryotic nature of microalgae, coupled with their photosynthetic characteristics, leads to a complex compartmentalization of metabolic processes. For example, there is specific production of coenzymes, such as NADPH in different organelles of the cell. This compartmentalization is also dependent on the species.

Not only the knowledge of the reactions taking place is necessary to understand the metabolism, but also computational tools such as flux balance analysis (FBA) can help us comprehend the overall functionality of the metabolism. Here, we will briefly mention the main pathways and their main reactions, although details of how metabolic fluxes operate are species-dependent. While we will briefly touch upon the main pathways and their primary reactions, the specific operation of metabolic reactions is contingent upon the particular microalgal species under consideration.

2.2.1 Photosynthesis

Microalgae exhibiting autotrophic or mixotrophic growth strategies employ carbon dioxide (CO_2) as an inorganic carbon source, while cellular energy is derived from light. Photosynthesis occurs within the chloroplast and encompasses a combination of light-dependent reactions and the Calvin cycle.

In the thylakoid lumen, energy derived from photons is utilized for the synthesis of adenosine triphosphate (ATP) and reduced nicotinamide adenine dinucleotide phosphate (NADPH), with oxygen (O_2) being produced as a byproduct. Within the Calvin cycle, CO_2 undergoes a reaction with ribulose biphosphate, facilitated by the enzyme ribulose-1,5-bisphosphate carboxylase/oxygenase (Rubisco), resulting in the production of two molecules of glyceraldehyde 3-phosphate (GAP). Subsequently, GAP is transported to the cytosol, where it integrates into other metabolic pathways, such as its reduction to glucose 6-phosphate (glucose 6-P) for carbohydrate production or its oxidation within the tricarboxylic acid (TCA) cycle. Notably, the Calvin cycle remains inactive in the absence of light (Tibocha-Bonilla *et al.*, 2018).

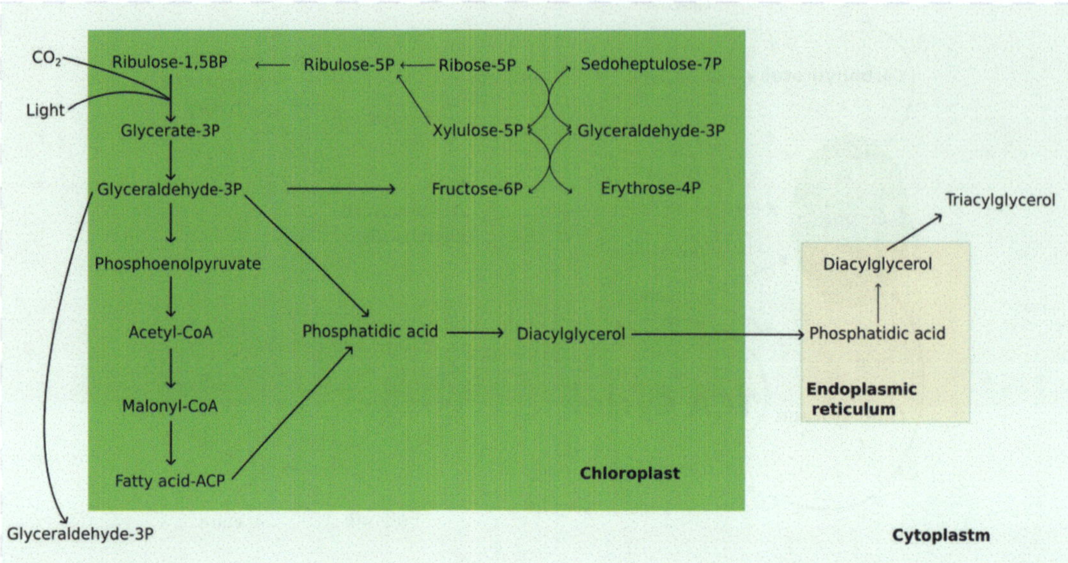

Figure 2.2 Metabolic reactions related to photosynthesis (Calvin cycle and light-dependent reactions) taking place in the chloroplast. Also, simplified pathway for the synthesis of lipids, takes place in the chloroplast and cytoplasm.

2.2.2 Glycolysis and pentose phosphate pathway

Glycolysis, a fundamental metabolic pathway in living organisms, occurs within the cytosol and plays an essential role in energy metabolism. In the case of heterotrophic growth external glucose is absorbed by phosphorylation producing glucose 6-P. Glucose 6-P can be utilized for the synthesis of carbohydrates or directed toward the pentose phosphate pathway (PPP), a parallel metabolic pathway to glycolysis, where it undergoes further transformations leading to the production of pentoses (5-carbon molecules). Most importantly, these pentoses are used for the synthesis of nucleic acid, but they are also used for the synthesis of many other biomass precursors. During the oxidative phase, the PPP produces NADPH.

2.2.3 Tricarboxylic acid cycle

The tricarboxylic acid (TCA) cycle, also known as the citric acid cycle or Krebs cycle, plays a crucial role in energy production and the synthesis of biosynthetic precursors. In the context of autotrophic or mixotrophic growth, the active reactions within the TCA cycle are primarily focused on the generation of biosynthetic precursors, whereas during heterotrophic growth, the emphasis shifts in the direction of energy production. The TCA cycle is essential for its anaplerotic reactions, which replenish intermediary metabolites. Therefore, it is considered a central axis in the core metabolism. Typically, at the entrance of the cycle there is acetyl-CoA that reacts with oxaloacetate-producing citrate. This metabolic pathway leads to the production of NADH, $FADH_2$, and GTP, thereby increasing cellular energy levels. However, there is carbon loss via the excretion of CO_2.

It is worth noting that not all microalgae possess the complete TCA cycle, as the presence of specific enzymes is dependent on the species. Furthermore, depending on certain environmental conditions, bypass variations may also take place. In those cases, for example, we can have pyruvate at the entrance of the cycle, producing oxaloacetate regulated by the enzyme phosphoenol pyruvate carboxylase (Fachet *et al.*, 2020).

2.2.4 Glyoxylate shunt

During the process of heterotrophic growth, microalgae have the capability to utilize acetate and, to a lesser extent, butyrate – because of its lower affinity – as carbon sources through the operation of the glyoxylate shunt. This metabolic pathway occurs within the glyoxysome, a specialized cellular organelle. Here, acetate and butyrate are converted first to acetyl-CoA, although additional enzymatic steps are required for the metabolism of butyrate. Subsequently, acetyl-CoA is used for the synthesis of succinate. Succinate can, then, be integrated into the core metabolism via the TCA cycle. By incorporating succinate into the TCA cycle, microalgae can further utilize the carbon and energy obtained from acetate and butyrate during heterotrophic growth.

2.2.5 Lipid biosynthesis

The synthesis of lipids has been an important topic in microalgae research for its potential use for the production of biofuels. There are three main classes of lipids: triacylglycerol (TAG), phospholipids, and glycolipids. Phospholipids and glycolipids have a higher polarity and form a stable bilayer, as a result, they are the major components of cell membranes and in stable growth conditions they form the majority of a cell's content of lipids. On the other hand, under stress conditions, the majority of cell's lipids are made of TAGs, since they serve as energy reserves. TAGs have a better yield for the production of biofuels and, therefore, are the focus of this section.

There are three main steps in the synthesis of lipids: (1) the production of malonyl-CoA from acetyl-CoA catalyzed by the enzyme ACCase, (2) the elongation of the acyl chain by the fatty acids synthase, both steps occur in the chloroplast, and (c) the formation of TAGs in the endoplasmatic reticulum. This process is illustrated in Figure 2.2. Following TAG synthesis, these lipids are stored in the form of lipid droplets. Furthermore, different specific pathways exist for the production of TAGs, for example acyl-CoA-dependent and -independent pathways. These pathways contribute to the diversification of lipid biosynthesis strategies (Chen & Wang, 2021; Huerlimann & Heimann, 2013).

2.3 GENOME-SCALE METABOLIC MODELS

Genome-scale metabolic models (GEMs) are stoichiometric representations of the complete metabolism of an organism, encompassing the connections between genes, proteins (enzymes), and reactions. Significant advances have been made concerning the mapping of metabolic reactions through the analysis of genomic data (Kim *et al.*, 2017). GEMs are constructed based on whole-genome sequencing, but the process of building a functional metabolic model involves several steps and iterations. Initially, it is necessary to identify functional roles in the genome and link them to enzyme complexes and reactions (Cuevas *et al.*, 2016).

New GEMs are regularly developed and more and more organisms have their proper GEM (Kim *et al.*, 2017). These models are continuously updated and refined, particularly for model organisms such as *Escherichia coli*, as knowledge regarding their genomes and expressed proteins become consolidated (Singh and Lercher, 2020). Experimental validation of metabolic models is crucial in light of this ongoing refinement. One of the first GEMs constructed for cyanobacteria predicted, through FBA, that photorespiration would allow for optimal growth rates (Knoop *et al.*, 2010). Analysis of GEMs helps to gain insights into possible metabolic engineering interventions and substrate allocation (Kim *et al.*, 2017).

Usually, the construction of these models is first focused on the carbon-core metabolic network. Later on, they are refined by accounting for more details, such as improved compartmentalization by including more organelles. Microalgae metabolic models require, at least, the reactions taking place in the chloroplast, cytosol, and mitochondria. GEMs are particularly valuable for identifying targets to modify strains. Despite the increasing availability of high-throughput analytical tools, the gathering and application of proteomic and metabolomic data for metabolic engineering in microalgae purposes remain limited. Currently, most studies focus primarily on lipid production, while other metabolites and pathways receive less attention.

Although genome sequences of numerous microalgal species have been resolved and made publicly available, the information provided by the genome and transcriptome alone offers only a limited view of the cell's metabolic pathways. However, the task of deciphering the nature and function of metabolic pathways in microalgae is challenging due to variations introduced through evolutionary processes. As for other eukaryotic organisms, compartmental complexity and intracellular transport further increase the difficulty in considering all aspects in the model. For these reasons, it is important to remember that current comprehension of the functioning of the cell in different conditions, although vast, will continue to evolve and expand in the coming years.

Figures 2.1 and 2.2 give an overall view of the metabolism of microalgae. They show the network and interconnectedness of some metabolism and how they are connected to the production of biomass. In a GEM, those reactions are all represented as coefficients of the stoichiometric matrix, linking the reactants and products of each reaction. One of the important features of GEMs is the biomass reaction, which describes the composition of the cell and therefore what metabolites and substrates are necessary for the growth of the cell. Therefore, the accurate determination of the macromolecular biomass composition is crucial for achieving accurate flux and growth rate simulations.

2.4 MODELLING METABOLIC NETWORKS

Metabolic networks are chains of reactions taking place inside the cell. The different metabolic pathways keep the cell functioning, for instance, the production of energy via ATP or the synthesis of macromolecules such as DNA, lipids, and proteins. Metabolic network models can be constructed based on the knowledge of biochemical processes, such as photosynthesis or glycolysis, or on the genomic knowledge of the organism (through the use of GEMs), which in general produces more accurate, though more complex models. The level of detail in the model can also be constrained by the objective of its use, and many reactions can be omitted.

In general, simplifications and assumptions to reduce the size of the system are necessary because the large number of states in standard models of metabolic networks makes optimization and control impracticable. In general, it is assumed that the system is in quasy steady state (QSS), known as the quasy steady state approximation (QSSA).

The ordinary differential equations (ODEs) representing the system in a continuous perfectly mixed stirred tank and can be written in the following general form:

$$\frac{\mathrm{d}C}{\mathrm{d}t} = \frac{\mathrm{d}cX}{\mathrm{d}t} = Nv - D.C$$

$$\frac{\mathrm{d}X}{\mathrm{d}t} = \mu X - DX$$

$$\frac{\mathrm{d}P}{\mathrm{d}t} = N_{\mathrm{P}}vX - DP$$

$$\frac{\mathrm{d}S}{\mathrm{d}t} = N_{\mathrm{S}}vX + D(S_{\mathrm{in}} - S)$$

where $C \in \mathbb{R}^{n_c}$, $S \in \mathbb{R}^{n_s}$, $P \in \mathbb{R}^{n_p}$ are concentration vectors of size n_c, n_s, n_p, respectively, representing the number of internal metabolites, substrates, and products. X is the biomass concentration. Substrates, products, and biomass are written as mass per volume of the reactor; $N \in \mathbb{R}^{n_c * n_r}$ is the matrix of stochiometric indices of the reactions in the metabolic network; $v \in \mathbb{R}^{n_r}$ is the vector of the reactions kinetics, giving the rate of all n_r reactions of the network; μ is the growth rate of the microalgae; D is the dilution rate; $S_{in} \in \mathbb{R}^{n_s}$ is the concentration vector of incoming substrates. This system of equations also describes a batch cultivation process when D equals zero.

When the metabolite concentrations are written per mass/volume of the cell, the ODE is written as:

$$\frac{dc}{dt} = Nv - c\mu$$

where c is the concentration of metabolites inside the cell written as a fraction, that is, mass of metabolites per total mass of the cell.

In the QSSA, internal metabolites are assumed to be in steady state, that is, the equilibrium is reached instantaneously, while only the concentration of external metabolites or substrates behaves dynamically. Mathematically, the QSSA is written as:

$$\frac{dc}{dt} = 0$$

or

$$Nv - c\mu = 0$$

The term $c\mu$, which describes the dilution of metabolites due to cellular growth is generally ignored because the dynamics of the chemical reactions are considerably greater than the loss of concentration due to the change in the cell mass (Provost & Bastin, 2004). In the end, we have the following equation:

$$Nv = 0$$

The QSSA is a necessary assumption for most frameworks and modeling of metabolic networks. The QSSA cannot always be applied, for example in cases where metabolites accumulate inside the organism, such as in microalgae. Due to dial variations of light intensity, microalgae accumulate different metabolites depending on light availability. Consequently, the classical frameworks face limitations when applied to the modeling of microalgae systems.

Constrained-based modeling techniques considering the QSSA are the most widely used when dealing with metabolic networks, enabling the estimation of intracellular fluxes at different conditions (Tibocha-Bonilla *et al.*, 2018). The two most important techniques are elementary flux modes (EFM) and FBA (Lotz *et al.*, 2014).

2.5 TOOLS FOR STEADY-STATE CONDITIONS

2.5.1 Elementary flux modes

2.5.1.1 Mathematical construction of EFMs

EFMs) are often described as a minimum set of pathways capable of representing the total of the network at the steady state. A flux mode is defined mathematically as a set M:

$$M = \left\{ v \epsilon \mathbb{R}^{n_r} \middle| v = \lambda v^*, \lambda > 0, Nv^* = 0 \right\}$$

where v^* is a vector respecting the steady-state condition $Nv^* = 0$, having a subset $v^{irr} \geq 0$, corresponding to the irreversible reactions, while the subset v^{rev} corresponding to the subset of reversible reactions has no sign restriction (Schuster *et al.*, 1999).

A representative v^* of M is an EFM if and only if it fulfills the simplicity condition: there is no couple of vectors v', v'' with the following properties:

- v^* is a non-negative linear combination of v' and v''
- v'' and v'' satisfy the conditions to be a flux mode
- v' and v'' contain at least the same number of zero elements as v^*, and at least one of them contains more zero elements than v^*.
- The elements at boundary reactions of v' and v'' have the same sign or one element is a zero (e.g. $v_i' = -1, v_i'' = 0$).

The vectors v satisfying the steady-state equation are necessarily non-negative belonging to the kernel of the stoichiometric matrix N. Therefore, the space generating these vectors is a polyhedral cone in the intersection between the kernel of N and the positive orthant. The vectors v can then be written as a non-negative linear combination of a set of vector e_k which forms the unique convex base of the polyhedral cone.

$$v = \sum \lambda_k e_k; \lambda_k \geq 0$$

The vectors e_k forming the convex basis are the EFMs, being the simplest pathway connecting substrates to products at a steady-state condition. The EFMs are useful to deduce macroscopic reactions (MR) or global reactions in the metabolic network. Because of the QSSA, the dynamics of internal metabolites can be ignored, simplifying the dynamic equation of the macroreaction.

As $e_k \in \mathbb{R}^{n_r}$, each position corresponds to a reaction participating in the elementary mode.

The macroreactions are easily deduced by multiplying by zero the components representing the internal reactions between metabolites, then keeping the substrates and products in the remaining reactions. Figure 2.3 has a visual representation of the EFMs for a toy network. For more complex metabolic networks, the determination of EFMs requires much more computational effort, with the number of EFMs increasing exponentially with the size of the network. An efficient algorithm to calculate all EFMs may be necessary to reduce computational time. Also, the existence of reversible reactions in the network might increase the difficulty of determining the set of EFM.

2.5.1.2 Minimal generating sets and EFM reduction

The presence or not of reversible reactions in the metabolic network change the algorithm necessary to compute the set of EFMs. The simplest case is when all reactions are irreversible. In this case, every

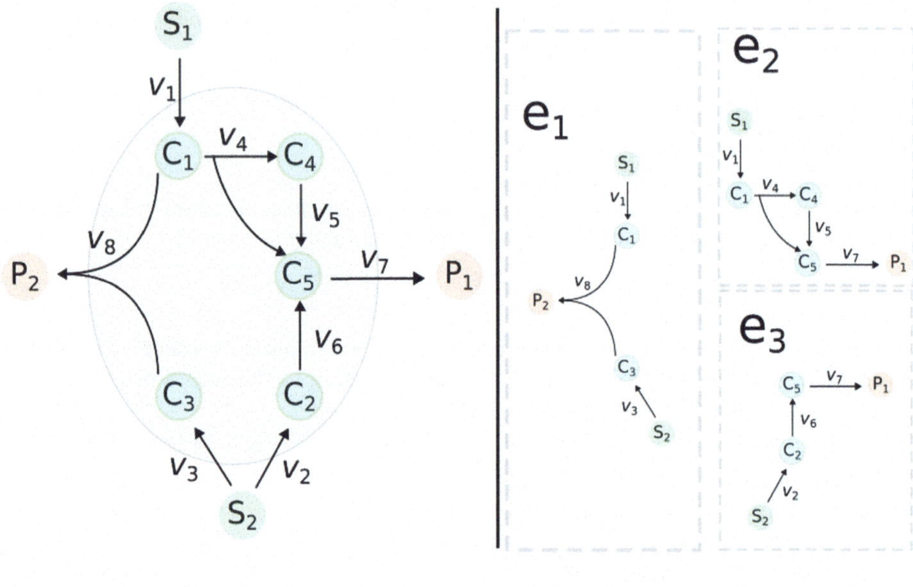

Figure 2.3 Example adapted from Provost and Bastin (2004) of the set of elementary flux modes being calculated for a toy-network. In this case, a linear combination of the three calculated elementary modes can reconstruct all possible steady states of the network.

component of all flux vectors v is non-negative and the cone representing the space of allowed flux vectors at steady state is a pointed cone. A convex cone K is pointed if:

$$K \cap -K = \{0\}$$

If a polyhedral cone is pointed there exists a unique minimal set of generating vectors and the elements of this set are the extreme rays of the cone. These vectors also serve as a complete set of representatives of elementary modes. In summary, when all reactions are irreversible, that is $v = v^{\text{irr}}$, there is a unique minimal generating set (MGS) which is equivalent to the set of EFMs.

There are three cases when reversible reactions are present in the network:

- The system has only irreversible elementary modes. Despite the presence of reversible reactions, no EFM can work in the reversible direction, that is, $\nexists e_k = -e_{k'}$
- The system has irreversible and reversible elementary modes
- The system has only reversible elementary modes.

In the first case, the polyhedral cone is still pointed, the MGS is unique, and it corresponds to the set of EFM. This is not the case anymore for the two remaining cases, where the cone is not pointed. Also, the MGS will not be unique and the set of EFM might be greater than the size of the MGS. Given this, the set of EFM will always be a superset of the MGS.

To understand the difference between the MGS and the EFMs getting deeper into convex analysis is necessary. In convex analysis it is shown that the space generating the solution of a linear homogenous system of equations is a convex polyhedral cone, C. Every point of such a cone is a non-negative combination of fundamental vectors, f, and basis vectors, b,

$$C = \left\{ v : v = \sum n_k f^k + \sum \lambda_m b^m; n_k, \lambda_m \geq 0 \right\}$$

The fundamental and basis vectors are also called the generating vectors. There is a minimum necessary number of generating vectors to span the cone. The basis vectors are the extreme rays of C for which the negative vector is also contained in C (Schuster *et al.*, 1999). The definition of C here is identical to the minimum set of elementary modes (MEMO) in Röhl and Bockmayr (2019), where every v vector in the steady-state cone is written as a non-negative linear combination:

$$v = \sum \lambda_e e + \sum \lambda_f f$$

where $f \in \left\{ U \cap E_N^{\text{irr}} \right\}, e \in \left\{ U \cap E_N^{\text{rev}} \right\}$, where U is an inclusion-minimal set which is a subset of E_N which is the set of all EFMs for the stoichiometric matrix N.

The basis vector b is then equivalent to the vector e corresponding to the reversible EFMs, while the fundamental vectors f are equivalent to the irreversible set of EFMs. This implies that when there are no reversible EFMs, that is, the cone is pointed, C has no basis vectors. By contrast, when there are only reversible EFMs, C has only basis vectors.

Jevremović and Boley (2013) and Röhl and Bockmayr (2019) provide algorithms to compute the MGS when there are reversible reactions in the metabolic network. Röhl and Bockmayr (2019) rely on a method of splitting reversible reactions, with a minimal number of splits, until no reversible EFM is left creating a pointed cone. While Jevremović and Boley (2013) divide the stoichiometric matrix based on the reversibility or not of the reaction, then they compute the null space of a modified matrix representing the reversible reactions and the MGS of the irreversible subnetwork.

The set of EFMs is in general much larger than the MGS. For example, the carbon metabolism of *E. coli* has 6421 EFMs while only 15 vectors are in the MGS/MEMO. The division of a network into subnetworks, as in the DRUM method (described below), also reduces the number of EFMs. The use of MGS may be another way to reduce the size of the system, though only the number of macroreactions is guaranteed to be reduced compared to the use of EFMs – the number of metabolites might still be the same. Furthermore, it is not guaranteed that the MGS is able to create meaningful macroreactions capable of accurately modeling the network, as it happens with the use of EFMs. The calculation of EFM becomes prohibitive when the metabolic network is too large, but the enumeration of EFMs is still possible by computing only a subset of the EFMs. Many methods have been developed in recent years to facilitate the computation of EFMs.

The method in Kaleta *et al.* (2009) is an example of subsystem analysis. This paper introduces the concept of elementary flux patterns, where instead of giving a stoichiometrical proportion to a reaction, it only considers the index. It means that it only calculates the list of reactions participating in an elementary mode. Oddsdóttir *et al.* (2015) use optimization in metabolic flux analysis to reduce the number of EFMs. The idea is to find the best-fitting EFMs to some measured external flux. There is an algorithm minimizing the difference between the measured flux and the EFMs to reproduce those fluxes.

Tabe-Bordbar and Marashi (2013) couple EFM with FBA. The proposed algorithm removes reactions by FBA, considering a random objective reaction. They select a list of reactions to remove, followed by FBA calculation, if the objective flux is non-zero, then they proceed with the deletion of the reactions. On the other hand, if the flux of the objective reaction is zero, then the reactions are kept. The goal is to find, at least, a subset of the EFMs of a genome-scale metabolic network by reducing the size of the total network. In a recent paper, Maton *et al.* (2022) calculate a reduced set of EFMs based on several steps, including geometrical criteria, optimization techniques, and also external observations to derive macroreactions for the system.

2.5.2 Flux balance analysis

Flux balance analysis (FBA) is one of the most common tools to analyze metabolic networks (Orth *et al.*, 2010). Together with EFMs, they can be used to identify feasible routes in the metabolic network and estimate internal metabolic fluxes based on substrate uptake and excretion rates (Lotz *et al.*, 2014). As in the case of EFMs, FBA also assumes the cell to be at steady state. However, instead of trying to determine the possible set of reactions constructing the steady state, the method consists of the maximization (or minimization) of an objective function:

$$\max Z = c^{\mathrm{T}} v$$

where $c \epsilon \mathbb{R}^{n_r}$ is a vector of weights, indicating how much a certain reaction influences the objective function, and v is the vector of fluxes of metabolic reactions. Besides the constraint of the steady state ($Nv = 0$), it considers boundaries for the vector of reaction fluxes v:

$$l_i \leq v_i \leq u_i$$

where l_i and u_i are the lower and upper boundary, respectively.

One of the most common cases is the maximization of biomass production, in this case c will be a vector containing zeros in every position, except the position for the reaction of biomass.

The system of equations and constraints of FBA leads to a linear programming problem, which can be computed with standard algorithms such as interior-point methods.

2.6 METABOLIC NETWORKS REDUCTION

The increasing size of metabolic networks makes it difficult to apply numerical analysis, especially when considering dynamical aspects. Even in the case of the steady state, computational power becomes limiting. For example, as discussed above, the number of EFMs grows exponentially as the metabolic network increases. As a consequence, calculation of EFMs for genome-scale models even for simple organisms such as *E. coli* may not be possible due to computational limitations. Methods to reduce the size of these genome-scale metabolic models become imperative to analyze steady state and dynamical behavior.

Here, we will briefly mention the current methods used to reduce metabolic networks, with emphasis on genome-scale networks.

Many methods have been released in the literature in recent years regarding the reduction of metabolic networks (Singh and Lercher 2020). They differentiate on the assumptions regarding the network in order to proceed with the model reduction, methods, and goals. While some techniques focus on keeping the same phenotype of the full network, others have a more greedy approach to the reduction, focusing on minimizing the most possible of the network and only keeping some desired reactions or phenotypes.

One of the first techniques used was the consideration of '*Enzymes Subsets*' (Pfeiffer *et al.*, 1999). An enzyme subset is a group of enzymes that work together in a metabolic pathway and can be considered as a unit structure catalyzing a series of reactions. Mathematically, a group of reactions (or enzymes) belongs to an enzyme subset if in all flux vectors v satisfying the steady-state condition, the ratio between the fluxes of the reactions in the enzyme subset, for example $v_n/v_{n'}$, has the same non-zero value and the directions of the reactions are not contradictory. It is possible therefore to reduce the network without losing the original information and capabilities, because it is considered that the enzymes belonging to such a subset are expressed coherently, regulating metabolism in the unit. Nevertheless, the drawback of this method is the limited capability in reducing the total size.

A method to further reduce the metabolic network was later implemented by Tabe-Bordbar and Marashi (2013), called minimal reaction sets. The method consists in solving a mixed-integer linear programming problem, where the objective is to minimize the number of reactions of the network, while still keeping a minimal flux of biomass production. A more recent method *NetworkReducer* has recently been published where the objective is to reduce the network while at the same time keeping certain protected phenotypes, metabolites, and reactions (Erdrich *et al.*, 2015). The algorithm functions in two major steps. First the pruning phase where reactions are iteratively removed until no more reaction can be deleted without breaking protected parts. Second, the compression phase is a loss-free simplification by the lumping of coupled reactions. An improvement of this method was made by Röhl and Bockmayr (2017), by including the minimization of the number of reactions as in Burgard *et al.* (2001). In Küken *et al.* (2021), a method of reduction based on the use of complexes (combination of the species participating in one side of the reaction), where the stoichiometric matrix is written as the product of two matrices, $N = Y.A$, where Y is a matrix having as columns the complexes and metabolites in the rows, A is a matrix having indices of -1, 0 or 1 with reactions represented on the columns and the complexes on the rows. Depending on the structural conditions and the balancing of the complexes, the network is reduced while keeping the phenotype of the original network.

2.6.1 The DRUM framework

DRUM (dynamic reduction of unbalanced metabolism) is a metabolic modeling framework created in order to circumvent the problem of inappropriate use of the QSSA to the whole metabolic network (Baroukh *et al.*, 2014). It was initially developed for organisms which dynamically accumulate and reuse some metabolites, such as microalgae under varying environmental conditions.

The idea of the DRUM framework is to divide the complete metabolic network into subnetworks, in which the QSSA is valid, also reducing the total number of state variables representing the system. After the division of the network, the EFMs are calculated for each subnetwork, generating MR, representing the result of all the internal reactions of the subnetwork with much simpler kinetics. See Figure 2.4 for a representation of the steps required in the DRUM framework.

After the application of the DRUM method, the dynamical equations of the system are reduced to the number of metabolites that are allowed to accumulate, external substrates and products. The form of the system of differential equations is the same as the original one, but the stoichiometric matrix is reduced and modified to represent the new MR deduced from the EFMs. The DRUM method is able

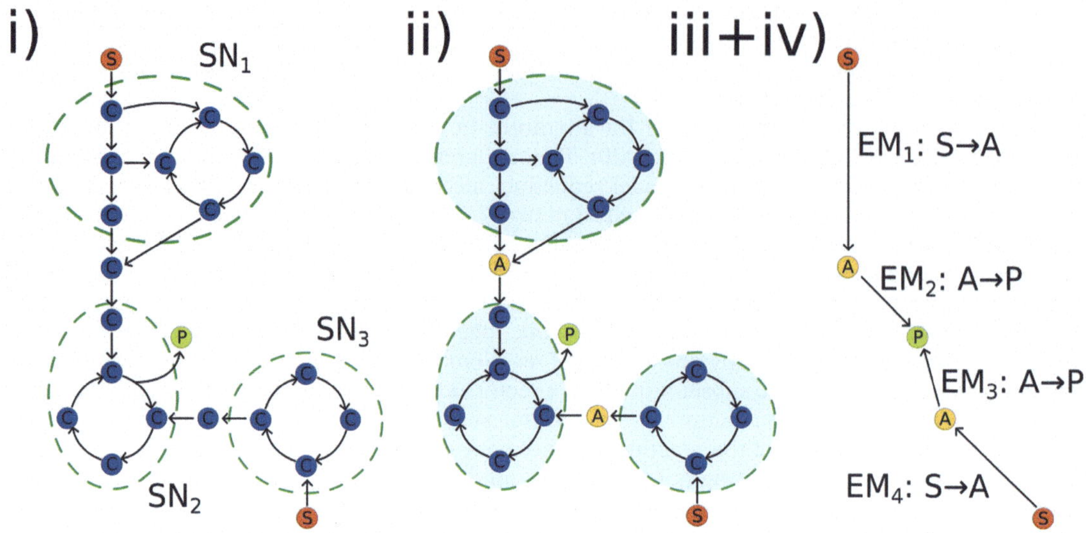

Figure 2.4 Example of application of the DRUM framework to a toy network. Initially, the network has two substrates (orange), one product (green), and 18 internal metabolites (blue). In step (i) the sub-networks are determined. In step (ii) there is the choice of accumulating metabolites connecting the SNs. In step (iii), the elementary flux modes are calculated for each of the SNs. In step (iv), the original network is reduced by the use of the macroreactions. Finally, the number of 18 internal metabolites is reduced to only 2.

to accurately represent empirical data, predicting for example the accumulation of carbohydrates and lipids during the day and its consumption during the night. Despite this, a more objective method to divide the subnetworks still needs to be defined. Finding new ways to split the metabolic network might reduce the size of the system even more, while still being able to predict the accumulation of metabolites.

The DRUM framework is grounded on the key concept and assumption of the different time scales characterizing metabolic reactions. This discrepancy in time scales gives rise to the accumulation of metabolites and consequential modifications in cellular composition. However, due to the present limitations in our knowledge of internal reaction rates, the selection of accumulating metabolites is currently non-deterministic and relies on prior knowledge or subjective preferences concerning the partitioning of the metabolic network. As discussed earlier, taking into consideration cellular compartments, intersection metabolites, and possible simplification strategies employed during the reduction process will influence the choice of accumulating metabolites. Nevertheless, with the advancement of knowledge regarding internal kinetic rates, future developments are expected to provide more rigorous approaches for determining accumulating metabolites. These refined methods will not only enhance the accuracy of phenotype approximation in reduced metabolic models, but also facilitate the evaluation of their proximity to the complete metabolic system.

In the DRUM framework, the metabolic network is represented by the following system of ODEs:

$$\frac{dM}{dt} = \frac{d}{dt}\begin{pmatrix} S \\ C \\ P \\ B \end{pmatrix} = \begin{pmatrix} N_S \\ N_C \\ N_P \\ N_B \end{pmatrix} .v(M).B - D.M + D.M_{in} = N.v(M).B - DM + DM_{in}$$

where M represents the vector of the concentrations of metabolites composed of substrate (S), intracellular metabolites (C), excreted products (P), and biomass (B). M_{in} is the influent concentration of these quantities. The dilution rate of the reactor (ratio of influent flow rate over the reactor volume) is D ($D=0$ for a batch process). All the concentrations are expressed as total concentrations in the solution. $v\epsilon\mathbb{R}^{n_r}$ is the reaction kinetic vector, while the matrices $N_S\epsilon\mathbb{R}^{n_s\times n_r}$, $N_C\epsilon\mathbb{R}^{n_c\times n_r}$, $N_P\epsilon\mathbb{R}^{n_p\times n_r}$, and $N_B\epsilon\mathbb{R}^{1\times n_r}$ correspond, respectively, to the stoichiometric matrices of substrates S, products P, intracellular metabolites C and biomass B ($n_s + n_c + n_p + 1 = n_m$).

The DRUM method consists in dividing the metabolic network into k quasi-stationary subnetworks, so the matrix N is rewritten in the following form:

$$N = [N_{SN_1}, N_{SN_2}, \ldots, N_{SN_k}]$$

where $N_{SN_i} \in \mathbb{R}^{n_m \times n_{SN_i}}$ and $\sum_{i=1}^{k} n_{SN_i} = n_r$. Each sub-network is assumed to be at steady state:

$$\forall i = 1, \ldots, k : N_{SN_i}.v_{SN_i} = 0$$

By considering the steady-state condition, it is possible to calculate the EFMs for each of these N_{SN_i} sub-networks, then construct macroreactions:

$$\forall i = 1, \ldots, k : v_{SN_i} = E_{SN_i}\alpha_{SN_i}, \alpha_{SN_i} \geq 0$$

$$\forall i = 1, \ldots, k : \to (N_{S_{SN_i}}.E_{SN_i}).S_{SN_i} \to (N_{P_{SN_i}}.E_{SN_i}).P_{SN_i}$$

where E_{SN_i} is the matrix of EFMs of the sub-network SN_i, and α_{SN_i} the kinetics of the MR described by the reduced stoichiometric matrix.

Following this step, we group all the sub-networks, and considering that only metabolites A are allowed to accumulate. Meaning that other metabolites have simple dynamics and their concentration is directly determined by the A metabolites. We obtain a reduced dynamic model, defined by the new metabolites vector $M'\epsilon\mathbb{R}^{n_m}$, the new stoichiometric matrix $N' \in \mathbb{R}^{n_m \times n_E}$ \$N' and α the kinetic vector associated with these MR:

$$\frac{\mathrm{d}M}{\mathrm{d}t} = \frac{\mathrm{d}}{\mathrm{d}t}\begin{pmatrix} S \\ A \\ P \\ B \end{pmatrix} = \begin{pmatrix} N_S \\ N_A \\ N_P \\ N_B \end{pmatrix}.\alpha.B - D.M + D.M_{\mathrm{in}} = N'.\alpha.B - DM' + DM'_{\mathrm{in}}$$

Furthermore, because of the consideration of the accumulating metabolites, there is a distinction between the dry weight biomass or total biomass (X) and the functional biomass. In this case, experimentally measured biomass is the sum of the functional biomass and the total mass of the accumulating metabolites:

$$X = B + \sum_{i=1}^{k} A_i$$

2.7 CASE STUDY: MICROALGAE CULTIVATION

2.7.1 Introduction: volatile fatty acid

In this section, the DRUM framework will be applied to a case study based on the research paper by Pessi *et al.* (2023). The objective of this case study is to construct a reduced metabolic model for the *Chlorella* sp. microalgae to address the treatment of volatile fatty acid (VFA) waste with the addition of organic carbon substrates, namely glucose and glycerol. The case study explores the concept of integrating wastewater treatment with biofuel production, where the VFA waste from dark fermentation, produced by bacteria, serves as the substrate for microalgae cultivation.

The metabolic network utilized in this study is based on the study by Pessi *et al.* (2023), initially with 188 reactions and 173 metabolites. Noticeably, this metabolic network is relatively small, since it is not a genome-scale metabolic network, but it was constructed based on core known metabolic reactions with some other reactions being included to enhance its coverage of metabolic possibilities. This means that in the case of using GEMs, as mentioned above, it may be harder to apply the DRUM method. Although it would be encouraged to apply it to a GEM since a more representative reduced model would be obtained, and after reduction, the model could be used again for other applications. The composition of VFAs coming from dark fermentation varies depending on the conditions, usually determined by the composition and characteristics of the substrate coming for treatment via dark fermentation. The VFAs present in the waste are primarily butyrate and acetate. Frequently, there is also production of lactate in this kind of process, but since *Chlorella* does not consume lactate, it is not included in the model.

The model encompasses four organic carbon substrates: butyrate (BUT), acetate (ACE), glucose (GLC), and glycerol (GLY), along with one inorganic carbon dioxide (CO_2). Additionally, the choice of the nitrogen source is an important consideration, as it influences the stoichiometric production of biomass. NH_4^+ is known to have a better yield for biomass production compared to other nitrogen sources, assuming the same carbon source.

The inclusion of glucose and glycerol as additional carbon sources in the treatment of wastewater containing VFAs is motivated by an optimization approach. Previous research has highlighted the inhibitory effects of butyrate on algal growth, manifested in both heterotrophic and mixotrophic growth conditions. Additionally, the presence of acetate has been observed to hinder the absorption of butyrate, leading to the occurrence of diauxic growth. Consequently, it is imperative to devise a strategy to mitigate this inhibition and enhance the consumption of butyrate by microalgae. The strategy proposed in this study involves supplementing the wastewater with glucose and glycerol, which serve as readily assimilated carbon sources for the microalgae. This supplementation aims to accelerate algal growth, enabling them to overcome the inhibitory effects of butyrate by attaining a higher biomass concentration that can more efficiently consume the remaining butyrate.

2.7.2 Determination of the subnetworks and accumulating metabolites

Upon determining the substrates to be represented within the model, the subsequent step involves the delineation of subnetworks and the identification of accumulating metabolites within the DRUM framework. This task resides under the discretion of the modeler, who may opt for a straightforward approach by considering cellular compartments. In the present case, the chloroplast and glyoxysome, previously discussed, are designated as initial compartments. Moreover, glyceraldehyde 3-phosphate (GAP) and succinate (SUC) are chosen as accumulating metabolites due to their pivotal role as intersection metabolites, facilitating the interconnection of diverse subnetworks within the model (Figure 2.5).

Succinate assumes paramount importance as an essential precursor, as it possesses the ability to enter the mitochondrial tricarboxylic acid (TCA) cycle, subsequently undergoing conversion to oxaloacetate and eventually participating in the generation of phosphoenolpyruvate. Likewise, GAP emerges as an important precursor, with its synthesis occurring, for example, in the chloroplast during photosynthesis and also in the cytosol during glycolysis, thus serving as a bridging molecule across multiple metabolic pathways.

It is important to bear in mind that these choices are devised to simplify the process of reducing the metabolic network. However, alternative strategies for partitioning may be deemed more appropriate as further knowledge is acquired. For instance, if the incorporation of lipid content in the model becomes essential, it may be prudent to exclude it from the biomass synthesis subnetwork and instead establish a separate subnetwork specifically dedicated to lipid synthesis. Nonetheless, this approach introduces an additional dynamic variable that will have to be simulated and entails the acquisition of more experimental data to calibrate the model.

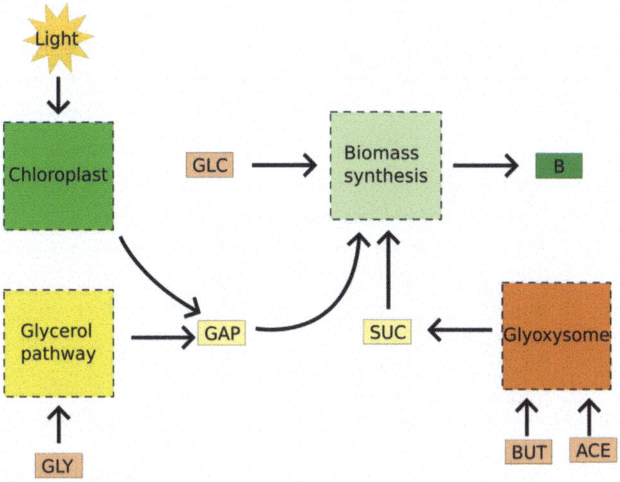

Figure 2.5 Simplified representation of the metabolic network of microalgae *Chlorella* after reduction with the DRUM framework. Sub-networks (squares with dashed lines), substrates (orange rectangles), accumulating metabolites (yellow rectangles), and biomass (green rectangles).

The first determination of subnetworks and selection of accumulating metabolites, by consequence, is able to pre-determine further subnetworks from still non-connected substrates. Given the prior selection of GAP, a subnetwork connecting GAP and glycerol (GLY) is naturally formed. The pathway created is, furthermore, called glycerol pathway. As for the remaining metabolic pathways, such as glycolysis, the TCA cycle, and protein and lipid synthesis, they are amalgamated into a comprehensive global network named here as the biomass synthesis pathway.

As for glucose, we establish a direct linkage to the biomass synthesis pathway, owing to the immediate synthesis of glucose-6P subsequent to glucose uptake by the cellular system. An alternative approach could involve the segregation of the glycolysis pathway, with the 'upper glycolysis' pathway serving as a distinct subnetwork. Such an arrangement would give rise to an elementary flux mode, encompassing a macroreaction that connects glucose (GLC) to glyceraldehyde 3-phosphate (GAP). It is worth noting that the construction of subnetworks may undergo iterations as the modeller refines the composition of the reduced network, ultimately culminating in the selection of a definitive configuration.

2.7.3 Derivation of MR

Following the completion of subnetwork and accumulating metabolite selection, the subsequent phase entails the determination of EFMs to derive the MR. This computational step can be accomplished using numerical tools such as efmtool (Terzer & Stelling, 2008) or COBRA methods (Ebrahim *et al.*, 2013). In situations where multiple EFMs are calculated for a given pair of subnetwork and accumulating metabolite, a selection process becomes necessary. Generally, the optimal approach for selection involves identifying the reaction that has the highest yield for the product. A list of the chosen EFMs and their corresponding MR for each subnetwork and accumulating metabolite/substrate is presented in Table 2.1.

2.7.4 Choice of kinetic model

After the reduction of the metabolic network and the determination of the final set of variables required for simulating the system dynamics, it is necessary to determine the appropriate kinetic models for

Table 2.1 List of the seven elementary flux modes selected to represent MR of the reduced network.

	Subnetwork	Macroscopic Reaction
MR_1	Glyoxysome	$2ACE + 3.5H + 0.5O_2 \rightarrow SUC + 0.5H_2O$
MR_2	Glyoxysome	$BUT + 7H + 1.5O_2 \rightarrow SUC + 5H_2O$
MR_3	Chloroplast	$Light + 3CO_2 + 2H_2O + Pi \rightarrow GAP + 3O_2$
MR_4	Glycerol pathway	$GLY + Pi \rightarrow GAP + H_2O$
MR_5	Biomass synthesis	$4.64GAP + 2.04O_2 + 0.99NO_3 + 0.98H + 0.02SO_4 + 0.01Mg_2 \rightarrow B + 5.39CO_2 + 2.90H_2O + 4.51Pi$
MR_6	Biomass synthesis	$4.90SUC + 5.28O_2 + 0.99NO_3 + 0.12Pi + 10.78H + 0.02SO_4 + 0.01Mg_2 \rightarrow B + 11.07CO_2 + 8.31H_2O$
MR_7	Biomass synthesis	$2.34GLC + 2.14O_2 + 0.99NO_3 + 0.12Pi + 0.98H + 0.02SO_4 + 0.01\ Mg_2 \rightarrow B + 5.49CO_2 + 7.63H_2O$

each MR. The selection of these kinetic models is crucial in ensuring accurate dynamical simulations. Even if the underlining representation of the metabolism is correct, inadequately estimated parameters or ill-fitted kinetic models will result in poor results.

A classical approach to model the kinetics of biochemical reactions is the use of Monod-like functions. In the case at hand, where acetate consumption is the MR of interest, the rate of the reaction increases with the concentration of the substrate, but reaches a maximum rate when the concentration saturates the quantity of enzymes catalyzing the reaction. Thus, the selected function is represented as follows:

$$\alpha_1 = \alpha_{1\max} \frac{ACE}{KS_1 + ACE}$$

As previously mentioned, there is inhibition in the consumption of butyrate coupled with a diauxic effect involving acetate. To model both of these effects, we first consider a Haldane-like function which describes the inhibition, wherein an optimal concentration exists at which the rate of the reaction is maximized, but after this concentration the rate is reduced. The Haldane function is then multiplied by a function that decreases with acetate concentration to account for diauxic growth. Hence, the model for the MR involving butyrate is expressed as follows:

$$\alpha_2 = \alpha_{2\max} \frac{BUT}{BUT + \frac{\alpha_{2\max}}{\beta_2}\left(\frac{BUT}{S_{2opt}} - 1\right)^2} \frac{k_d}{(ACE + k_d)}$$

Macroscopic reaction 3 (MR_3 in Table 2.1) describes the autotrophic growth, and as such, it is dependent on light intensity. Numerous functions can be employed to model autotrophic growth, taking into account phenomena such as photoacclimation, which introduces a dependency on past light intensities and necessitates the consideration of an additional dynamical variable. However, one important aspect is the consideration of light absorption due to medium turbidity, which increases with biomass concentration. In this case, light absorption is modeled by the Beer–Lambert equation, where light intensity depends on the light at the top of the reactor $(I)_0$, the extinction coefficient (σ), which depends on biomass concentration, and the depth of the reactor (L). See Martínez *et al.* (2018), for a thorough discussion on modeling light absorption.

$$I = I_0 \exp(-X.\sigma.L)$$

$$\sigma = aX^{1-b}$$

$$\alpha_3 = \frac{\alpha_{3\max}}{\sigma.X.L} \ln\left(\frac{k_I + \sigma I_0}{k_I + \sigma I_0 \exp(-\sigma.X.L)}\right)$$

The model for glycerol utilizes a Haldane-like function. However, if only low or moderate concentrations of glycerol are under consideration, a Monod-like function may also be appropriate.

$$\alpha_4 = \alpha_{4\max} \frac{GLY}{GLY + \frac{\alpha_{4\max}}{\beta_4}\left(\frac{GLY}{S_{4\mathrm{opt}}} - 1\right)^2}$$

For internal accumulating metabolites, because of the lack of experimental data of their dynamic concentrations, it is necessary to minimize the number of model parameters to facilitate the calibration process. For these reasons, linear kinetics is used to model the reactions with GAP and SUC. Furthermore, it is important to note that the Monod function is approximated to a linear function in low substrate concentrations, which also justify the choice of a linear equation. It is important to emphasize that, in the context of internal metabolites, the reaction rate relies on the internal concentration rather than the total concentration within the reactor.

$$\alpha_5 = \alpha_{5\max} \cdot \frac{GAP}{X}$$

$$\alpha_6 = \alpha_{6\max} \cdot \frac{SUC}{X}$$

Glucose, like glycerol, is also modeled by a Haldane-like function. Equally, if only low concentrations of glucose are being considered, a Monod-like function would also be fitting.

$$\alpha_7 = \alpha_{7\max} \frac{GLC}{GLC + \frac{\alpha_{7\max}}{\beta_7}\left(\frac{GLC}{S_{7\mathrm{opt}}} - 1\right)^2}$$

Finally, at the end of the process of applying the DRUM framework, we are going to have the following system of seven ODEs when considering a continuous reactor:

$$\frac{dACE}{dt} = -2\alpha_1.B + D(ACE_{\mathrm{in}} - ACE)$$

$$\frac{dBUT}{dt} = -\alpha_2.B + D(BUT_{\mathrm{in}} - BUT)$$

$$\frac{dSUC}{dt} = (\alpha_1 + \alpha_2 - 4.9\alpha_6)B - D.SUC$$

$$\frac{dGLY}{dt} = -\alpha_4.B + D(GLY_{\mathrm{in}} - GLY)$$

$$\frac{dGLC}{dt} = -2.34.\alpha_7.B + D(GLC_{\mathrm{in}} - GLC)$$

$$\frac{dGAP}{dt} = (\alpha_3 + \alpha_4 - 4.64\alpha_5)B - D.GAP$$

$$\frac{dB}{dt} = (\alpha_5 + \alpha_6 + \alpha_7)B - D.B$$

This set of equations can then be used to simulate the system, for process control and optimization.

2.7.5 Model calibration and validation

With the mathematical structure of the model at hand, it is now possible to simulate the dynamics of metabolite and biomass concentrations in the reactor. To achieve this, we need to determine the parameters governing the rates of the MR. The calibration process involves adjusting these parameters to align the model's predictions with experimental data, typically obtained from laboratory experiments.

Several methods can be employed for parameter estimation, and the choice of method depends on the available data and the complexity of the model. Some common approaches include least-squares fitting, maximum likelihood estimation, and Bayesian parameter estimation. In Pessi *et al.* (2023), a combination of methods is used. First, a global optimization method called differential evolution (Storn & Price, 1997) is used to identify initial parameter values and avoid local minima in the objective function. Subsequently, a Markov chain Monte Carlo method is applied to identify parameter values within a range of uncertainty (Foreman-Mackey *et al.*, 2013). Alternatively, if the modeler possesses a good estimation of the parameter range, local optimization with Markov chain Monte Carlo might suffice.

During the calibration process, it is essential to consider the uncertainties in the experimental data and the model structure. The use of statistical tools to quantify parameter uncertainty and confidence intervals can aid in this process. Uncertainty in the experimental data can be included during calibration with the Markov chain Monte Carlo method and the confidence interval for the parameters is also obtained at the end of this process.

Furthermore, to facilitate calibration it is possible to divide the set of parameters to calibrate, following a 'divide and conquer' strategy (Mairet & Bernard, 2019). In the case of our system of equations, it is possible to calibrate the model in multiple ways. For instance, kinetics of glucose, glycerol, and acetate consumption rely solely on their respective concentrations, allowing for separate calibration of the relevant parameters. Only butyrate consumption also depends on the acetate concentration, the relation described by the parameter k_d. The parameters of internal metabolites (GAP and SUC) have to be calibrated with at least one substrate, or even in autotrophic conditions for GAP. Ideally, α_5 and α_6 should be calibrated using data of multiple external substrates.

Once calibration is completed, the model should be tested against other independent experimental data for validation. This step is crucial to assess the model's predictive capability and involves comparing the calibrated model's predictions with additional experimental data that were not used during the calibration phase. Successful validation ensures the reliability of the model and the accuracy of the parameter values.

Following model calibration and validation, it can be used for various simulation scenarios to gain insights into the system's behavior under different conditions. One important aspect is the optimization of the process, where optimal operating conditions can be found.

2.7.6 Example of application: optimization of waste treatment time

When optimizing a process, multiple targets for optimization can be considered, such as minimizing costs, maximizing profits, or enhancing product production. In the context of butyrate inhibition, the model can be used to minimize treatment time, by overcoming the slow consumption of butyrate. The initial concentrations of acetate and butyrate are fixed, as they are the product of dark fermentation. Only the optimal addition of glucose and glycerol is found using numerical optimization techniques. When considering continuous or fed-batch cultivation, the objective function is as follows:

$$\min_{D,y} t_f : S(t_f) \leq \bar{S}$$

Here, $\bar{S}$ is a vector of the regulation threshold for the external substrates, indicating the maximal allowed concentration of external substrates at the end of the process. t_f corresponds to the time of process completion, when all substrate concentrations are below the defined threshold. The ratio y denotes the proportion between glucose and glycerol added, and D is the dilution rate (or flux) for the added substrates.

In the case of a batch system, the optimization problem consists in finding the optimal addition of glucose or glycerol at the beginning of the process. The objective function is formulated as follows:

$$\min_{S_{in}} t_f : S(t_f) \leq \bar{S}$$

Here, S_{in} is a vector containing the initial concentrations of the added substrates.

Solving these optimization problems employs similar algorithms used during the calibration process since they consist of function minimization. The major difficult arises in case of fed-batch or continuous cultivation, where the optimal D needs to be found. D can be approximated as a function of time ($D = f(t)$), specified by the modeler, with the parameters of this function obtained during optimization. In other words, the problem of optimization of the process would become a calibration for the D function. Otherwise, more advance techniques of optimal control would be necessary.

Finally, the obtained optimal results could be tested experimentally. If experimental results are coherent with the optimal conditions predicted by the model, it validates the model's performance. Usually, some deviations between model prediction and experimental data will be encountered. In such cases, the newly acquired information can be used as a feedback for the model to check if parameters must be re-calibrated or if the structure of the model should be modified.

2.8 CONCLUSION

In this chapter, we reviewed the current state of the art of metabolic modeling for microalgae, exploring various aspects such as metabolic network regulation, reduction methods, and dynamic simulations. A case study on microalgae-based wastewater treatment with a focus on VFAs provided practical insights into the application of these modeling techniques.

The exploration of metabolic network modeling and its application to microalgae-based wastewater treatment and biofuel production has provided valuable insights into the potential of sustainable biotechnology applications. Throughout this chapter, the DRUM framework has been demonstrated as a powerful approach for reducing complex metabolic networks, facilitating the creation of simplified ODEs that describe the dynamics of the system. Following model calibration and validation using experimental data, the resulting model becomes a robust tool for simulating microalgae cultivation processes under various conditions.

Future research in the field should continue to focus on advancing modeling techniques and experimental data collection, in particular of internal metabolites. Additionally, a more in-depth understanding of microalgae metabolic pathways and their interactions with environmental factors, such as temperature, can lead to more accurate models. Additionally, integrating the metabolism of multiple organisms into future models is essential, as it has been demonstrated that the microbial community plays a crucial role in such processes.

While the current method of metabolic network reduction and dynamical simulation shows promising results, modelers must stay attuned to the rapid developments in this field. The continuous advancement in biological understanding and modeling techniques related to computational biology is inevitable. As a result, we can anticipate significant improvements in sustainable biotechnology applications using microalgae.

REFERENCES

Baroukh C., Muñoz-Tamayo R., Steyer J.-P. and Bernard O. (2014). DRUM: a new framework for metabolic modeling under non-balanced growth. Application to the carbon metabolism of unicellular microalgae. *PLOS ONE*, **9**(8), e104499, https://doi.org/10.1371/journal.pone.0104499

Barsanti L. and Gualtieri P. (2018). Is exploitation of microalgae economically and energetically sustainable? *Algal Research*, **31**, 107–115, https://doi.org/10.1016/j.algal.2018.02.001

Burgard A. P., Vaidyaraman S. and Maranas C. D. (2001). Minimal reaction sets for *Escherichia coli* metabolism under different growth requirements and uptake environments. *Biotechnology Progress*, **17**(5), 791–797, https://doi.org/10.1021/bp0100880

Castillo T., Ramos D., García-Beltrán T., Brito-Bazan M. and Galindo E. (2021). Mixotrophic cultivation of microalgae: An alternative to produce high-value metabolites. *Biochemical Engineering Journal*, **176**, 108183, https://doi.org/10.1016/j.bej.2021.108183

Chen H. and Wang Q. (2021). Regulatory mechanisms of lipid biosynthesis in microalgae. *Biological Reviews*, **96**(5), 2373–2391, https://doi.org/10.1111/brv.12759

Cuevas D. A., Edirisinghe J., Henry C. S., Overbeek R., O'Connell T. G. and Edwards R. A. (2016). From DNA to FBA: How to build your own genome-scale metabolic model. *Frontiers in Microbiology*, **7**, 907, https://doi.org/10.3389/FMICB.2016.00907

Ebrahim A., Lerman J. A., Palsson B. O. and Hyduke D. R. (2013). COBRAPY: constraints-based reconstruction and analysis for Python. *BMC Systems Biology*, **7**(1), 74, https://doi.org/10.1186/1752-0509-7-74

Erdrich P., Steuer R. and Klamt S. (2015). An algorithm for the reduction of genome-scale metabolic network models to meaningful core models. *BMC Systems Biology*, **9**(1), 48, https://doi.org/10.1186/s12918-015-0191-x

Fachet M., Witte C., Flassig R. J., Rihko-Struckmann L. K., McKie-Krisberg Z., Polle J. E. W. and Sundmacher K. (2020). Reconstruction and analysis of a carbon-core metabolic network for *Dunaliella salina*. *BMC Bioinformatics*, **21**(1), 1, https://doi.org/10.1186/s12859-019-3325-0

Foreman-Mackey D., Hogg D. W., Lang D. and Goodman J. (2013). Emcee: The MCMC hammer. *Publications of the Astronomical Society of the Pacific*, **125**(925), 306–312, https://doi.org/10.1086/670067

Huerlimann R. and Heimann K. (2013). Comprehensive guide to acetyl-carboxylases in algae. *Critical Reviews in Biotechnology*, **33**(1), 49–65, https://doi.org/10.3109/07388551.2012.668671

Jevremović D. and Boley D. (2013). Finding minimal generating set for metabolic network with reversible pathways. *Biosystems*, **112**(1), 31–36, https://doi.org/10.1016/j.biosystems.2013.02.003

Kaleta C., de Figueiredo L. F. and Schuster S. (2009). Can the whole be less than the sum of its parts? Pathway analysis in genome-scale metabolic networks using elementary flux patterns. *Genome Research*, **19**(10), 1872–1883, https://doi.org/10.1101/gr.090639.108

Kim W. J., Kim H. U. and Lee S. Y. (2017). Current state and applications of microbial genome-scale metabolic models. *Current Opinion in Systems Biology*, **2**, 10–18, https://doi.org/10.1016/j.coisb.2017.03.001

Knoop H., Zilliges Y., Lockau W. and Steuer R. (2010). The metabolic network of *Synechocystis* sp. PCC 6803: systemic properties of autotrophic growth. *Plant Physiology*, **154**(1), 410–422, https://doi.org/10.1104/pp.110.157198

Küken A., Wendering P., Langary D. and Nikoloski Z. (2021). A structural property for reduction of biochemical networks. *Scientific Reports*, **11**(1), 17415, https://doi.org/10.1038/s41598-021-96835-1

Lotz K., Hartmann A., Grafahrend-Belau E., Schreiber F. and Junker B. H. (2014). Elementary flux modes, flux balance analysis, and their application to plant metabolism. *Methods in Molecular Biology*, **1083**, 231–252, https://doi.org/10.1007/978-1-62703-661-0_14

Mairet F. and Bernard O. (2019). Twelve quick tips for designing sound dynamical models for bioprocesses. *PLOS Computational Biology*, **15**(8), e1007222, https://doi.org/10.1371/journal.pcbi.1007222

Martínez C., Mairet F. and Bernard O. (2018). Theory of turbid microalgae cultures. *Journal of Theoretical Biology*, **456**, 190–200, https://doi.org/10.1016/j.jtbi.2018.07.016

Mata T. M., Martins A. A. and Caetano N. S. (2010). Microalgae for biodiesel production and other applications: a review. *Renewable and Sustainable Energy Reviews*, **14**(1), 217–232, https://doi.org/10.1016/j.rser.2009.07.020

Maton M., Bogaerts P. and Van de Wouwer A. (2022). A systematic elementary flux mode selection procedure for deriving macroscopic bioreaction models from metabolic networks. *Journal of Process Control*, **118**, 170–184, https://doi.org/10.1016/j.jprocont.2022.09.002

Oddsdóttir HÆ, Hagrot E., Chotteau V. and Forsgren A. (2015). On dynamically generating relevant elementary flux modes in a metabolic network using optimization. *Journal of Mathematical Biology*, **71**(4), 903–920, https://doi.org/10.1007/s00285-014-0844-1

Orth J. D., Thiele I. and Palsson B. Ø. (2010). What is flux balance analysis? *Nature Biotechnology*, **28**(3), 245–248, https://doi.org/10.1038/nbt.1614

Pessi B. A., Baroukh C., Bacquet A. and Bernard O. (2023). A universal dynamical metabolic model representing mixotrophic growth of *Chlorella* sp. on wastes. *Water Research*, **229**, 119388, https://doi.org/10.1016/j.watres.2022.119388

Pfeiffer T., Sánchez-Valdenebro I., Nuño J. C., Montero F. and Schuster S. (1999). METATOOL: for studying metabolic networks. *Bioinformatics*, **15**(3), 251–257, https://doi.org/10.1093/bioinformatics/15.3.251

Provost A. and Bastin G. (2004). Dynamic metabolic modelling under the balanced growth condition. *Journal of Process Control*, **14**(7), 717–728, https://doi.org/10.1016/j.jprocont.2003.12.004

Röhl A. and Bockmayr A. (2017). A mixed-integer linear programming approach to the reduction of genome-scale metabolic networks. *BMC Bioinformatics*, **18**(1), 2, https://doi.org/10.1186/s12859-016-1412-z

Röhl A. and Bockmayr A. (2019). Finding MEMo: minimum sets of elementary flux modes. *Journal of Mathematical Biology*, **79**(5), 1749–1777, https://doi.org/10.1007/s00285-019-01409-5

Schade S. and Meier T. (2019). A comparative analysis of the environmental impacts of cultivating microalgae in different production systems and climatic zones: {A} systematic review and meta-analysis. *Algal Research*, **40**, 101485, https://doi.org/10.1016/j.algal.2019.101485

Schuster S., Dandekar T. and Fell D. A. (1999). Detection of elementary flux modes in biochemical networks: a promising tool for pathway analysis and metabolic engineering. *Trends in Biotechnology*, **17**(2), 53–60, https://doi.org/10.1016/S0167-7799(98)01290-6

Singh D. and Lercher M. J. (2020). Network reduction methods for genome-scale metabolic models. *Cellular and Molecular Life Sciences*, **77**(3), 481–488, https://doi.org/10.1007/s00018-019-03383-z

Storn R. and Price K. (1997). Differential evolution – a simple and efficient heuristic for global optimization over continuous spaces. *Journal of Global Optimization*, **11**(4), 341–359, https://doi.org/10.1023/A:1008202821328

Tabe-Bordbar S. and Marashi S.-A. (2013). Finding elementary flux modes in metabolic networks based on flux balance analysis and flux coupling analysis: application to the analysis of *Escherichia coli* metabolism. *Biotechnology Letters*, **35**(12), 2039–2044, https://doi.org/10.1007/s10529-013-1328-x

Tan X. B., Lam M. K., Uemura Y., Lim J. W., Wong C. Y. and Lee K. T. (2018). Cultivation of microalgae for biodiesel production: a review on upstream and downstream processing. *Chinese Journal of Chemical Engineering*, **26**(1), 17–30, https://doi.org/10.1016/j.cjche.2017.08.010

Terzer M. and Stelling J. (2008). Large-scale computation of elementary flux modes with bit pattern trees. *Bioinformatics*, **24**(19), 2229, https://doi.org/10.1093/bioinformatics/btn401

Tibocha-Bonilla J. D., Zuñiga C., Godoy-Silva R. D. and Zengler K. (2018). Advances in metabolic modeling of oleaginous microalgae. *Biotechnology for Biofuels*, **11**(1), 241, https://doi.org/10.1186/s13068-018-1244-3

Williams P. J. L. B. and Laurens L. M. (2010). Microalgae as biodiesel & biomass feedstocks: review & analysis of the biochemistry, energetics & economics. *Energy & Environmental Science*, **3**(5), 554–590, https://doi:10.1039/B924978H

doi: 10.2166/9781789063547_0031

Chapter 3

Wastewater treatment using microalgal–bacterial consortia in the photo-activated sludge process

A. Rada-Ariza, C. Lopez-Vazquez, N. P. Van der Steen and P. N. L. Lens*

IHE Delft Institute for Water Education, Westvest 7, 2601, DA Delft, the Netherlands
**Corresponding author: p.vandersteen@un-ihe.org*

ABSTRACT

Nitrogen-rich wastewaters (10–400 mg N/L) are produced by municipal, industrial and agricultural wastes, including effluents from anaerobic treatment processes. These represent a risk to the environment due to the high nutrient concentrations (nitrogen and phosphorous), which can cause eutrophication of water bodies, deteriorating the quality of the ecosystems. As a solution, the nitrogen removal capacity of a novel bio-treatment system, the photo-activated sludge (PAS), composed of microalgae and bacteria consortia can be applied. Photobioreactors used for the simultaneous cultivation of microalgae and bacteria under sequencing batch conditions showed that microalgal–bacterial consortia can remove ammonium 50% faster than solely microalgal consortia. The increase in ammonium removal rates is due to the action of nitrifying bacteria, supplied with oxygen produced by the algae. The microalgal–bacterial system offers the possibility of reducing the hydraulic retention time, which can decrease the large area requirements often demanded by algal systems. The SRT is the main parameter to control the efficiency of the technology. The control of the suspended solids concentration, by adjusting the SRT, influences the light penetration within the reactor, which can limit or enhance the oxygen production of the algae. The photo-activated sludge system using microalgal–bacterial consortia is a sustainable treatment option for ammonium-rich wastewaters, providing clean effluents and opening reuse options for the biomass.

3.1 MICROALGAL–BACTERIAL CONSORTIA

3.1.1 Use of microalgal–bacterial consortia in environmental technologies

Microalgae and bacteria co-habit in freshwater, wastewater and marine systems. Symbiosis among aerobic bacteria and microalgae for treatment of wastewater was first reported by Oswald *et al.* (1953) in oxidation ponds. One of the interactions reported is the exchange of oxygen: the oxygen produced by the microalgae, through photosynthesis, is used by aerobic bacteria (heterotrophic and nitrifiers) to oxidize organic matter and ammonium (Figure 3.1). Heterotrophic bacteria produce carbon dioxide through respiration and oxidation of organic matter, which can be taken up as a carbon source by the microalgae. In the case of nitrogen, after nitrate is produced, it can be taken up by microalgae as a source of nitrogen, or further denitrified by bacteria when anoxic conditions are met, usually during dark periods, or dark zones within the reactor. These interactions create a synergistic

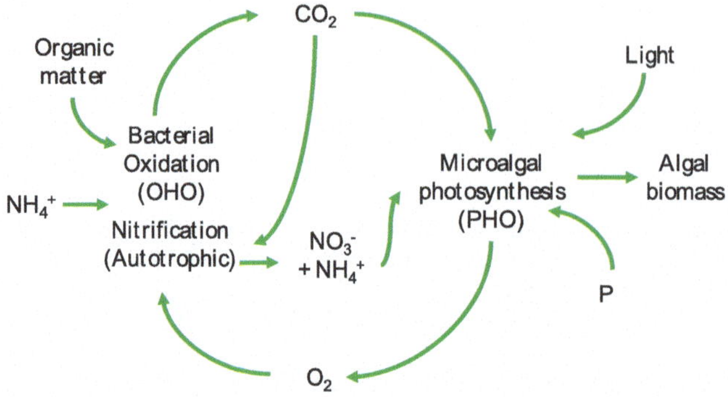

Figure 3.1 Microalgae and bacterial oxidation interactions in microalgal–bacterial consortia. (Source: adapted from Muñoz and Guieysse, 2006). OHO: heterotrophic organisms, PHO: phototrophic organisms and P: phosphorous.

relationship between microalgae, heterotrophs and nitrifiers in which the required oxygen is supplied by microalgae. The aeration supplied by microalgae is defined as photosynthetic oxygenation. The term was first defined by Oswald *et al.* (1953) as '*production of oxygen through the action of light on the chloroplastic tissue of microscopic green plants, growing dispersed in the aqueous medium*'.

The symbiosis occurs in waste stabilization ponds, oxidation ponds and high-rate algae ponds (HRAP). Zhou *et al.* (2006) reported removal of nutrients through nitrification/denitrification in HRAP treating rural domestic wastewater. About 50% of the nitrogen was removed through nitrification/denitrification, followed by algae assimilation and sedimentation. In the case of phosphorus, the main removal mechanisms were through algae assimilation followed by chemical precipitation.

Additional to the removal of nutrients, a consortium of algae and bacteria is able to remove hazardous pollutants, as reviewed by Muñoz and Guieysse (2006). Pollutants such as acetonitrile were found to be removed at a rate of 2300 mg/L/d by a consortium of *Chlorella sorokiniana* and a bacterial consortium suspended in a stirred tank reactor. Safonova *et al.* (2004) reported the removal of different xenobiotic compounds through a consortium of algae and bacteria. They observed different removal efficiencies for phenols (85%), anionic surfactants such as secondary alkane sulfonates (73%), oil spills (96%), copper (62%), nickel (62%), zinc (90%), manganese (70%) and iron (64%). The consortia used consisted of the algal strains *Chlorella* sp., *Scenedesmus obliquus*, *Stichococcus* and *Phormidium* sp. and of bacterial strains such as *Rhodococcus* sp., *Kibdelosporangium aridium* and two other unidentified bacterial strains. The removal mechanisms were the association between the oil-degrading bacteria and the algal strains, the ability of algae to supply oxygen and at the same time the ability of aerobic bacteria to degrade hydrocarbons.

3.1.2 Interactions within microalgal–bacterial consortia

The interactions between algae and bacteria are not limited to the exchange of carbon dioxide and oxygen. On the opposite, the interactions can be either mutualism, parasitism or commensalism (Ramanan *et al.*, 2016). As a result, algae and bacteria are able to change their physiology and metabolism (Ramanan *et al.*, 2016).

There are several studies showing the benefits and negative effects of bacteria and algae when present in consortia (Unnithan *et al.*, 2014). Algae can either promote bacterial growth through the release of organic exudates (Abed *et al.*, 2007), nutrient exchange as result of algal lysis (Unnithan *et al.*, 2014), or decreased algal growth through the release of algicidal substances by bacteria (Fukami *et al.*, 1997) and/or pH fluctuations as a result of the photosynthesis. Kirkwood *et al.* (2006) reported how the production of exudates by cyanobacteria did not completely inhibit bacterial growth, but instead were used as substrate in a consortium of heterotrophic bacteria and cyanobacteria treating pulp and paper wastewater. In addition, the study revealed that the exudates also enhanced the removal of dichloroacetate and at the same time affected the removal of phenolic compounds.

Choi *et al.* (2010) reported the negative effect of cyanobacteria on the nitrification rates in a bioreactor growing only nitrifiers. The presence of algae and cyanobacteria in the autotrophic bioreactor inhibited the maximum nitrification by a factor of 4, however, the ammonium was still efficiently removed (Choi *et al.*, 2010). Other negative effects of microalgae on bacteria are the increase in pH due to the photosynthetic activity and high dissolved oxygen concentration. The fast growth rate of microalgae can create a high density in the culture that led to the increase of dark zones, in which microalgae can perform respiration and diminish the amount of oxygen for bacteria (Muñoz & Guieysse, 2006).

On the contrary, there are also microalgae growth-promoting bacteria. As the name states, these bacteria enhance the growth of microalgae. de Bashan *et al.* (2004) demonstrated how the bacterium *A. brasilense* boosted the growth of *Chlorella sorokiniana*, which lead to an effluent with less nitrogen and phosphorus. Additionally, the consumption of oxygen by the aerobic bacteria helps to prevent oxygen saturation conditions.

The presence of bacteria in microalgal cultures improves the flocculation of suspended algae. Some studies have reported the improvement in the settling characteristics of the biomass in microalgal–bacterial cultures through the formation of granules or aggregates (Gutzeit *et al.*, 2005; Lee *et al.*, 2013; Van Den Hende *et al.*, 2014). The formation of flocs in an algal–bacterial consortium is promoted by the bacterial exopolymers, increasing the aggregation and stabilizing the already existing aggregates, while increasing settleability (Subashchandrabose *et al.*, 2011). Algal–bacterial flocs vary from 50 μm to 1 mm, but the predominant size is between 400 and 800 μm (Gutzeit *et al.*, 2005). Tiron *et al.* (2017) reported the development of granules or as the author calls them 'activated algae flocs', for this already formed algal flocs and the bacterial population already present in the raw dairy wastewater were used as inoculum. The developed activated algae granules had a size between 600 and 2000 μm and a settling velocity of 21.6 (±0.9) m/h (Tiron *et al.*, 2017). Figure 3.2 presents an example of an activated algae granule. This positive effect tackles one of the drawbacks of solely algal systems: efficient biomass harvesting. Tiron *et al.* (2017) show that the formation of the granules was achieved

Figure 3.2 Algae granules containing the algae strains: *Chlorella* sp. and *Phormidium* sp. (Source: Tiron *et al.*, 2017).

in a 1.5 L photobioreactor operated as sequencing batch using diluted pretreated dairy wastewater (15.3–21.8 mg NH_4^+-N/L) with an HRT between 96 and 24 hours.

Despite some of the negative interactions, a consortium of microalgae and bacteria enhances the removal of nutrients and other pollutants. The synergistic relationship provides sturdiness to overcome extreme environmental conditions and fluctuations due to operational changes. The complexity of these interactions needs to be understood in order to maximize the positive effects to develop culture conditions that enhance wastewater treatment.

3.1.3 Nutrient removal by microalgal–bacterial consortia

The main difference between an algal system and a microalgal-bacterial consortium in terms of nitrogen removal is the removal pathway. In algal systems, assimilation into the biomass and ammonium volatilization due to pH fluctuations are the two main removal mechanisms. In microalgal-bacterial consortia these are not the only removal mechanisms, but another important pathway of nitrogen removal is nitrification, as nitrifiers can make use of the oxygen produced by the microalgae (Karya *et al.*, 2013). The exchange of oxygen and carbon dioxide allows the efficient removal of organic matter and nitrogen by heterotrophic and nitrifying bacteria. Furthermore, open and closed photobioreactors contain dark zones in which anoxic conditions allow denitrification by anoxic heterotrophic (denitrifying) bacteria.

Phosphorus can be removed from the water either by chemical or microbiological mechanisms. Like nitrogen, phosphorus is an essential nutrient for microalgae. Phosphorus is taken up by algae preferably in the forms of $H_2PO_4^-$ and HPO_4^{2-} and incorporated into the cell through phosphorylation (transformation into high energy organic compounds) (Martínez *et al.*, 1999). However, there is no clear description in the literature about how the phosphorous removal is achieved in waste stabilization ponds, as the reasons are not well understood (Powell *et al.*, 2008). The chemical mechanism of phosphorus removal is through precipitation. This mechanism depends on the pH and the dissolved oxygen concentration in the bulk liquid. At high pH and dissolved oxygen concentrations, phosphorus will precipitate (Cai *et al.*, 2013). de Bashan and Bashan (2004) presented a review of the different forms of phosphorus precipitation. Usually it can occur at pH higher than 9, depending on the concentrations of the different ions and phosphorus. Due to the fact that phosphorus does not exist in gaseous form (like atmospheric nitrogen which eventually could be fixed by algae) and that it can be easily bound with other ions, it is the most important growth limiting factor in microalgae cultivation, besides light (Grobbelaar, 2008). Phosphorus assimilation is the main biological mechanism of removal in algal systems. Di Termini *et al.* (2011) achieved phosphorus removal between 80 and 90% in outdoor and indoor closed photobioreactors through microalgae assimilation.

Several authors have reported the use of microalgal–bacterial consortia for nutrient (nitrogen and phosphorous) removal from real or synthetic wastewater using different types of photobioreactors (Subashchandrabose *et al.*, 2011; Zhang *et al.*, 2018). The different studies showed nitrogen removal efficiencies were between 100% and 15%, whereas the phosphorous removal efficiencies were between 90% and 31.5% (Subashchandrabose *et al.*, 2011).

The symbiosis between microalgae and bacteria offers a large potential for the treatment of nutrient-rich wastewaters, although some aspects need to be taken into account, as they determine the nutrient removal efficiencies or the nutrient removal pathways. The selection of a particular strain for wastewater treatment is a decisive step when engineering a consortium of microalgae and bacteria. In open ponds, there is a natural selection of the microalgae species, which depends on the organic load of the wastewater, species interactions, seasonal environmental conditions, competition and interactions among the microorganisms present in the culture (Riaño *et al.*, 2012). Natural selection of microalgae within a microalgal–bacterial consortium allows to achieve higher efficiencies as there are no inhibitory effects by the source of the wastewater.

González-Fernández *et al.* (2011) compared the removal efficiency of four ponds using microalgal–bacterial consortia for the treatment of pig slurry. The ponds differed in terms of operational conditions (optimal and real conditions) and source of the slurry (anaerobically digested or fresh). The three reported removal mechanisms were nitrification/denitrification, stripping and biomass uptake. Among these three, the main driving force of removal depended on the substrate source. The NH_4^+-N/COD ratio of the substrates was responsible for the different removal rates and the main removal pathway. The anaerobic digested slurry had a ratio of 0.46 NH_4^+-N/COD, whereas the fresh slurry had a NH_4^+-N/COD ratio of 0.13. Since the organic matter in the anaerobically digested slurry is more recalcitrant, the oxygen is more likely taken up for nitrification, the reason why nitrification rates were higher for ponds fed with anaerobically digested slurry (González-Fernández *et al.*, 2011).

Molinuevo-Salces *et al.* (2010) compared open and close configurations and the results showed that even though ammonium was completely removed, the removal mechanisms were different. In the open configuration the biomass uptake was between 38 and 47%, while 52–29% was nitrified/denitrified. In the closed reactor 10.5% was volatilized and 11.3% nitrified, 41% nitrified/denitrified and 31.3% taken up by algae (Molinuevo-Salces *et al.*, 2010). About 80% of the phosphorous was removed regardless of the configuration.

Ammonium removal through nitrification/denitrification as the main removal mechanism in microalgal–bacterial systems has the advantage of achieving faster removal rates in comparison with solely algal systems, especially for high concentrated effluents from industrial sectors. Wang *et al.* (2015) used microalgal–bacterial consortia to treat anaerobically digested swine manure with ammonium concentrations up to 297 (±29) mg NH_4^+-N/L (value after 3 times dilution) in a sequencing batch photobioreactor (4 days hydraulic retention time), achieving a 90% total nitrogen (TN) removal efficiency, from which 80% was removed through nitritation/denitritation without any external aeration. Furthermore, Manser *et al.* (2016) reported the successful combination of microalgae, ammonium-oxidizing bacteria (AOB) and anammox in a sequencing batch photobioreactor achieving ammonium oxidation to nitrite at a rate of 7.0 mg NH_4^+-N/L/h in the light periods and during the night periods in which anoxic conditions were achieved, about 82% of the nitrite was reduced by anammox bacteria (Table 3.1).

Table 3.1 Nutrient removal using microalgal–bacterial consortia for different types of wastewater and using different types of reactors.

Cyanobacterium/ Microalga	Bacterium	Source of Waste Water	Nutrients and Removal Efficiency	System-Reactor Used
Spirulina platensis	Sulphate-reducing bacteria	Tannery effluent	Sulphate 80% (2000 mg/L)	HRAP
Chlorella vulgaris	*Azospirillum brasilense*	Synthetic wastewater	Ammonia 91% (21 mg/L)	Chemostat
			Phosphorous 75% (15 mg/L)	
Chlorella vulgaris	Wastewater bacteria	Pretreated sewage	DOC 93% (230 mg C/L	Photobioreactor pilot-scale
			Nitrogen 15% (78.5 mg/L)	
			Phosphorous 47% (10.8 mg/L)	
Chlorella vulgaris	*Alcaligenes* sp.	Coke factory wastewater	NH_4^+ 45% (500 mg/L)	Continuous photobioreactor with sludge recirculation
			Phenol 100% (325 mg/L)	

(Continued)

Table 3.1 Nutrient removal using microalgal–bacterial consortia for different types of wastewater and using different types of reactors (*Continued*).

Cyanobacterium/ Microalga	Bacterium	Source of Waste Water	Nutrients and Removal Efficiency	System-Reactor Used
Chlorella vulgaris	*A. brasilense*	Synthetic wastewater	Phosphorous 31.5% (50 mg/L) Nitrogen 22% (50 mg/L)	Inverted conical glass bioreactor
Chlorella sorokiniana	Mixed bacterial culture from an activated sludge process	Synthetic wastewater	Phosphorous 86% (15 mg/L) Nitrogen 99% (180 mg/L)	Tubular biofilm photobioreactor
Chlorella sorokiniana	Activated sludge bacteria	Pretreated piggery wastewater	TOC 86% (645 mg/L) Nitrogen 87% (373 mg/L)	Glass bottle
Chlorella sorokiniana	Activated sludge bacteria	Pretreated swine slurry	TOC 9–61% (1247 mg/L) Nitrogen 94–100% (656 mg/L) Phosphorous 70–90% (117 mg/L)	Tubular biofilm photobioreactor
Chlorella sorokiniana	Activated sludge bacteria	Piggery wastewater	TOC 47% (550 mg/L) Phosphorous 54% (19.4 mg/L) NH_4^+ 21% (350 mg/L)	Jacketed glass tank photobioreactor
Euglena viridis	Activated sludge bacteria	Piggery wastewater	TOC 51% (450 mg/L) Phosphorous 53% (19.4 mg/L) NH_4^+ 34% (320 mg/L)	Jacketed glass tank photobioreactor
Microalgae present in tertiary stabilization pond treating domestic wastewater	Bacteria present in tertiary stabilization pond treating domestic wastewater	Piggery wastewater	COD 58.7% (526 mg/L) Total Kjeldahl nitrogen 78% (59 mg/L)	HRAP

Source: Subashchandrabose *et al.* (2011).

3.1.4 Microalgal–bacterial systems and configurations

Algal wastewater treatment systems can be divided into open and clos photobioreactors. According to the reactor geometry, closed photobioreactors can be divided into: (1) vertical columns, (2) tubular reactors and (3) flat panel reactors (Wang *et al.*, 2012). Open reactors can be listed into: (a) waste stabilization ponds (WSP), (b) raceway ponds and (c) HRAP. Figure 3.3 presents a scheme of the three most used photobioreactors for algal cultivations. Currently, open systems are the most used type for wastewater treatment and biomass cultivation using microalgae (Carvalho *et al.*, 2006;

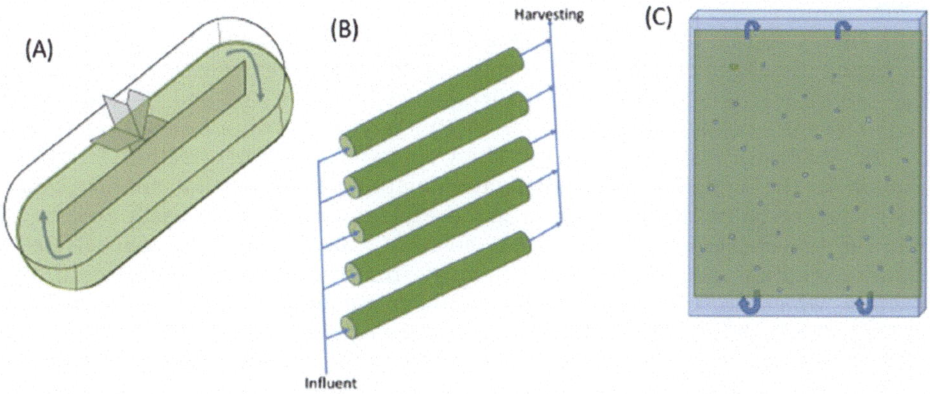

Figure 3.3 The three most used algal system configurations. (a) high-rate algae pond, (b) closed tubular photobioreactor and (c) flat panel airlift reactor. (Source: Wang *et al.*, 2018).

Wang *et al.*, 2012) due to their low investment and maintenance cost and easiness to scale up (Cai *et al.*, 2013). Closed systems are mostly used for sensitive microalgae strains, products vulnerable to microbial degradation or when the harvested biomass is aimed at direct human consumption such as for cosmetics or nutritional supplements (Carvalho *et al.*, 2006). Closed systems have a higher light harvesting, thus biomass production can achieve a higher population density; however, the investment and maintenance costs are higher compared with open systems (Carvalho *et al.*, 2006).

HRAP are the most efficient open systems as they are operated with a higher depth in comparison with the other options. HRAP are raceway-type ponds with depths between 0.2 and 1 m. They can treat up to 35 g BOD/m^2/d compared with 5–10 BOD/m^2/d in waste stabilization ponds (Muñoz & Guieysse, 2006). However, light penetration in such reactors is limited by the depth or solids concentration. Furthermore, open and close systems both require large areas for operation in order to either efficiently remove the contaminants or to achieve high biomass production. Therefore, the reactor selection and the growth medium composition depend on the objective of the system.

3.1.5 Limiting and operational conditions of microalgal–bacterial photobioreactors

There are several factors that can affect the growth of algae and bacteria, especially when using wastewater as growth medium, since there are many substances, compounds and factors to take into account. In open and close photobioreactors there are physical, chemical, biological and operational factors that can limit the growth of microalgae (Borowitzka, 1998). Among those, the parameters that have a strong effect on the efficiency of microalgae and bacteria when treating wastewater are: pH, light intensity, temperature, dissolved carbon dioxide, nutrients, mixing, dilution and algae harvesting (Borowitzka, 1998; Rawat *et al.*, 2011).

In terms of operation, different operational parameters have an effect on the cultivation of microalgae and bacteria separately. Therefore, special attention should be given when combining these two groups of microorganisms. One of the most critical operational parameters is the biomass retention time, which in the case of a consortium can be determined by the influent flow rate and whether there is biomass recirculation. Solid retention time (SRT) and hydraulic retention time (HRT) influence the biomass concentration and the overall productivity of the microalgal–bacterial systems (Valigore *et al.*, 2012). This PhD research study focused on open photobioreactors such as HRAP. For this reason, the implications of some of the factors limiting algal and bacterial growth in high rate open algal ponds are described below.

3.1.5.1 Light

Light is the energy source to perform photosynthesis, allowing microalgae growth. Hence, the uptake efficiency of light is crucial for the productivity of algal biomass and photo-oxygenation. Microalgae can absorb only a fraction of the irradiance, between 400 and 700 nm. This range is called the photosynthetically active radiation (PAR). Open ponds obtain this irradiance from the sun hence, the ponds are shallow in order to allow a maximal light penetration. Height is not the only limitation for the light irradiance, attenuation by the biomass itself is another factor, which can increase when co-cultured with bacteria, and the fact that light can be easily absorbed by other materials or substances (Fernández *et al.*, 2013; Jeon *et al.*, 2005). Dense and concentrated cultures present mutual shading, reducing the light intensity from the illuminated surface to the centre of the reactors, which increase the dark zones and consequently microalgal respiration (Chen *et al.*, 2011; Fernández *et al.*, 2013). Due to this, microalgae are exposed to light/dark zones. For instance, in open ponds except for the upmost thin layer, the irradiance in the pond is below the photo-compensation point for algal growth (Barbosa *et al.*, 2003), as a result of this photosynthetic rates decrease, as well as algal growth. This effect can be compensated by a good mixing which allows the cells to be exposed to a sufficient amount of irradiance (Chen *et al.*, 2011). In open ponds, usually the mixing is provided by a paddle wheel, while aeration is usually applied in closed photobioreactors.

Indoor cultures and closed photobioreactors use other sources of light different from sunlight. For instance, high-pressure sodium lamps, tungsten–halogen lamps, fluorescent tubes and light-emitting diodes (LED lights). Although, these lamps provide a reliable source of energy, the disadvantages are the high power consumption and high operational costs, and they do not contain the full spectrum of light energy (Chen *et al.*, 2011). On the contrary, sunlight is free and holds the full spectrum of light energy.

3.1.5.2 pH

pH is one of the most important parameters in microalgal cultures, as it determines the solubility of carbon dioxide, removal of other nutrients like P and N, and most importantly it affects the metabolism of the microalgae (Becker, 1994). Furthermore, pH fluctuations can inhibit bacterial activity such as autotrophic and heterotrophic bacteria. Fluctuations of pH in microalgae cultures are a consequence of the processes of photosynthesis and respiration during the light and dark periods, respectively. During the day, the pH increases due to the assimilation of CO_2 and the release of OH^-. pH values of up to 10 have been reported after the depletion of NO_3^- and CO_2 (Becker, 1994). Increments of the pH are limited in some cases by the respiration of the different microorganisms. Additionally, nitrogen removal through nitrification has an effect on the pH fluctuations, since the pH decreases during this process due to the release of H^+. Therefore, the addition of ammonium can help to reduce the pH increment (Larsdotter, 2006), making it a good option for pH control in open ponds. Also, the addition of CO_2 can help to control the pH as shown by Park and Craggs (2010).

pH values can affect the growth of microalgae and therefore the removal of nutrients, this can vary for the different strains. Some algae such as *Microcystis aeruginosa* and *Anabena spiroides* have growth limitations and inhibition when exposed to a pH below 6 (Wang *et al.*, 2011). pH fluctuations can also determine the removal of N and P, as higher pH causes ammonium volatilization and phosphorus precipitation. When this occurs faster than the uptake by algae, it leads to algal growth limitation due to the lack of nutrients. Therefore, pH control strategies must be developed in order to avoid possible negative effects caused by drastic pH fluctuations.

In the case of nitrifiers, the growth is suppressed when the pH is not within the 7–8 range (Ekama & Wentzel, 2008a). Nitrification performed by aerobic bacteria release hydrogen ions, reducing the alkalinity of the bulk liquid. Stoichiometrically, for every 1 mg free and saline ammonia (FSA) nitrified, 7.14 mg alkalinity ($CaCO_3$) is consumed (Ekama & Wentzel, 2008a). When alkalinity is lower than 40 mg/L in activated sludge systems, the pH decreases to low values, compromising the nitrification rates and settleability characteristics of the sludge (Ekama & Wentzel, 2008a). In systems working with algae and bacteria, the pH drop by nitrification can be counterbalanced by photosynthetic activity. Also denitrification recovers alkalinity, which occurs under anoxic conditions. In algal–bacterial systems, dark conditions guarantee the absence of oxygen production by algae, instead algae respire releasing CO_2, which helps to decrease the pH. Based on this, it is evident that the balance in terms of alkalinity between microalgae and bacteria is important.

3.1.5.3 Hydraulic retention time

Hydraulic retention time controls the nutrient loading rates, which at the same time will control the productivity and nutrient removal rate of an algae system. In an open pond with well mixed and steady-state conditions, the productivity is governed by the dilution rate and the depth of the pond. The HRT corresponds to the reciprocal of the dilution rate. In algal ponds and HRAP, the HRT is the same as the solids retention time (SRT), since it is not common to recirculate the biomass, as the harvesting of algal biomass is one of the biggest challenges due to their low cell size (Lee *et al.*, 2013). Therefore, in order to achieve complete removal rates of pollutants, it is common practice to operate algal systems at a HRT between 2 and 8 days and depths between 0.2 and 0.5 m (Shilton, 2006). Due to seasonal variations, it is recommended to vary the HRT, as the temperature changes limit or enhance the growth rates.

Furthermore, shorter HRT in algal systems enhance the biomass production (Oswald *et al.*, 1953; Takabe *et al.*, 2016). Valigore *et al.* (2012) compared different HRT (from 8 to 1.4 days) in a microalgal–bacterial culture, concluding that a shorter HRT enhanced the biomass productivity. However, a shorter HRT can decrease the nutrient removal rates in microalgal–bacterial systems, especially when it can promote wash out of the biomass. An optimum HRT enhances nutrient removal by allowing the proper growth of algal–bacterial populations, which will promote faster nitrification rates, especially since the growth rate of nitrifying microorganisms is low, that is $\mu_m = 0.45$ per d at 20°C (Ekama & Wentzel, 2008a). Therefore, the HRT must be chosen depending on the objective, whether the maximization of the biomass production or the treatment of wastewater. Also, it must be taken into account that due to the depth of the HRAP, a longer HRT will result in larger areas, therefore, optimization of this parameter is crucial for algal systems.

3.1.5.4 Solid retention time

When working with a consortium of microalgae and activated sludge bacteria for nutrient and organic matter removal through photo-oxygenation, the sludge retention time plays an important role within the operational parameters. In fact, it is the most fundamental and important decision for the design of activated sludge systems (Ekama & Wentzel, 2008b). Sludge retention time controls the growth of the microorganisms and corresponds to the relation between the volume of the reactor and the waste biomass flow from the reactor. Therefore, the sludge production in activated sludge systems decreases with the increase of the SRT (Ekama & Wentzel, 2008b). On the other hand, for suspended algae systems, the algae biomass production is controlled by the HRT. This parameter controls the biomass concentrations, which will affect the light utilization by microalgae (Lambeert–Beer law).

3.1.5.4.1 Biomass

Figure 3.4 presents the productivity curve for a flat panel reactor for different biomass concentrations and light intensity. The optimal concentration ($C_{x,opt}$), where the biomass production is at the maximum, will depend on the efficient use of light. This is achieved when the light at the back of the reactor equals the compensation point for microalgae growth. For lower concentrations, the light will pass through the reactor un-used, whereas for higher values, the light will not be able to reach the bottom/back of the photobioreactor (Janssen & Lamers, 2013). Therefore, there is a need for optimum SRT and HRT combinations to achieve a microalgal–bacterial biomass concentration that allows complete nitrification by ensuring sufficient oxygen without biomass wash-out.

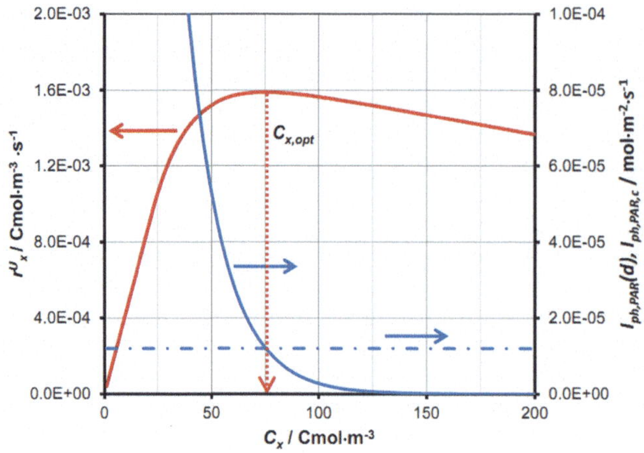

Figure 3.4 Volumetric productivity of a photobioreactor r_x^u as a function of the biomass concentration C_x. Light intensity at the back of the reactor $I_{ph,PAR}$ (d) and the compensation light intensity $I_{ph,PARc}$, are also shown. (Source: Janssen and Lamers, 2013).

3.1.5.4.2 Nitrogen removal

Valigore *et al.* (2012) concluded that biomass recycling at a SRT higher than the HRT reduces the wash-out of the microorganisms present in the reactor. Therefore, an appropriate SRT will ensure the successful growth of nitrifiers (slower growing microorganisms in activated sludge) and in addition guarantees light availability for photo-oxygenation. The recommended ranges of SRT values for complete nitrification are divided in two: (1) intermediate, between 10 and 15 days, this range ensures complete nitrification and (2) long sludge age refers to more than 20 days, for which the production of sludge is low with a rather inactive sludge (Ekama & Wentzel, 2008b).

Rada-Ariza *et al.* (2017) showed that the uncoupling of the SRT and HRT is imperative for the development of a steady nitrifying microalgal–bacterial consortium. Furthermore, Arashiro *et al.* (2016) and Rada-Ariza *et al.* (2019) showed the effects of the SRT on the removal mechanism of microalgal–bacterial consortia, still the ammonium removal efficiency was 100% under the different operational conditions tested. In both studies, volumetric and specific ammonium removal rates were higher at shorter SRTs (17 days SRT for Arashiro *et al.* (2016) and 7 days SRT for Rada-Ariza *et al.* (2019)). Furthermore, the ammonium removal mechanisms differ at different durations of the SRT. In Arashiro *et al.* (2016), at a longer SRT of 52 days, ammonium removal by algal uptake represented up to 38% of the total ammonium removal, while it decreased up to 11% at an SRT of 17 days. In both cases, the main ammonium removal mechanism was nitrification/denitrification.

One of the most important operational parameters to control the efficiency and rates of ammonium removal in microalgal–bacterial consortia is the SRT. The SRT controls the amount of solids in the reactor, which will have a high impact on the light penetration used for algal growth and consequently oxygen production. Longer SRTs in activated sludge increase the concentration of endogenous residues, which reduce the active fraction of the biomass and increase the oxygen consumption through respiration of the bacterial biomass (Ekama & Wentzel, 2008b). In addition, longer SRTs increase the solids concentration in the reactor, hence the dark zones within the reactor increase, which will also increase the oxygen consumption by algal respiration. As a result, oxygen is less available for the aerobic processes such as organic carbon oxidation and nitrification, resulting in a shift in the removal mechanism from nitrification to algal uptake. However, if the HRT is not long enough and the ammonium concentration in the influent is high, the efficiency of the system could be hindered and both high concentrations of nitrite and ammonium (partial nitrification and no denitritation) and organic carbon can end up in the effluent.

The uncoupling of the SRT from the HRT permits to select an optimum SRT that allows enough light penetration to maximize the nitrification rates and reduce the solids concentration. This will decrease the endogenous residue by the bacterial biomass, while at the same time increase the growth rate of the nitrifiers (Ekama & Wentzel, 2008a). Decreasing of the SRTs and increasing the ammonium removal rates can help to further decrease the HRT, which would as well offer the possibility to reduce the area requirement of the technology as stated above. However, HRTs shorter than 0.5 days have, to the best of our knowledge, not yet been tested. Therefore, further research studies are required to demonstrate the feasibility of this low HRT. Furthermore, it is imperative to not fall below the SRT_{min} for nitrifiers, since below this value, nitrifiers would be washed out of the system and the system would collapse. Finally, based on the experiments conducted by Arashiro *et al.* (2016) and Rada-Ariza *et al.* (2019), the optimum SRTs for microalgal–bacterial reactors is between 5 and 10 days.

3.1.5.4.3 Biomass retention

The sludge retention time also plays a role in the floc formation, since longer SRT and biomass recirculation enhances the biomass settleability and floc formation (Gutzeit *et al.*, 2005; Medina & Neis, 2007; Valigore *et al.*, 2012). It was reported that settleability of algal–bacterial biomass increased from 13 to 93% when the SRT increases up to 40 days (Valigore *et al.*, 2012). Additionally, Gutzeit *et al.* (2005) achieved during a period of 18 months a flocculent algal–bacterial biomass with excellent sedimentation characteristics, using a SRT between 20 and 25 days. On the contrary, longer SRT promotes algal death due to high solids concentrations, which limits the light penetration and creates higher dark zones increasing the respiration activity (Oswald *et al.*, 1953). Since HRT and SRT can operationally define the removal rate, biomass characteristics and productivity, it is essential to further investigate different conditions of these two in order to define the operational conditions for novel algal–bacterial-based wastewater treatment systems.

3.2 ADVANTAGES OF MICROALGAL–BACTERIAL CONSORTIA FOR AMMONIUM REMOVAL

3.2.1 Advantages on ammonium removal rates

Microalgal–bacterial consortia removed ammonium 50% times faster than in a solely microalgal system, which ultimately increases the efficiency of the system. Rada-Ariza *et al.* (2017) achieved the highest ammonium removal rate and specific ammonium removal rate in comparison with their other studies (Table 3.2). The main removal mechanism that contributed to the increase in the ammonium removal rates was nitrification. Furthermore, other studies have also reported the successful treatment of high strength wastewater using microalgal–bacterial cultures (de Godos *et al.*, 2010; González *et al.*, 2008; Wang *et al.*, 2015; Zhao *et al.*, 2014). The removal rates reported in Table 3.2 based on the research study of Rada-Ariza (2018) are higher than those reported by solely algal cultures treating a diverse range of ammonium concentrations in the influent (Abou-Shanab *et al.*, 2013; Aslan & Kapdan, 2006; Cabanelas *et al.*, 2013). Furthermore, the algae strains used as inoculum were a combination between eukaryotic algae and prokaryotic cyanobacteria (Rada-Ariza *et al.*, 2017). Yet, once the reactors reached steady state, the most predominant algal strain was *Chlorella*. In the literature, it can be found that the most used strains of microalgae for wastewater treatment are *Chlorella* sp. (Cabanelas *et al.*, 2013; Ruiz *et al.*, 2011), *Scenedesmus* sp. (Kim *et al.*, 2013; Park *et al.*, 2010) and *Spirulina* sp. (Olguín, 2003).

The presence of nitrifiers in the microalgal culture increased the volumetric and specific ammonium removal rates. The oxidation of ammonium by nitrifiers is faster than the algal uptake (Arashiro *et al.*, 2016). Therefore, the presence of nitrifiers in the biomass has a strong impact on the removal of ammonium despite they have a low content in the total biomass composition, between 1.8 and 17%

Table 3.2 Volumetric and specific ammonium removal rates of algal bacterial reactors under the different operational conditions tested by Rada-Ariza (2018).

Influent (mg NH_4^+/L)	r_{Am_T} (mgNH_4^+-N/L/h)	k_{Am_T} (mgNH_4^+-N mg/VSS/d)	SRT (d) and HRT (d)	Light intensity (μmol/m²/s)	Reference
297.3	4.16 ± 0.75	1.84 ± 0.12	SRT: 4.2 ± 0.3 HRT: 1	700	Rada-Ariza *et al.* (2017)
23	2.12	0.063 ± 0.009	SRT: 17 HRT: 0.5	25.9	Rada-Ariza *et al.* (2019)
264 ± 10	2.4 ± 0.17	0.033 ± 0.002	SRT: 7 HRT: 4	84 ± 3	Arashiro *et al.* (2016)
45.36 ± 5.52	3.21 ± 0.24	0.063 ± 0.012	SRT: 10 HRT: 1	766.5 ± 154.1	Rada-Ariza (2018)

r_{Am_T}: Volumetric ammonium removal rate; k_{Am_T}: specific ammonium removal rate.

(Arashiro *et al.*, 2016; Rada-Ariza *et al.*, 2017, 2019). Also, the presence of other microorganisms played an important role in the total nitrogen removal. For instance, heterotrophic bacteria not just removed the organic carbon present in the influent, but also removed ammonium for their biomass growth (Arashiro *et al.*, 2016; Rada-Ariza *et al.*, 2019). In addition, during anoxic periods, heterotrophic bacteria, when sufficient organic carbon is present, could denitrify the nitrate or nitrite produced by nitrification (Arashiro *et al.*, 2016; Rada-Ariza *et al.*, 2019).

3.2.2 Operational conditions and area requirement

The ammonium removal rate by a reactor containing just microalgae was 1.84 (±0.66) mg NH_4^+-N/L/h and the specific ammonium removal rate was 0.025 (±0.009) mg NH_4^+-N mg/VSS/d (Rada-Ariza *et al.*, 2017). These values are significantly lower than those for the microalgal–bacterial reactors tested (Table 3.2). Thus, for 100% ammonium removal in the microalgal reactor described in Rada-Ariza *et al.* (2017) and assuming that the volumetric ammonium removal would remain similar, the required HRT would be approximately 6.7 days, assuming all other macronutrients and micronutrients are sufficient. Alcántara *et al.* (2015) calculated that in a microalgae-based system, such as HRAP treating medium-strength domestic water, the necessary HRT would be 7.5 for complete nitrogen and phosphorous removal. Higher nitrogen uptake by algae would result in a higher concentration of solids, which limits the light penetration and thus reduces the growth rate of algae. Noteworthy, HRT values in HRAP could be reduced when carbon dioxide is sparged to avoid inorganic carbon limitation. This can also help as a pH control to maintain an optimum pH. Park and Craggs (2011) obtained ammonium removal efficiencies of up to 83.3% at a HRT of 4 days with CO_2 addition in a high rate algae pond treating an effluent from anaerobic digestion. However, in HARPs with CO_2 supply, the growth of nitrifiers can be enhanced, especially when inorganic carbon is not limiting and in most cases when the HRT is not long enough for nitrifiers to grow (de Godos *et al.*, 2016; Park & Craggs, 2011). The latter occurs in conventional HRAPs where the HRT and the SRT are not uncoupled and therefore the HRT corresponds to the SRT.

The high ammonium removal rates (volumetric and specific) by microalgal–bacterial consortia can further help to reduce the HRT of the system. This can be done by ensuring that the main ammonium removal mechanism within the microalgal–bacterial system is through nitrification. Comparing the oxygen production by algae with the oxygen consumption by nitrification, the yield of oxygen on ammonium consumed is 16.85 gO_2 gNH_4^+-N^{-1} consumed (Mara, 2004). This is significantly higher than the 4.57 gO_2 gNH_4^+-N^{-1} required for complete nitrification (Ekama & Wentzel, 2008a). Therefore, the design of a microalgal–bacterial system should ensure enough oxygen production by algae to support all aerobic processes. Another important condition that should be met is the retention of nitrifiers within the system. Thus, for the cultivation of a microalgal–bacterial consortium in which nitrification is envisioned as the main removal mechanism, there should be an uncoupling between the SRT and the HRT (Rada-Ariza *et al.*, 2017; Valigore *et al.*, 2012).

The possibility of reducing further the HRT by the uncoupling between the SRT and HRT in a microalgal–bacterial system has positive effects on the nitrification process and the objective of microalgae supplying the necessary oxygen to support the aerobic processes. Also, the reduction of the HRT contributes to the reduction of the large area requirements of algal systems. Since microalgae would not be the main removal mechanisms, the limitation of light by solids should be enough to support photo-oxygenation. Therefore, the designing depths of reactors using microalgal–bacterial consortia could be deeper. The microalgal–bacterial system of Rada-Ariza *et al.* (2017) had a surface removal rate of 10.2 g $NH_4^+ - N/m^2/d$, compared with 4.4 g $NH_4^+ - N/m^2/d$ for the microalgal consortia. Comparing these values with the study of Tuantet *et al.* (2014), who achieved a maximum removal rate of 54.1 mg $NH_4^+ - N/L/h$ using urine as growth medium, the surface ammonium removal rate calculated was 6.5 g $NH_4^+ - N/m^2/d$. This value is lower than for microalgal–bacterial systems and also the reactor used for cultivation by Tuantet *et al.* (2014) had a short light path of 5 mm, which avoided any light limitation in the culture.

In practice, HRAP are designed with a HRT between 2 and 8 days and depths between 0.2 and 0.5 m (Shilton, 2006). Using the information reported by Park and Craggs (2011) in a HRAP treating domestic wastewater, the surface removal rate was estimated to be 1.1 g $NH_4^+ - N/m^2/d$, which is considerably lower than the values found in this thesis. In summary, the uncoupling of the HRT and SRT allows to develop a higher settleable biomass. Consequently, both SRT and HRT can be further shortened, which has a positive result on the light limitation by solids and on the nutrient removal rates. As a result, the depth (light path) of the reactors using microalgal–bacterial consortia, in which the main ammonium removal mechanism is through nitrification, can be further decreased, which would help reduce area requirements. Rada-Ariza *et al.* (2017) showed that the area requirements for microalgal–bacterial consortia can be reduced up to 50% in comparison with solely algal systems. Nonetheless, the rates presented in the above study were calculated based on laboratory-scale experiments and more research is required at pilot- and full-scale levels in order to define minimum depths that are able to meet the necessary oxygen production and at the same time maintain the nutrient removal efficiency of the system.

3.2.3 Photo-oxygenation and algal harvesting

Another important advantage of the use of microalgal–bacterial consortia over other technologies are the economic costs. Especially on two aspects: the cost of aeration when comparing this technology with activated sludge and the cost of harvesting when comparing with algal systems. Comparing this technology with activated sludge systems, the oxygen required for nitrification and COD oxidation is fully supported by microalgae (Rada-Ariza *et al.*, 2017, 2019). Operational costs by aeration can represent up to 60–80% (Holenda *et al.*, 2008) of the total operational costs in activated sludge plants. The energy consumption is on average between 0.33 and 0.60 kWh/m^3 in activated sludge plants in the United States (Plappally & Lienhard, 2012), while for HRAP the power consumption for mixing, calculated by Alcántara *et al.* (2015), was 0.023 kWh/m^3. Therefore, the energy needed for removal of ammonium in high strength wastewater using an activated sludge process would be considerably higher when compared with a microalgal–bacterial system.

Another advantage of the microalgal–bacterial systems is the improvement in the settling characteristics of the biomass (Arashiro *et al.*, 2016; Rada-Ariza *et al.*, 2019) when compared with algal systems. The uncoupling of the SRT and HRT, and the operation in sequencing batch creates a selective environment for fast settleable microalgae and furthermore promoted the formation of algal–bacterial aggregates. This positive effect on biomass harvesting by the presence of bacteria in algal systems has been reported by other studies as well (Gutzeit *et al.*, 2005; Park & Craggs, 2011; Van Den Hende, 2014). Furthermore, the increase in settleability reduces the cost of operation in these systems and so no extra energy is required for solids separation, such as centrifugation or dissolved air flotation. In addition, the bioflocculation avoids contamination of the biomass, since no chemicals are needed to promote flocculation (Su *et al.*, 2011).

Several studies found ways to improve this positive effect of algae and bacteria aggregation. Tiron *et al.* (2017) published an approach to develop activated algae granules which have sedimentation velocities of 21.6 (±0.9) m/h and in terms of the separation of the algal biomass from the bulk liquid, the biomass recoveries were up to 99%. Zhang *et al.* (2022a, 2022b) investigated the granulation process of algae/bacteria granules, starting from aerobic granular sludge growing on acetate-based synthetic domestic wastewater. The inoculum aerobic granular sludge size greatly affected the characteristics of the photo-granule and the optimal inoculum aerobic granular sludge size for the start-up of photo-granule process was 0.8–1.4 mm (Zhang *et al.*, 2022a). Furthermore, the granulation process could be accelerated by applying algal–mycelial pellets as nuclei for the rapid development of the symbiotic algal–bacterial granular sludge (Zhang *et al.*, 2022b).

3.3 MICROALGAL–BACTERIAL MODELLING

Modelling of processes in wastewater treatment has the advantage of getting insight into the performance of the technology, evaluation of possible scenarios for upgrading, evaluation of new plant design, support to the decision making related with operational conditions and personal training (van Loosdrecht *et al.*, 2008). Modelling of microalgae systems, more specifically for open ponds, has to take into account several factors, such as light, wind, stripping of ammonia and carbon dioxide, as well as biological and hydrodynamic processes (Gehring *et al.*, 2010). There are several models which focus on different microalgae processes, for instance on the net growth of microalgae (Decostere *et al.*, 2013; Solimeno *et al.*, 2015; Wágner *et al.*, 2016), models dealing with light limitation and photosynthesis rates (Yun & Park, 2003), kinetics of nutrient removal (Kapdan & Aslan, 2008), pigments dynamics and respiration (Bernard, 2011) and dissolved oxygen rates (Kayombo *et al.*, 2000).

In the case of activated sludge, bacteria are mostly modelled by a set of models (ASM1, ASM2 and ASM3, ASM3, ASM2d, ASM3-bio-P) developed by task groups of the International Water Association (IWA) and the metabolic model developed at Delft University of Technology (Gernaey *et al.*, 2004). The activated sludge model no. 1 (ASM1) (Henze, 2000) is considered the reference model. It describes the removal of organic carbon compounds and nitrogen, while consuming oxygen and nitrate as electron acceptors. Additionally, it describes the sludge production and has adopted the chemical oxygen demand (COD) as measurement unit for organic matter (Gernaey *et al.*, 2004). Furthermore, similar to ASM1, ASM3 was developed to correct the deficiencies of the ASM1 model. The main difference of the ASM3 model is the inclusion of the intracellular storage process of readily biodegradable COD, for the slower conversion from readily biodegradable into slowly biodegradable organic matter (Gernaey *et al.*, 2004; van Loosdrecht *et al.*, 2008). Other models include biological phosphorus removal, i.e. ASM2d and the TUDelft model (van Loosdrecht *et al.*, 2008).

As mentioned in previous sections, usually in open ponds that are treating wastewater, not only microalgae play a role in the removal of nutrients and biomass production, but at the same time, heterotrophic and nitrifying bacteria carry out different processes like oxidation of organic matter, nitrification, denitrification and respiration (Figure 3.1). Therefore, they make the system more complex as those microorganisms and their associated parameters and variables should be taken into account. Furthermore, models describing these complex relationships should be based on the microalgae models and activated sludge models. Models describing the relationships of algal–bacterial consortia in open ponds have been reported at first by Buhr and Miller (1983). Their objective was to develop a mathematical model for high-rate algal–bacterial wastewater treatment systems. This model takes into account the algal and bacterial growth, light limitation and solution equilibrium related with the pH and mass balances. The variations of pH, DO and substrate concentrations along the pond length were evaluated under different feed loads and hydraulic residence times. Later on,

Gehring *et al.* (2010) developed a model to simulate the processes in a waste stabilization pond. The activated sludge model no. 3 (ASM3) was used as a basis. The new components were the integration of algae biomass and gas transfer processes for oxygen, carbon dioxide and ammonia depending on wind velocity. Furthermore, it had the possibility to model the algae concentrations based on measured chlorophyll-*a*, light intensity and total suspended solids (TSS) measurements (Gehring *et al.*, 2010). However, modelling of nitrification and denitrification was not considered in the simulations carried out by Gehring *et al.* (2010) because the experimental data did not show any nitrification or denitrification rates. Therefore, the model was not evaluated under the two conditions of nitrification and algal growth.

In the literature some models focused on algal–bacterial consortia (Solimeno *et al.*, 2017; van der Steen *et al.*, 2015; Wolf *et al.*, 2007; Zambrano *et al.*, 2016). Solimeno *et al.* (2017) developed the BIO-ALGAE model for suspended microalgal–bacterial biomass, which was an updated version of the algal model proposed by the same author (Solimeno *et al.*, 2015). The model was calibrated and validated, reporting good results on the prediction of biomass characterization. Furthermore, it identified the light factor as one of the most sensitive parameters for microalgal growth. The model takes into account the algal growth on carbon and nutrients, gas transfer to the atmosphere, photorespiration and photoinhibition.

The PHOBIA model was developed by Wolf *et al.* (2007) for microalgal–bacterial biofilms. It includes the modelling of different kinetic mechanisms of phototrophic microorganisms, such as internal polyglucose storage, growth in darkness, photoadaptation and photoinhibition, as well as nitrogen preference (Wolf *et al.*, 2007). These models can serve as a basis for the development of further models whose aim is to explain and describe the microalgae–bacteria symbiosis for their cultivation for wastewater treatment in suspended cultures. For this reason, there is still a need for models calibrated and validated with longer data sets or at different operational conditions treating diverse types of wastewaters.

3.4 INTEGRATION OF PHOTOACTIVATED SLUDGE IN WASTEWATER TREATMENT CONCEPTS

The photo-activated sludge (PAS) system could fit within a holistic approach for wastewater treatment consisting of an anaerobic digester coupled with a microalgal–bacterial photobioreactor (Figure 3.5). The anaerobic digester is used for bioenergy production through a combined heat and power (CHP) system and the high nutrient strength centrate is further treated in a microalgal–bacterial photobioreactor. The biomass produced in the photobioreactor can be returned to the anaerobic digester to increase biogas production by co-digestion with the main waste(water) streams (Wang & Park, 2015). Part of the stabilized solids from the anaerobic digester and the microalgal–bacterial reactor could be used as biosolids for fertilizer replacement, promoting a circular economy within the treatment of wastewater.

At full scale and using sunlight as energy source, it is important to take into account the feeding conditions of the medium. However, this also depends on the final objective of the water reclamation of the treated effluent. For instance, effluents with high concentrations of nitrate, when just nitrification is performed in the microalgal–bacterial system, can support irrigation for crop growth (Taylor *et al.*, 2018). In case that due to the prior treatment there is a lack of micronutrients or other nutrients such as phosphorous, the effluent can be mixed in a certain ratio with the influent from the anaerobic digester to supply all the compounds needed. When the objective of the microalgal–bacterial system is the treatment of the wastewater to negligible ammonium and total nitrogen concentrations, the system should support nitrification and denitrification as shown by Arashiro *et al.* (2016) and Rada-Ariza *et al.* (2019). Then, during a HRT of 1 day, nitrification can be performed during the daylight and denitrification can be supported at night when there is no longer oxygen production. Therefore, it is recommended that the influent is fed during the dark conditions, then some of the oxygen

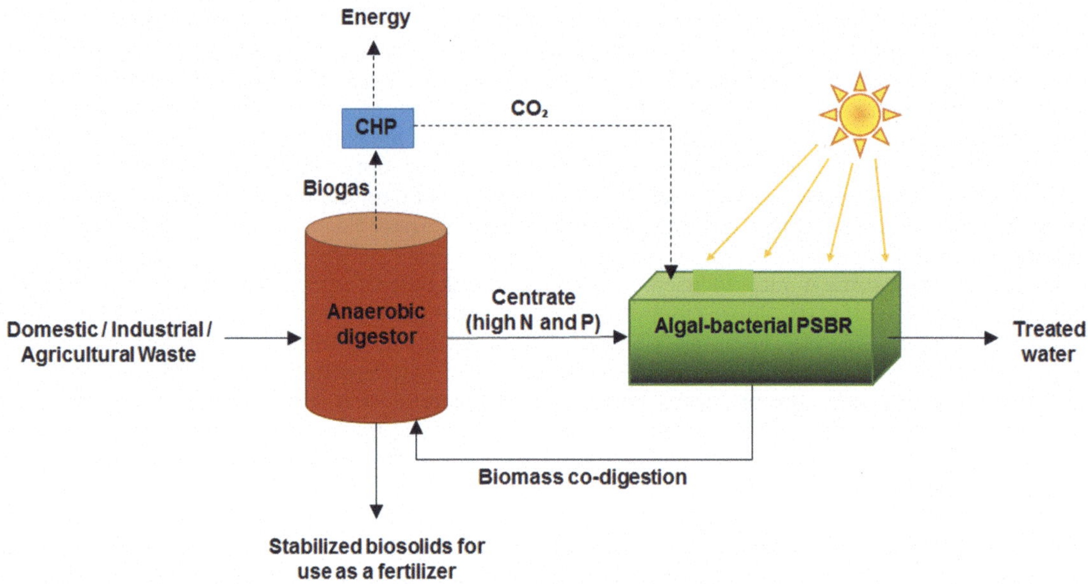

Figure 3.5 Scheme of the proposed holistic approach for treatment of domestic, industrial and agricultural wastes. CHP: combined heat and power system, N: nitrogen and P: phosphorous.

still present from the light phase would be consumed for organic matter oxidation and part of the ammonium would be oxidized or taken up by algae. The rest of the organic matter would be used for denitrification and the remaining ammonium that is not nitrified or taken up in the dark phase would be nitrified in the next light phase.

3.5 CONCLUSIONS

Microalgal–bacterial consortia are able to effectively remove nitrogen at shorter SRTs and HRTs than usually used in algal systems, showing high ammonium removal efficiencies. Furthermore, the co-cultivation of microalgae and bacteria offers advantages such as higher ammonium removal rates through nitrification/denitrification and consequently reduction of the area requirements in the implementation of the technology. Also the development of a bioflocculant algal–bacterial biomass without the addition of chemicals or energy input is an advantage. The symbiosis of microalgae and bacteria has shown promising results not just for nutrient and organic carbon removal, but for the elimination of other pollutants and contaminants from different industries as well (Rawat *et al.*, 2011). This offers new directions for research on microalgal–bacterial consortia. New studies on the co-culturing of different microorganisms for treatment of wastewater have already been reported (Manser *et al.*, 2016; Mukarunyana *et al.*, 2018). This shows the ability of algae to be resilient and adapt to different microbial populations and environments, and can help to further develop microalgal–bacterial consortia as sustainable approach to today's and tomorrow's wastewater problems.

REFERENCES

Abed R. M. M., Zein B., Al-Thukair A. and de Beer D. (2007). Phylogenetic diversity and activity of aerobic heterotrophic bacteria from a hypersaline oil-polluted microbial mat. *Systematic and Applied Microbiology*, **30**, 319–330, https://doi.org/10.1016/j.syapm.2006.09.001

Abou-Shanab R. A. I., Ji M.-K., Kim H.-C., Paeng K.-J. and Jeon B.-H. (2013). Microalgal species growing on piggery wastewater as a valuable candidate for nutrient removal and biodiesel production. *Journal of Environmental Management*, **115**, 257–264, https://doi.org/10.1016/j.jenvman.2012.11.022

Alcántara C., Posadas E., Guieysse B. and Muñoz R. (2015). Chapter 29 – microalgae-based wastewater treatment. In: Handbook of Marine Microalgae, S.-K. Kim (ed.), Academic Press, Boston, pp. 439–455.

Arashiro L. T., Rada-Ariza A. M., Wang M., van der Steen P. and Ergas S. J. (2016). Modelling shortcut nitrogen removal from wastewater using an algal-bacterial consortium. *Water Science and Technology*, **75**, 782–792, https://doi.org/10.2166/wst.2016.561

Aslan S. and Kapdan I. K. (2006). Batch kinetics of nitrogen and phosphorus removal from synthetic wastewater by algae. *Ecological Engineering*, **28**, 64–70, https://doi.org/10.1016/j.ecoleng.2006.04.003

Barbosa M. J., Janssen M., Ham N., Tramper J. and Wijffels R. H. (2003). Microalgae cultivation in air-lift reactors: modeling biomass yield and growth rate as a function of mixing frequency. *Biotechnology and Bioengineering*, **82**, 170–179, https://doi.org/10.1002/bit.10563

Becker E. W. (1994). Microalgae: Biotechnology and Microbiology. Cambridge University Press, Cambridge, U.K, p. 230.

Bernard O. (2011). Hurdles and challenges for modelling and control of microalgae for CO_2 mitigation and biofuel production mitigation and biofuel production. *Journal of Process Control*, **21**, 1378–1389, https://doi.org/10.1016/j.jprocont.2011.07.012

Borowitzka M. A. (1998). Limits to growth. In: Wastewater Treatment with Algae, Biotechnology Intelligence Unit, Y.-S. Wong and N. F. Y. Tam (eds.), Springer, Berlin, Heidelberg, pp. 203–226.

Buhr H. O. and Miller S. B. (1983). A dynamic model of the high-rate algal-bacterial wastewater treatment pond. *Water Research*, **17**, 29–37, https://doi.org/10.1016/0043-1354(83)90283-X

Cabanelas I. T. D., Ruiz J., Arbib Z., Chinalia F. A., Garrido-Pérez C., Rogalla F., Nascimento I. A. and Perales J. A. (2013). Comparing the use of different domestic wastewaters for coupling microalgal production and nutrient removal. *Bioresource Technology*, **131**, 429–436, https://doi.org/10.1016/j.biortech.2012.12.152

Cai T., Park S. Y. and Li Y. (2013). Nutrient recovery from wastewater streams by microalgae: status and prospects. *Renewable and Sustainable Energy Reviews*, **19**, 360–369, https://doi.org/10.1016/j.rser.2012.11.030

Carvalho A. P., Meireles L. A. and Malcata F. X. (2006). Microalgal reactors: a review of enclosed system designs and performances. *Biotechnology Progress*, **22**, 1490–1506, https://doi.org/10.1021/bp060065r

Chen C.-Y., Yeh K.-L., Aisyah R., Lee D.-J. and Chang J.-S. (2011). Cultivation, photobioreactor design and harvesting of microalgae for biodiesel production: a critical review. *Bioresource Technology*, **102**, 71–81, https://doi.org/10.1016/j.biortech.2010.06.159

Choi O., Das A., Yu C.-P. and Hu Z. (2010). Nitrifying bacterial growth inhibition in the presence of algae and cyanobacteria. *Biotechnology and Bioengineering*, **107**, 1004–1011, https://doi.org/10.1002/bit.22860

de Bashan L. E. and Bashan Y. (2004). Recent advances in removing phosphorus from wastewater and its future use as fertilizer (1997–2003). *Water Research*, **38**, 4222–4246, https://doi.org/10.1016/j.watres.2004.07.014

de Bashan L. E., Hernandez J.-P., Morey T. and Bashan Y. (2004). Microalgae growth-promoting bacteria as "helpers" for microalgae: a novel approach for removing ammonium and phosphorus from municipal wastewater. *Water Research*, **38**, 466–474, https://doi.org/10.1016/j.watres.2003.09.022

Decostere B., Janssens N., Alvarado A., Maere T., Goethals P., Van Hulle S. W. H. and Nopens I. (2013). A combined respirometer–titrimeter for the determination of microalgae kinetics: experimental data collection and modelling. *Chemical Engineering Journal*, **222**, 85–93, https://doi.org/10.1016/j.cej.2013.01.103

de Godos I., Vargas V. A., Blanco S., González M. C. G., Soto R., García-Encina P. A., Becares E. and Muñoz R. (2010). A comparative evaluation of microalgae for the degradation of piggery wastewater under photosynthetic oxygenation. *Bioresource Technology*, **101**, 5150–5158, https://doi.org/10.1016/j.biortech.2010.02.010

de Godos I., Arbib Z., Lara E. and Rogalla F. (2016). Evaluation of high rate algae ponds for treatment of anaerobically digested wastewater: effect of CO_2 addition and modification of dilution rate. *Bioresource Technology*, **220**, 253–261, https://doi.org/10.1016/j.biortech.2016.08.056

Di Termini I., Prassone A., Cattaneo C. and Rovatti M. (2011). On the nitrogen and phosphorus removal in algal photobioreactors. *Ecological Engineering*, **37**, 976–980, https://doi.org/10.1016/j.ecoleng.2011.01.006

Ekama G. A. and Wentzel M. C. (2008a). Nitrogen removal. In: Biological Wastewater Treatment Principles, Modelling and Design, M.M. Henze, M.C. van Loosdrecht, G.A. Ekama and D. Brdjanovic (eds), IWA Publishing, London, UK, pp. 87–139.

Ekama G. A. and Wentzel M. C. (2008b). Organic material removal. In: Biological Wastewater Treatment Principles, Modelling and Design, M.M. Henze, M.C. van Loosdrecht, G.A. Ekama and D. Brdjanovic (eds), IWA Publishing, London, UK, pp. 53–86.

Fernández F. G. A., Sevilla J. M. F. and Grima E. M. (2013). Photobioreactors for the production of microalgae. *Reviews in Environmental Science and Bio/Technology*, **12**, 131–151, https://doi.org/10.1007/s11157-012-9307-6

Fukami K., Nishijima T. and Ishida Y. (1997). Stimulative and inhibitory effects of bacteria on the growth of microalgae. *Hydrobiologia*, **358**, 185–191, https://doi.org/10.1023/A:1003139402315

Gehring T., Silva J. D., Kehl O., Castilhos A. B., Costa R. H. R., Uhlenhut F., Alex J., Horn H. and Wichern M. (2010). Modelling waste stabilisation ponds with an extended version of ASM3. *Water Science and Technology*, **61**, 713–720, https://doi.org/10.2166/wst.2010.954

Gernaey K. V., van Loosdrecht M. C., Henze M., Lind M. and Jørgensen S. B. (2004). Activated sludge wastewater treatment plant modelling and simulation: state of the art. *Environmental Modelling & Software*, **19**, 763–783, https://doi.org/10.1016/j.envsoft.2003.03.005

González C., Marciniak J., Villaverde S., León C., García P. A. and Muñoz R. (2008). Efficient nutrient removal from swine manure in a tubular biofilm photo-bioreactor using algae-bacteria consortia. *Water Science and Technology*, **58**, 95–102, https://doi.org/10.2166/wst.2008.655

González-Fernández C., Molinuevo-Salces B. and García-González M. C. (2011). Nitrogen transformations under different conditions in open ponds by means of microalgae–bacteria consortium treating pig slurry. *Bioresource Technology*, **102**, 960–966, https://doi.org/10.1016/j.biortech.2010.09.052

Grobbelaar J. U. (2008). Algal nutrition – mineral nutrition. In: Handbook of Microalgal Culture: Biotechnology and Applied Phycology, A. Richmond (ed.), Blackwell Science Ltd, Oxford, UK, pp. 97–115.

Gutzeit G., Lorch D., Weber A., Engels M. and Neis U. (2005). Bioflocculent algal-bacterial biomass improves low-cost wastewater treatment. *Water Science and Technology*, **52**, 9–18, https://doi.org/10.2166/wst.2005.0415

Henze M. (2000). Activated Sludge Models ASM1, ASM2, ASM2d and ASM3. IWA Publishing, London, UK.

Holenda B., Domokos E., Rédey Á. and Fazakas J. (2008). Dissolved oxygen control of the activated sludge wastewater treatment process using model predictive control. *Computers & Chemical Engineering*, **32**, 1270–1278, https://doi.org/10.1016/j.compchemeng.2007.06.008

Janssen M. and Lamers P. (2013). Microalgae Biotechnology Notes.

Jeon Y.-C., Cho C.-W. and Yun Y.-S. (2005). Measurement of microalgal photosynthetic activity depending on light intensity and quality. *Biochemical Engineering Journal*, **27**, 127–131, https://doi.org/10.1016/j.bej.2005.08.017

Kapdan I. K. and Aslan S. (2008). Application of the Stover–Kincannon kinetic model to nitrogen removal by *Chlorella vulgaris* in a continuously operated immobilized photobioreactor system. *Journal of Chemical Technology & Biotechnology*, **83**, 998–1005, https://doi.org/10.1002/jctb.1905

Karya N. G. A. I., van der Steen N. P. and Lens P. N. L. (2013). Photo-oxygenation to support nitrification in an algal-bacterial consortium treating artificial wastewater. *Bioresource Technology*, **134**, 244–250, https://doi.org/10.1016/j.biortech.2013.02.005

Kayombo S., Mbwette T. S., Mayo A., Katima J. H. and Jorgensen S. (2000). Modelling diurnal variation of dissolved oxygen in waste stabilization ponds. *Ecological Modelling*, **127**, 21–31, https://doi.org/10.1016/S0304-3800(99)00196-9

Kim T.-H., Lee Y., Han S.-H. and Hwang S.-J. (2013). The effects of wavelength and wavelength mixing ratios on microalgae growth and nitrogen, phosphorus removal using *Scenedesmus* sp. for wastewater treatment. *Bioresource Technology*, **130**, 75–80, https://doi.org/10.1016/j.biortech.2012.11.134

Kirkwood A. E., Nalewajko C. and Fulthorpe R. R. (2006). The effects of cyanobacterial exudates on bacterial growth and biodegradation of organic contaminants. *Microbial Ecology*, **51**, 4–12, https://doi.org/10.1007/s00248-004-0058-y

Larsdotter K. (2006). Wastewater treatment with microalgae – a literature review. *Vatten*, **62**, 31–38.

Lee J., Cho D.-H., Ramanan R., Kim B.-H., Oh H.-M. and Kim H.-S. (2013). Microalgae-associated bacteria play a key role in the flocculation of *Chlorella vulgaris*. *Bioresource Technology*, **131**, 195–201, https://doi.org/10.1016/j.biortech.2012.11.130

Manser N. D., Wang M., Ergas S. J., Mihelcic J. R., Mulder A., van de Vossenberg J., van Lier J. B. and van der Steen P. (2016). Biological nitrogen removal in a photosequencing batch reactor with an algal-nitrifying bacterial consortium and anammox granules. *Environmental Science & Technology Letters*, **3**, 175–179, https://doi.org/10.1021/acs.estlett.6b00034

Mara D. (2004). Domestic Wastewater Treatment in Developing Countries. Earthscan, London.

Martínez M. E., Jiménez J. M. and El Yousfi F. (1999). Influence of phosphorus concentration and temperature on growth and phosphorus uptake by the microalga *Scenedesmus obliquus*. *Bioresource Technology*, **67**, 233–240, https://doi.org/10.1016/S0960-8524(98)00120-5

Medina M. and Neis U. (2007). Symbiotic algal bacterial wastewater treatment: effect of food to microorganism ratio and hydraulic retention time on the process performance. *Water Science and Technology*, **55**, 165–171, https://doi.org/10.2166/wst.2007.351

Molinuevo-Salces B., García-González M. C. and González-Fernández C. (2010). Performance comparison of two photobioreactors configurations (open and closed to the atmosphere) treating anaerobically degraded swine slurry. *Bioresource Technology*, **101**, 5144–5149, https://doi.org/10.1016/j.biortech.2010.02.006

Mukarunyana B., van de Vossenberg J., van Lier J. B. and van der Steen P. (2018). Photo-oxygenation for nitritation and the effect of dissolved oxygen concentrations on anaerobic ammonium oxidation. *Science of the Total Environment*, **634**, 868–874, https://doi.org/10.1016/j.scitotenv.2018.04.082

Muñoz R. and Guieysse B. (2006). Algal–bacterial processes for the treatment of hazardous contaminants: a review. *Water Research*, **40**, 2799–2815, https://doi.org/10.1016/j.watres.2006.06.011

Olguín E. J. (2003). Phycoremediation: key issues for cost-effective nutrient removal processes. *Biotechnology Advances*, **22**, 81–91, https://doi.org/10.1016/S0734-9750(03)00130-7

Oswald W. J., Gotaas H. B., Ludwig H. F. and Lynch V. (1953). Algae symbiosis in oxidation ponds: III. Photosynthetic oxygenation. *Sewage and Industrial Wastes*, **25**, 692–705.

Park J. B. K. and Craggs R. J. (2010). Wastewater treatment and algal production in high rate algal ponds with carbon dioxide addition. *Water Science and Technology*, **61**, 633–639, https://doi.org/10.2166/wst.2010.951

Park J. B. K. and Craggs R. J. (2011). Nutrient removal in wastewater treatment high rate algal ponds with carbon dioxide addition. *Water Science and Technology*, **63**, 1758–1764, https://doi.org/10.2166/wst.2011.114

Park J., Jin H.-F., Lim B.-R., Park K.-Y. and Lee K. (2010). Ammonia removal from anaerobic digestion effluent of livestock waste using green alga *Scenedesmus* sp. *Bioresource Technology*, **101**, 8649–8657, https://doi.org/10.1016/j.biortech.2010.06.142

Plappally A. K. and Lienhard J. H. (2012). Energy requirements for water production, treatment, end use, reclamation, and disposal. *Renewable and Sustainable Energy Reviews*, **16**, 4818–4848, https://doi.org/10.1016/j.rser.2012.05.022

Powell N., Shilton A. N., Pratt S. and Chisti Y. (2008). Factors influencing luxury uptake of phosphorus by microalgae in waste stabilization ponds. *Environmental Science & Technology*, **42**, 5958–5962, https://doi.org/10.1021/es703118s

Rada-Ariza M. A. (2018). Photo-Activated Sludge; a Novel Algal-Bacterial Biotreatment for Nitrogen Removal from Wastewater. Dissertation, IHE Delft, the Netherlands.

Rada-Ariza M. A., Lopez-Vazquez C. M., Van der Steen N. P. and Lens P. N. L. (2017). Nitrification by microalgal-bacterial consortia for ammonium removal in flat panel sequencing batch photo-bioreactors. *Bioresource Technology*, **245**, 81–89, https://doi.org/10.1016/j.biortech.2017.08.019

Rada-Ariza M. A., Lopez-Vazquez C. M., Van der Steen N. P. and Lens P. N. L. (2019). Ammonium removal mechanisms in a microalgal-bacterial sequencing-batch photobioreactor at different SRT. *Algal Research*, **39**, 101468, https://doi.org/10.1016/j.algal.2019.101468

Ramanan R., Kim B.-H., Cho D.-H., Oh H.-M. and Kim H.-S. (2016). Algae–bacteria interactions: evolution, ecology and emerging applications. *Biotechnology Advances*, **34**, 14–29, https://doi.org/10.1016/j.biotechadv.2015.12.003

Rawat I., Ranjith Kumar R., Mutanda T. and Bux F. (2011). Dual role of microalgae: phycoremediation of domestic wastewater and biomass production for sustainable biofuels production. *Applied Energy*, **88**, 3411–3424, https://doi.org/10.1016/j.apenergy.2010.11.025

Riaño B., Hernández D. and García-González M. C. (2012). Microalgal-based systems for wastewater treatment: effect of applied organic and nutrient loading rate on biomass composition. *Ecological Engineering*, **49**, 112–117, https://doi.org/10.1016/j.ecoleng.2012.08.021

Ruiz J., Álvarez P., Arbib Z., Garrido C., Barragán J. and Perales J. A. (2011). Effect of nitrogen and phosphorus concentration on their removal kinetic in treated urban wastewater by *Chlorella vulgaris*. *International Journal of Phytoremediation*, **13**, 884–896, https://doi.org/10.1080/15226514.2011.573823

Safonova E., Kvitko K. V., Iankevitch M. I., Surgko L. F., Afti I. A. and Reisser W. (2004). Biotreatment of industrial wastewater by selected algal-bacterial consortia. *Engineering in Life Sciences*, **4**, 347–353, https://doi.org/10.1002/elsc.200420039

Shilton A. (2006). Pond Treatment Technology. IWA Publishing, London, United Kingdom.

Solimeno A., Samsó R., Uggetti E., Sialve B., Steyer J.-P., Gabarró A. and García J. (2015). New mechanistic model to simulate microalgae growth. *Algal Research*, **12**, 350–358, https://doi.org/10.1016/j.algal.2015.09.008

Solimeno A., Parker L., Lundquist T. and García J. (2017). Integral microalgae-bacteria model (BIO_ALGAE): application to wastewater high rate algal ponds. *Science of The Total Environment*, **601–602**, 646–657, https://doi.org/10.1016/j.scitotenv.2017.05.215

Su Y., Mennerich A. and Urban B. (2011). Municipal wastewater treatment and biomass accumulation with a wastewater-born and settleable algal-bacterial culture. *Water Research*, **45**, 3351–3358, https://doi.org/10.1016/j.watres.2011.03.046

Subashchandrabose S. R., Ramakrishnan B., Megharaj M., Venkateswarlu K. and Naidu R. (2011). Consortia of cyanobacteria/microalgae and bacteria: biotechnological potential. *Biotechnology Advances*, **29**, 896–907, https://doi.org/10.1016/j.biotechadv.2011.07.009

Takabe Y., Hidaka T., Tsumori J. and Minamiyama M. (2016). Effects of hydraulic retention time on cultivation of indigenous microalgae as a renewable energy source using secondary effluent. *Bioresource Technology*, **207**, 399–408, https://doi.org/10.1016/j.biortech.2016.01.132

Taylor R. P., Jones C. L. W., Laing M. and Dames J. (2018). The potential use of treated brewery effluent as a water and nutrient source in irrigated crop production. *Water Resources and Industry*, **19**, 47–60, https://doi.org/10.1016/j.wri.2018.02.001

Tiron O., Bumbac C., Manea E., Stefanescu M. and Lazar M. N. (2017). Overcoming microalgae harvesting barrier by activated algae granules. *Scientific Reports*, **7**, 4646, https://doi.org/10.1038/s41598-017-05027-3

Tuantet K., Temmink H., Zeeman G., Janssen M., Wijffels R. H. and Buisman C. J. N. (2014). Nutrient removal and microalgal biomass production on urine in a short light-path photobioreactor. *Water Research*, **55**, 162–174, https://doi.org/10.1016/j.watres.2014.02.027

Unnithan V. V., Unc A. and Smith G. B. (2014). Mini-review: a priori considerations for bacteria–algae interactions in algal biofuel systems receiving municipal wastewaters. *Algal Research*, **4**, 35–40, https://doi.org/10.1016/j.algal.2013.11.009

Valigore J. M., Gostomski P. A., Wareham D. G. and O'Sullivan A. D. (2012). Effects of hydraulic and solids retention times on productivity and settleability of microbial (microalgal–bacterial) biomass grown on primary treated wastewater as a biofuel feedstock. *Water Research*, **46**, 2957–2964, https://doi.org/10.1016/j.watres.2012.03.023

Van Den Hende S. (2014). Microalgal Bacterial Flocs for Wastewater Treatment: From Concept to Pilot Scale. Dissertation, Ghent University, Belgium.

Van Den Hende S., Carré E., Cocaud E., Beelen V., Boon N. and Vervaeren H. (2014). Treatment of industrial wastewaters by microalgal bacterial flocs in sequencing batch reactors. *Bioresource Technology*, **161**, 245–254, https://doi.org/10.1016/j.biortech.2014.03.057

van der Steen P., Rahsilawati K., Rada-Ariza A. M., Lopez-Vazquez C. M. and Lens P. N. L. (2015). A new photo-activated sludge system for nitrification by an algal-bacterial consortium in a photo-bioreactor with biomass recycle. *Water Science and Technology*, **72**, 443–450, https://doi.org/10.2166/wst.2015.205

van Loosdrecht M. C. M., Ekama G. A., Wentzel M. C., Brdjanovic D. and Hooijmans C. M. (2008). Modelling activated sludge processes. In: Biological Wastewater Treatment Principles, Modelling and Design, M.M. Henze, M.C. van Loosdrecht, G.A. Ekama, D. Brdjanovic (eds.), IWA Publishing, London, UK, pp. 361–392.

Wágner D. S., Valverde-Pérez B., Sæbø M., Bregua de la Sotilla M., Van Wagenen J., Smets B. F. and Plósz B. G. (2016). Towards a consensus-based biokinetic model for green microalgae – The ASM-A. *Water Research*, **103**, 485–499, https://doi.org/10.1016/j.watres.2016.07.026

Wang M. and Park C. (2015). Investigation of anaerobic digestion of *Chlorella* sp. and *Micractinium* sp. grown in high-nitrogen wastewater and their co-digestion with waste activated sludge. *Biomass and Bioenergy*, **80**, 30–37, https://doi.org/10.1016/j.biombioe.2015.04.028

Wang X., Hao C., Zhang F., Feng C. and Yang Y. (2011). Inhibition of the growth of two blue-green algae species (*Microsystis aruginosa* and *Anabaena spiroides)* by acidification treatments using carbon dioxide. *Bioresource Technology*, **102**, 5742–5748, https://doi.org/10.1016/j.biortech.2011.03.015

Wang B., Lan C. Q. and Horsman M. (2012). Closed photobioreactors for production of microalgal biomasses. *Biotechnology Advances*, **30**, 904–912, https://doi.org/10.1016/j.biotechadv.2012.01.019

Wang M., Yang H., Ergas S. J. and van der Steen P. (2015). A novel shortcut nitrogen removal process using an algal-bacterial consortium in a photo-sequencing batch reactor (PSBR). *Water Research*, **87**, 38–48, https://doi.org/10.1016/j.watres.2015.09.016

Wang M., Keeley R., Zalivina N., Halfhide T., Scott K., Zhang Q., van der Steen P. and Ergas S. J. (2018). Advances in algal-prokaryotic wastewater treatment: a review of nitrogen transformations, reactor configurations and molecular tools. *Journal of Environmental Management*, **217**, 845–857, https://doi.org/10.1016/j.jenvman.2018.04.021

Wolf G., Picioreanu C. and van Loosdrecht M. C. M. (2007). Kinetic modeling of phototrophic biofilms: the PHOBIA model. *Biotechnology and Bioengineering*, **97**, 1064–1079, https://doi.org/10.1002/bit.21306

Yun Y.-S. and Park J. M. (2003). Kinetic modeling of the light-dependent photosynthetic activity of the green microalga *Chlorella vulgaris*. *Biotechnology and Bioengineering*, **83**, 303–311, https://doi.org/10.1002/bit.10669

Zambrano J., Krustok I., Nehrenheim E. and Carlsson B. (2016). A simple model for algae-bacteria interaction in photo-bioreactors. *Algal Research*, **19**, 155–161, https://doi.org/10.1016/j.algal.2016.07.022

Zhang B., Lens P. N. L., Shi W., Zhang R., Zhang Z., Guo Y., Bao X. and Cui F. (2018). Enhancement of aerobic granulation and nutrient removal by an algal–bacterial consortium in a lab-scale photobioreactor. *Chemical Engineering Journal*, **334**, 2373–2382, https://doi.org/10.1016/j.cej.2017.11.151

Zhang B., Li W., Wu L., Shi W. and Lens P. (2022a). Rapid start-up of photo-granule process in a photo-sequencing batch reactor under low aeration conditions: effect of inoculum AGS size. *Science of the Total Environment*, **820**, 153204, https://doi.org/10.1016/j.scitotenv.2022.153204

Zhang B., Wu L., Shi W., Zhang Z. and Lens P. (2022b). A novel strategy for rapid development of a self-sustaining symbiotic algal-bacterial granular sludge: applying algal-mycelial pellets as nuclei. *Water Research*, **214**, 118210, https://doi.org/10.1016/j.watres.2022.118210

Zhao X., Zhou Y., Huang S., Qiu D., Schideman L., Chai X. and Zhao Y. (2014). Characterization of microalgae-bacteria consortium cultured in landfill leachate for carbon fixation and lipid production. *Bioresource Technology*, **156**, 322–328, https://doi.org/10.1016/j.biortech.2013.12.112

Zhou Q., He S. L., He X. J., Huang X. F., Picot B., Li X. D. and Chen G. (2006). Nutrients removal mechanisms in high rate algal pond treating rural domestic sewage in East China. *Water Supply*, **6**, 43–50, https://doi.org/10.2166/ws.2006.956

doi: 10.2166/9781789063547_0053

Chapter 4

Macroalgae biorefinery and its role in achieving a circular economy

James Lawrence[1], Armando Oliva[1,2], Jerry D. Murphy[3] and Piet N. L. Lens[1]*

[1]University of Galway, H91 TK33, Galway, Ireland

[2]Department of Civil, Architectural and Environmental Engineering, University of Naples Federico II, Via Claudio 21, 80125, Naples, Italy

[3]Civil, Structural and Environmental Engineering, School of Engineering and Architecture, University College Cork, Cork, Ireland

**Corresponding author: j.lawrence4@nuigalway.ie*

ABSTRACT

Availability of fossil fuels and feedstocks is a major problem currently faced by a variety of sectors thus highlighting the importance of transitioning towards a circular economy. Increased pollution, fossil fuel availability and other adverse effects are just some of the reasons that have prompted a need to find additional resources for fuel. One such feedstock that has shown to be both promising and viable is macroalgae. This chapter focuses on the latest scientific literature related to the development of macroalgae biorefineries, focusing on the different biological processes and how the resulting generated bioproducts can positively impact the global bioeconomy. The fundamental biological processes are explained while also providing details on specific problems the sector currently faces. Potential areas of further development and recent scientific discoveries of a variety of macroalgal species are also discussed.

4.1 INTRODUCTION

The world energy consumption was recorded at 488 EJ (exajoule) in 2005, 580 EJ in 2018 (Kober *et al.*, 2020) and is expected to exceed 650 EJ by 2025; 86% of this can be attributed to fossil fuel energy (Drapcho *et al.*, 2008). These figures indicate a clear overreliance on the use of fossil fuels across many different industries. One such sector which plays a huge role involving this energy consumption is the transportation sector. The current use of high-powered vehicles in the transportation sector makes it difficult to promote decarbonization. For this reason, researchers have focused on the promotion of biofuel usage and production in achieving a more sustainable future involving transportation. According to the International Energy Agency (IEA) (2014), one third of the final energy consumption is associated with transport-related liquid fuels such as petrol and diesel. This clear overreliance on a non-renewable energy source has led to many organizations working towards development of a plan to transition to a mode of cleaner energy consumption. The need for this transition is further highlighted by directives from both world and European environmental agencies in highlighting responsibilities regarding energy admission and consumption. One such directive is the terms of the European Union (EU) Renewable Energy Directive stating a new, legally binding aim for the EU's use of renewable energy for 2030 of at least 32%, with a provision for a potential modification to the higher level in 2023. This objective builds on the 20% renewable energy supply goal for 2020 (EEA, 2023).

To fulfil such directives, the production and utilization of biofuels is of utmost importance. Electrification of transport through battery electric vehicles (BEV) is a viable option for light vehicles and short-distance heavier transport (Forrest *et al.*, 2020), but to decarbonize long-distance heavy vehicles we will need renewable hydrocarbon fuels either in gaseous or liquid form (Gray *et al.*, 2021). Biofuels, such as biogas, biomethanol, bioethanol and biobutanol, are considered an alternative to fossil fuels going forward because they can reduce transport emissions and increase the security of supply (Nigam & Singh, 2011). Another biofuel which has been shown to have potential consists of a hydrogen (H_2) and methane (CH_4) blend known as biohythane (Lay *et al.*, 2020). Biohythane consists of a blend of 70–90% v/v methane and 10–30% v/v hydrogen (Bolzonella *et al.*, 2018). Research has shown that this biofuel exhibits major potential in terms of application in the transport sector. By harnessing this potential, this approach can contribute to decarbonizing and fuelling maritime ferries (Dahlgren *et al.*, 2022), along with specific elements within the broader transportation sector (long-distance haulage, coaches and ships). One source of biomass that is effective in the production of these biofuels (biohythane (Keskin *et al.*, 2019) and biogas (Saqib *et al.*, 2013)) is macroalgae (seaweed). Interest in this area has been constantly growing due to the increase in energy demand as well as the potential shown by microalgae in wastewater treatment (Chapter 5) and biofuel production (Chapter 9).

4.2 MACROALGAE SPECIES

4.2.1 Green algae

One species of green algae considered to holster much potential in terms of energy is *Ulva lactuca* (Figure 4.1a). Utilization of this species is appealing due to its high potential growth rate and high content of carbohydrates. Nutritional composition studies have shown that carbohydrates are the major component of *U. lactuca*, nearing 60% (Rasyid, 2017). Pre-treatment, saccharification, fermentation, and distillation are all steps in the conversion of macroalgae to bioethanol. Korzen *et al.* (2015) demonstrated the use of sonication as a pretreatment method in bioethanol production from *Ulva* sp. Since macroalgae contain low lignin (5.11% according to Allouache *et al.*, 2021), they can be easily depolymerized. Enzymatic hydrolysis of the resultant polysaccharides, followed by the addition of microbes such as *Saccharomyces cerevisiae*, can convert them into ethanol.

Currently, most of the naturally produced *U. lactuca* biomass is an unused resource ending up in a landfill due to the waste problems it poses to beaches and ultimately not being used efficiently for energy conversion. This build-up has also resulted in beach waste problems (Figure 4.1b, d and f) in countries such as Spain (Madejón *et al.*, 2022), Brazil (Harb & Chow, 2022) and Korea (Sunwoo *et al.*, 2017). This highlights a worldwide issue in that there is a need for better utilization for algal bloom waste management. Utilization of seaweed such as *U. lactuca* as a potential source for biofuel production dates back as far as the 'aquatic species programme' that was run in the United States from 1978 to 1996. The conclusion of this study stated that *U. lactuca* usage as a source of energy was not economically feasible (Ryther *et al.*, 1984). While this study may have demonstrated a lack of sustainability going forward, the need to revisit utilizing aquatic energy crops for biofuel production has resurfaced in recent times. Due to issues around climate change and growing opportunities in renewable energy production, traditional biomass availability has plummeted. For this reason, macroalgae are back on the radar due to their offering as an alternative and sustainable resource in terms of production of bioenergy (Lehahn *et al.*, 2016).

U. lactuca growth is commonly found worldwide although the strains vary among regions due to the influence played by different climates. Studies have shown the species has been harvested from shallow coastal areas (Cecchi *et al.*, 1996) or else land-built systems. *Ulva* blooms occur mainly in shallow waters with surplus of nutrients and the decomposition of this alga can produce acidic vapours, which highlights the importance of controlling and cultivating the biomass.

Figure 4.1 Major species of macroalgae with potential for biorefinery applications: (a) *Ulva* sp., (c) *Sargassum* sp., and (e) *Laminaria* sp. as well as their respective blooms (b), (d), and (f).

Growth conditions such as season are the predominant factors which affect the chemical composition of macroalgae (Thorsteinsson *et al.*, 2023). One of the main points of attraction for using *Ulva* sp. in biofuel production is attributed to its high carbohydrate content. This is illustrated by Ortiz *et al.* (2006), who highlighted the total solid carbohydrate content to be close to 60%. This carbohydrate content is predominantly in the form of the complex hydrocolloid ulvan, see Section 4.3. This sulphated polysaccharide is a structural component of the cell wall alongside cellulose (Lahaye & Robic, 2007). The unique chemical properties of ulvan make it an attractive prospect to be used as an active polymer for the pharmaceutical and agricultural sector.

4.2.2 Brown algae

4.2.2.1 Laminaria sp.

Promising macroalgal species used within the bioenergy and bioproducts industry also include *Laminaria* sp. (Figure 4.1e) and *Sargassum* sp. (Figure 4.1c), both of which belong to the brown algae family. Brown macroalgae, referred to as *Phaeophyceae* are the second largest group of macroalgae with over 2000 species identified to date (Guiry, 2023). *Laminaria* contains many structural and functional polysaccharides with compositions as high as 60% (Holdt & Kraan, 2011) as well as its unique alginate composition makes this species an ideal candidate for alginate production. Alongside this, *Laminaria* is also a source of a range of high-value products which are the precursors to biofuels and biochemicals (Bojorges *et al.*, 2022).

To maximize the potential of this species in obtaining these valuable products it is pivotal to select an appropriate pretreatment method. One study demonstrated that hydrothermal treatment is an effective means of improving biohydrogen and methane yield (showing an increase of 26.7%) via two-stage dark fermentation of the species *Laminaria* (Ding *et al.*, 2020). Further appeal in the utilization of this species is illustrated in its wide-ranging polyphenol content featuring both low-weight phenolic acids and sulphated phenolic compounds (Wekre *et al.*, 2022). Phenolics have shown potential in terms of bioactivity features such as acting as an antioxidant, with antidiabetic and anti-cancer properties making them highly desirable for the medical industry (Wekre *et al.*, 2023). One method to extract phenolic compounds from *Laminaria* sp. is an ionic liquid-based extraction which uses three kinds of 1-alkyl-3-methylimidazolium with different cations and anions coupled with ultrasonic treatment (Han *et al.*, 2011). High phenolic compound concentrations have also been proven to function as an inhibiting factor in terms of the digestion process and produce a lower biomethane potential yield (BMP) in brown seaweeds (Hierholtzer *et al.*, 2013). Whereas work conducted on the brown seaweed *Ascophylum nodosum* detailed how seasonal variation during the summer months increases polyphenolic content of the seaweed and in turn adversely affects BMP yield (Tabassum *et al.*, 2016).

4.2.2.2 *Sargassum sp.*

Sargassum sp. (Figure 4.1c) also belongs to the family Phaeophyceae. This macroalgae often floats on the ocean's surface in large quantities which results in the formation of *Sargassum* blooms (Figure 4.1d). Pelagic *Sargassum* blooms, linked to rising sea temperatures and nutrient discharge from the Amazon basin (Thompson *et al.*, 2020), have caused a huge waste management problem for tropical Atlantic countries since 2011 with the costs attributed to beach cleanup rising to US$0.3–1.5 million per kilometre (Rodríguez-Martínez *et al.*, 2023). While this highlights a clear economic problem for the countries affected by more frequent blooming events, it has a detrimental impact in terms of both ecology and human health-related problems. Ecological impacts include the smothering of coral reefs causing fish deaths due to hypoxia and the alteration of pH in coastal waters. One study illustrates this impact on the sea urchin species *Diadema antillarum* (Cabanillas-Terán *et al.*, 2019). The hypoxic conditions generated by the leachates released from the decomposition of *Sargassum* led to reduced taxonomic diversity of the macroalgal food sources. Further findings saw that these changes impacted the trophic characteristics of *D. antillarum*, which highlights the need for this ongoing *Sargassum* problem to be addressed before further impacts into the functioning of coastal ecosystems and alterations in biodiversity arise.

While much attention has been given to the environmental impacts of *Sargassum*, it is highly important to focus on the health hazards it can pose to humans and animals. Following the decomposition of *Sargassum* onshore, large amounts of toxic gases such as hydrogen sulphide (H_2S) and ammonia (NH_3) are produced which is a problem also associated with *U. Lactuca* in both France (Loret *et al.*, 2020) and Ireland (Murphy *et al.*, 2015). Human exposure to such gases can have health consequences such as hypoxic pulmonary, neurological, and cardiovascular lesions. Across an 8-month spell in 2018, it was found that exposure to such toxic gases reached case numbers of 3341 in Guadeloupe and 8061 in Martinique (Resiere *et al.*, 2018). Alongside these health threats, *Sargassum* blooms have also been shown to impact the economy and loss of income due to many of the impacted countries relying heavily on tourism. To combat these ongoing problems government agencies have developed ecological briefs detailing best practices and methods of remediation of the waste generated (Hinds *et al.*, 2016) as well as providing funding to the affected countries (Oxenford *et al.*, 2021).

4.3 BIOMATERIALS AND BIOPRODUCTS FROM MACROALGAE

Macroalgae exhibit many advantages over alternative biofuel feedstocks. Unlike feedstocks used for second-generation biofuels high in lignocellulosic materials, macroalgae are easier to biologically degrade. Subsequently, the digestion of algae may be shown to be cost-effective in comparison to feedstocks derived from lignocellulosic crops. Moreover, macroalgae do not compete with food sources for land usage or irrigation by freshwater (Smith *et al.*, 2010); though they are a significant resource for food in Asian countries (Pereira, 2021). Macroalgae can take advantage of the nutrients present in wastewater and seawater to promote growth. In addition, macroalgae also boast a faster growth rate with higher biomass yields in comparison to other terrestrial plants (Dutta *et al.*, 2014).

Ulva sp. provide a potential in terms of extraction of its high-value product, that is Ulvan, and utilization of the leftover biomass in terms of biofuel production. Figure 4.2 details the possible routes in which *Ulva* sp. may be utilized within a biorefinery concept. Ulvan is a cell wall polysaccharide found in *Ulva* species and its percentage composition in dry-weight biomass shows a variance from species to species of 8–29% (Lahaye and Robic, 2017), and 9–36% (Lakshmi *et al.*, 2020). Ulvan is a value-added product which is used in the pharmaceutical industry as a biomaterial. It can be harvested prior to *Ulva* biomass use in anaerobic processes for biofuel production to boost the overall efficiency and profitability of the process, as ulvan within a reactor is a potential precursor for high sulphide levels that is an inhibiting compound for anaerobic bacteria (Chen *et al.*, 2008). Ulvan is used in hydrogels (Morelli & Chiellini, 2010), membranes and films, and, particularly, in food packaging due to its antioxidant properties (Ganesan *et al.*, 2018). Pharmaceutically it is currently being investigated for anticancer properties although there are thus far no human trials (Kidgell *et al.*, 2019) and similar investigations are being undertaken regarding its immunomodulatory effects (Kidgell *et al.*, 2020).

Ulvan can be extracted in several different ways including acid extraction, combined enzymatic and chemical extraction (Yaich *et al.*, 2017) and Soxhlet extraction (Ben Amor *et al.*, 2021). Acid extraction is particularly cost-effective and eco-friendly as citric acid can be utilized (Manikandan & Lens, 2022a, 2022b). The principle relies on the hydrophobic nature of rhamnose causing ulvan to fold into a neutral pH that will then aggregate in the presence of NaCl allowing for easy removal from the solution (Kidgell *et al.*, 2019).

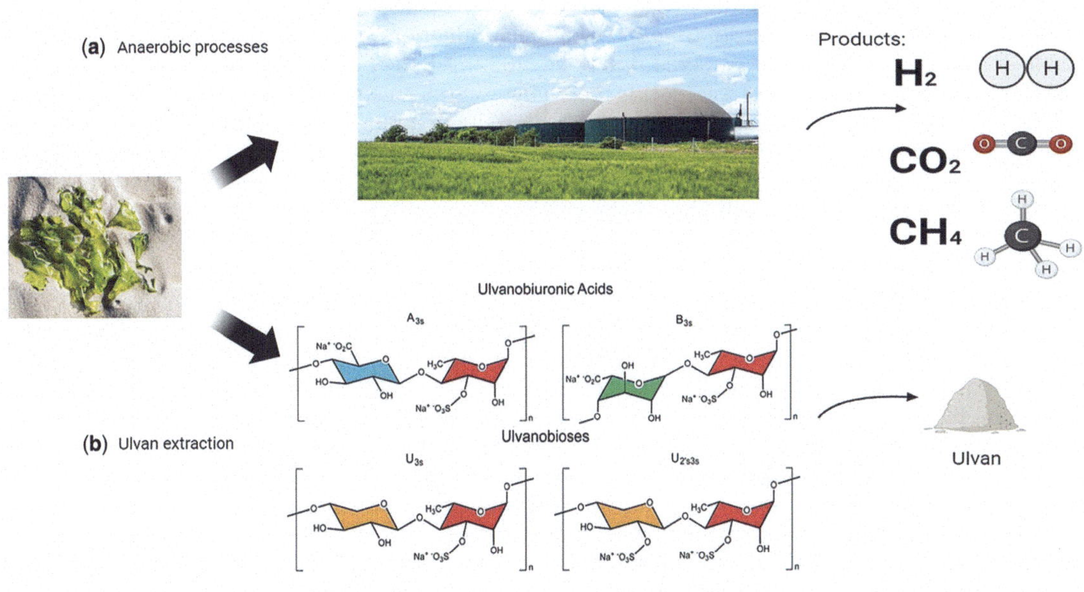

Figure 4.2 Potential biofuel and bioproducts produced from *Ulva* sp. (a) Biofuel production – H_2, CO_2, CH_4 and (b) ulvan – a cell wall polysaccharide utilized within the biopharmaceutical industry.

4.4 BIOFUELS FROM MACROALGAE

The transportation sector is one of the largest and fastest-growing energy consumers in today's world, while also being difficult to fully decarbonize (Papadis & Tsatsaronis, 2020). For this reason, it is pivotal to commit to a future that consists of lowering carbon consumption and increasing sustainable energy. To allow for this transition to occur, it is important to find means of maximizing energy efficiency, discovering renewable energy supplies, and optimizing energy systems from source to end use. Transitioning to a cleaner energy future using biofuel energy will bring inflated costs due to the need for robust investment in research and technological development. For this reason, it is pivotal that innovative and cost-saving technologies are used. The process of anaerobic digestion has been demonstrated to be an effective and feasible way of producing biofuels from the digestion of various feedstocks such as seaweed (Tabassum *et al.*, 2017a, 2017b).

Due to the ongoing fuel and climate crisis and efforts being made in reaching a circular economy, there is a newfound importance placed on maximizing bioprocesses to generate multiple products from the same biomass. This is not different to the seaweed industry with growing interest due to its potential use in creation of a variety of bioproducts and biofuels. One of the most promising products in the seafood sector, the commercial seaweed market is anticipated to rise from $15.01 billion in 2021 to $24.92 billion in 2028 at a compound annual growth rate (CAGR) of 7.51% (CBI, 2023). Furthermore, the compound annual growth rate of the industry is estimated at 9.7% for the years 2020–2025 (Mordor Intelligence, 2022).

4.4.1 Biogas

An advantage of utilizing algae is that algal tissue which is used to produce biofuels may potentially be a waste product from other industries. Chemical compounds and components of many algal species are used in food and livestock feed production. The extraction of value-added products from biofuel feedstock creates a sustainable cyclical system, especially considering the potential for leftover biomass after fuel production as fertilizer for crops or substrate for generating other types of biofuels. *Ulva lactuca* has been identified as having potential use to produce biofuels (Bikker *et al.*, 2016), such as methane (Bruhn *et al.*, 2011) and hydrogen (Dogmaz & Cavas, 2023).

Previous research into the role of *Ulva* sp. in biofuel production suggested that it is not economically viable or sustainable (Ryther *et al.*, 1984; Liu *et al.*, 2022). Allen *et al.* (2013) indicated that despite a low C:N ratio in *Ulva* sp., if pretreated, this macroalgae can be a suitable source for third-generation biofuel production. *Ulva* sp. is favourable for biofuel production due to its composition being enriched in polysaccharides, starch, and cellulose which are vital components required for microbes to feed on in producing clean biofuels like H_2 (Olsson *et al.*, 2020). Table 4.1 details various biomethane yields obtained from a range of seaweeds featuring a variety of pretreatment methods.

4.4.2 Biohydrogen

Biohydrogen production from marine macroalgal biomass is considered a clean energy technology with a high caloric value produced via dark fermentation. In comparison to the complex and extremely variable cell wall of lignocellulosic biomass (Oliva *et al.*, 2022), macroalgae features a much simpler carbohydrate cell wall which allows for a variety of biomass pretreatment methods to be applied in enhancing biohydrogen production. Various pretreatment technologies feature physical, chemical, biological, and combinational that allow for the breakdown of algal biomass into simpler compounds and releasing fermentable sugars efficiently. Table 4.2 details various biohydrogen yields obtained from a range of seaweeds featuring a variety of combined pretreatment methods.

Issues surrounding the use of algal biomass in biohydrogen production may be attributed to factors such as its high ammonium, sodium, and sulphate content (Xia *et al.*, 2016). This high sulphur content can lead to increased levels of H_2S production, which is a foul smelling, toxic and corrosive harmful gas. Optimization of the carbon to sulphur ratio can overcome the bottleneck that comes with utilizing *Ulva* with a high sulphur content in a dark fermentation process (Allen *et al.*, 2014).

Table 4.1 Comparison of biomethane yields and varying pretreatment conditions obtained from *Laminaria, Sargassum* and *Ulva* sp.

Seaweed	Inoculum	Pretreatment	Biomethane Yield	Reference
Laminaria digitata	Digested slurry	Mechanical	282 L CH_4/kg VS	Tabassum *et al.* (2017a, 2017b)
Sargassum fulvellum	Digested slurry	Enzymatic	186.60 mL CH_4/g VS	Farghali *et al.* (2021)
Sargassum fulvellum	Digested slurry	Mechanical	142.91 ± 0.004 mL CH_4/g VS	Yuhendra *et al.* (2021)
Ulva lactuca	Cattle digestate	Biologically	408 ± 20.02 mL CH_4/g VS	Mhatre *et al.* (2019)
Ulva lactuca	Digested slurry	Drying	250 L CH_4/kg VS	Allen *et al.* (2013)

Table 4.2 Comparison of biohydrogen yields and varying pretreatment conditions obtained from *Laminaria, Sargassum and Ulva* sp.

Seaweed	Inoculum	Pretreatment	Biohydrogen Yield	Reference
Laminaria japonica	Seed sludge	Mechanical	71.4 mL H_2/g TS	Shi *et al.* (2011)
Laminaria japonica	Anaerobic sludge	Microwave – acid treatment	28 mL H_2/g TS	Yin and Wang (2018)
Sargassum tennerimum	Rumen fluid	Ultrasonic coupled treatment	86 mL H_2/g COD	Snehya *et al.* (2022)
Sargassum sp.	*C. saccharolyticus* DSM 8903	Mechanical	91.3 ± 3.3 L H_2/kg VS	Costa *et al.* (2015)
Ulva fasciata	Rumen fluid	Surfactant-coupled sonication	91.7 mL H_2/g COD	Snehya *et al.* (2021)
Ulva reticulata	Digested sludge	Surfactant-induced microwave disintegration	54.9 mL H_2/g COD	Kumar *et al.* (2022)

4.4.3 Biohythane

Biohythane – a H_2 and CH_4 blend, is produced in a two-stage fermentation process. The first stage (operated at a low pH and retention time with a corresponding relatively high organic loading rate with inhibited methanogenesis) involves H_2 production controlled by a diverse population of hydrolytic and acidogenic bacteria. The metabolism of hydrogen involves the oxidation of pyruvate to acetyl-CoA by the enzyme pyruvate-ferredoxin oxidoreductase by obligate anaerobes. Hydrogen is then formed due to the reduction of ferredoxin as it undergoes oxidation by the enzyme hydrogenase. Hydrogen can also be formed by facultative anaerobes which oxidize pyruvate to formate and acetyl CoA following the catalysis of the enzyme pyruvate formate lyase (Hallenbeck, 2013). Meanwhile, in the second stage (neutral pH, retention time typically 5 times higher and organic loading rate typically five times lower than that of the first stage), methanogenic archaea control methane generation with the enzyme methyl-coenzyme M reductase (MCR) playing a key role (Ghimire *et al.*, 2017). This role is key to the fact that methanogenic microorganisms have an energy metabolism which is controlled by the reduction of C1 transfer coenzymes, enzymes and activated C1 intermediates.

By combining a dark fermentation reactor alongside an AD reactor in a two-phase process, biohythane can be produced in a cost-effective and environmentally friendly way (Bolzonella *et al.*, 2018). Biohydrogen production via dark fermentation is typically carried out by anaerobic bacteria, such as *Clostridium* spp., *Thermoanaerobacterium* spp., *Enterobacter* and *Bacillus* (Reith *et al.*, 2003). This occurs due to the breakdown of glucose into pyruvate through the glycolytic pathway. The fate of pyruvate is then dependent on the microbes present as the pyruvate formate lyase (PFL) pathway is utilized by facultative anaerobes whereas the pyruvate : ferredoxin oxidoreductase (PFOR) pathway is for strict anaerobic microorganisms (Cao *et al.*, 2022). Alongside the presence of a suitable microbial community, environmental conditions such as pH (6.0) (Ding *et al.*, 2020), temperature (20°C–45°C) (Qu *et al.*, 2022) and HRT of 72 h (Soares *et al.*, 2020) are favourable in maintaining bacterial cooperation and in turn enhancing the dark fermentation process for hydrogen production. Meanwhile, biomethane production is produced by microorganisms such as *Methanosarcina barkeri* and *Methanococcus*, which require a more stable temperature and pH as well as less vigorous agitation (Battista *et al.*, 2016).

In addition, numerous by-products are formed because of the above biological processes involved in producing gas from macroalgae. Volatile fatty acids (VFAs) are a by-product of hydrogen production via dark fermentation and are a value-added product because of the demand for VFAs in industries such as cosmetics, food, bioenergy, and pharmaceuticals. The most well-known VFA is acetic acid, which is often used in food preservation. Butyric acid may be used in bioenergy production as a precursor in the form of ethyl butyrate or butyl butyrate. Butyric acid is also valued in pharmaceuticals as an intermediate in the production of drugs for the treatment of cancers such as leukaemia and colorectal cancer (Pouillart, 1998).

4.4.4 Bioethanol and biobutanol

4.4.4.1 Acetone–butanol–ethanol fermentation

Numerous studies have been completed on the conversion of lipids from algal species into alcohols by a variety of different methods. This fermentation strategy is known as acetone–butanol–ethanol (ABE) fermentation (Figure 4.3). The strategy usually utilizes *Clostridium* sp. to ferment sugars to form acetone, butanol, and ethanol in a ratio of 3:6:1 (Awang *et al.*, 1988). Several different microorganisms from the *Clostridium* genus can be used in ABE fermentation with all having slightly different product distributions, nutrient requirements, and carbon source preferences. Such organisms are from the *Clostridia* species such as *Clostridium acetobutylicum*, *Clostridium beijerinckii*, *Clostridium saccharobutylicum*, and *Clostridium saccharoperbutylacetonicum* (Patakova *et al.*, 2013).

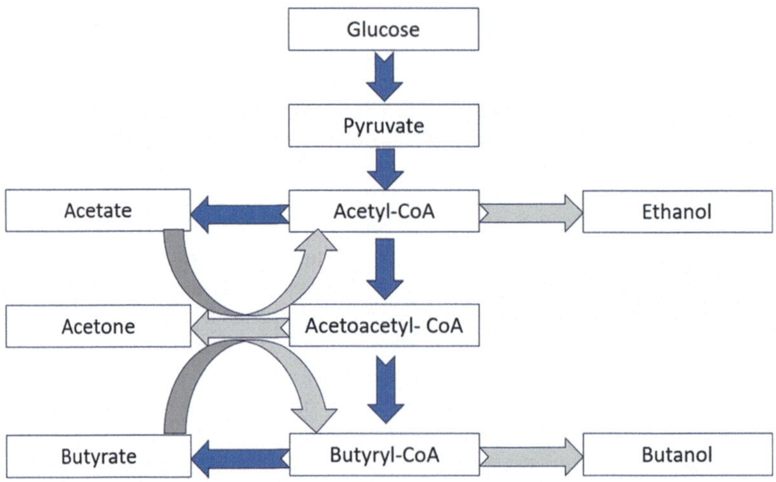

Figure 4.3 Schematic diagram of the metabolic pathway in ABE fermentation – arrows classifying an enzymatic conversion. Notched arrows (⇨), as seen first from glucose to pyruvate, indicate multiple steps shown as one. Arrows in blue represent the steps in acidogenesis, and arrows in grey represent reactions during the solventogenesis phase.

Typically, ABE solvents are produced during two designated time-based phases (Potts *et al.*, 2018). First bacteria produce organic acids such as lactic acid, acetic acid, and butyric acid, which is followed by bacteria converting the acids to their corresponding solvents. In this two-stage process, stage one is known as acidogenesis, while stage two is referred to as the solventogenesis phase. The move from acidogenesis to solventogenesis is triggered as cell growth slows down following the rapid production of acetate and butyrate that occurs during acidogenesis (Amador-Noguez *et al.*, 2011). Typically, a change in pH of the fermentation broth allows metabolism to transition between acidogenesis and solventogenesis (Richter *et al.*, 2016).

4.4.4.2 Biobutanol

Like ethanol, butanol is also typically a biomass-based renewable fuel that can be produced by alcoholic fermentation of a range of different feedstocks with one being macroalgae, as reported by Potts *et al.* (2012). By comparison of carbon structures, butanol (C_4H_9OH) possesses a four-carbon structure whereas methanol (CH_3OH) and ethanol (CH_3CH_2OH) have a one- and two-carbon structure, respectively. A benefit of butanol is its ability to blend with gasoline efficiently while studies have also demonstrated potential of blending with diesel (Yusri *et al.*, 2019). Due to butanol possessing a higher oxygen content than biodiesel, there is a reduction in the amount of soot produced. A further advantage of using butanol over ethanol and petrol blends (Sanap *et al.*, 2023), is related to the fact NO_x emissions can be reduced due to its higher heat evaporation, thus resulting in a lower combustion temperature (Rakopoulos *et al.*, 2010). The main disadvantage centred around the use of butanol is related to its low production rates and end-product toxicity and for this reason often ethanol production was favoured over that of butanol. Nevertheless, thanks to recent advancements in technology and the development of butanol fermentative techniques, the production rates of butanol have been improved. One study indicated that the production cost of butanol from wheat straw stands at $1.37/kg (Wang *et al.*, 2023). Meanwhile, a second study also detailed through a technoeconomic analysis of the production of butanol alongside further biorefinery products from the macroalgae *Ulva rigida* is also economically feasible. Results from the modelling indicated an internal rate of return (IRR) of 37% (Llano *et al.*, 2023).

4.4.4.3 Bioethanol

The utilization of biofuels in the transportation sector is constantly growing. According to preliminary European Environment Agency (EEA) statistics, in 2021, the proportion of renewable energy utilized for transportation in the EU stabilized at 10.2% (EEA, 2021a, 2021b). Two products which have demonstrated their potential for use in this sector are butanol and ethanol. Ethanol is a biomass-based renewable fuel that is commonly produced by the fermentation of sugar from a range of different substrates one being macroalgae (Enquist-Newman *et al.*, 2014). It is often considered an alternative fuel for internal combustion engines (Li *et al.*, 2019). The adoption of ethanol blended fuel (E85, 85% ethanol and 15% fossil fuels) vehicles together with electric and compressed natural gas vehicles is expected to make up 34% of all private vehicle stock by 2050 (Saraf & Shastri, 2023). Meanwhile, methanol has also been shown to be a promising fuel of the future with numerous technoeconomic studies available detailing its potential in decarbonization of the maritime industry (de Fournas & Wei, 2022; Shi *et al.*, 2023). Methanol is produced from coal or petrol-based products (Khalafalla *et al.*, 2020) but in future will be generated from reforming of biomethane or reaction of green hydrogen with biogenic CO_2 (Rinaldi & Visconti, 2023). Thus, in 2023 ethanol production is considered more favourable than methanol in industry due to the higher technology readiness of the decarbonized versions of the fuel; although some concerns are detailed in relation to its sustainability from the use of food crops (Kumar *et al.*, 2023). In this reasoning, bioethanol production from macroalgae offers a promising solution (Aslanbay Guler *et al.*, 2023).

While ethanol has clear benefits for use as an engine fuel, several shortfalls need to be addressed to favour its commercialization at a large scale. Due to ethanol being corrosive, problems can occur to the engine's pipelines. Ethanol is corrosive in three different ways: general corrosion, dry corrosion, and wet corrosion. Ionic impurities such as chloride ions and acetic acid are the main causes of general corrosion. Metals such as magnesium, lead and aluminium are often at risk of chemical attack due to dry corrosion, while wet corrosion is caused by ethanol absorbing moisture from the atmosphere leading to an oxidation of most metals (Jin *et al.*, 2011). One such method to overcome this is the use of an inhibitor such as ascorbyl palmitate that acts in protection against corrosion in C-steel in blended fuel (Deyab, 2016).

4.5 MACROALGAL BIOREFINERIES

4.5.1 Biorefinery concepts

All biorefinery concepts focus thoroughly on the maximum valorization of the algae biomass by the production of target compounds of increased value (see Chapter 10). This can be achieved by selection of the cell content and growth characteristics of macroalgae strains, which are often impacted by environmental growth conditions such as light intensity, growth habitat, seawater salinity and temperature (Biris-Dorhoi *et al.*, 2020). Meanwhile, it is also key to look at stimulating the main target compounds during macroalgae cultivation. In recent times, researchers have laid emphasis on the importance of finding multiple cascading approaches to biorefining different species of macroalgae for multiple product generation (Manikandan & Lens, 2023) (Figure 4.4).

Depending on the type of species used and the manner of cultivation, macroalgae can produce biofuels such as CH_4, CO_2, H_2, ethanol and butanol (see Section 4.4). Macroalgal biomass has several advantages over conventional energy crops. Although macroalgae are typically cultivated in the sea, land cultivation is also viable with a tumbling technique adopted. This sees a steady flow of air injected into the cultivation tank suspending the macroalgae and allowing for agitation (Titlyanov and Titlyanova, 2010). Higher production costs associated with land-based cultivation has resulted in this approach being far less common in comparison to offshore farming (Ghadiryanfar *et al.*, 2016). As of 2019, 97% of the global aquaculture output came from artificial

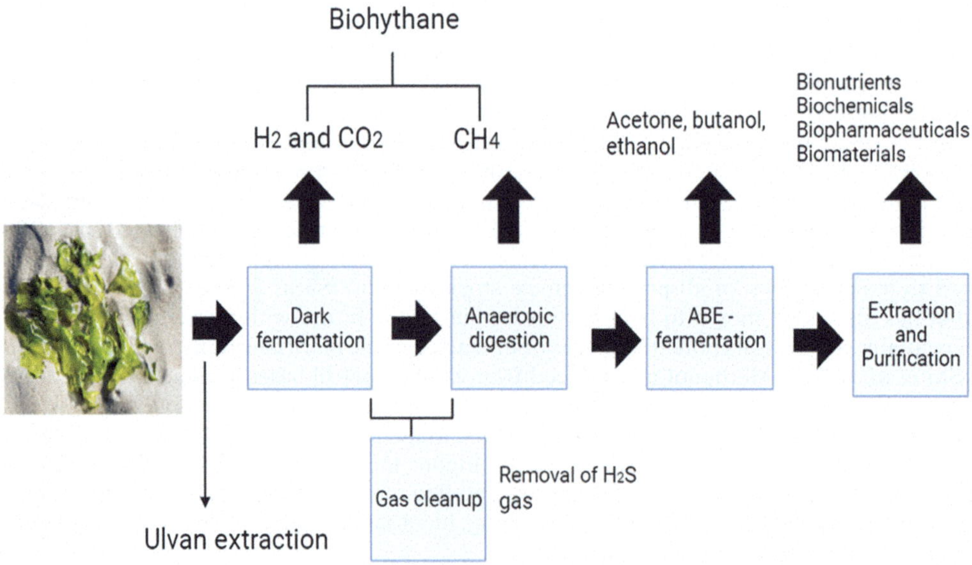

Figure 4.4 Potential biorefinery routes and products achievable utilizing macroalgae as a feedstock.

farming (Zhang *et al.*, 2022) with one study indicating that approximately 20.8 km^2 of the ocean is suitable for farming macroalgae (Liu *et al.*, 2023). As such, a major advantage of macroalgae is that it is cultivated either in the sea or on marginal non-fertile land which leads to a decrease in competition of land for human crop foods (McKennedy & Sherlock, 2015). A second important factor is related to the fact that macroalgae do not need freshwater to grow as seen by their capability to grow in salt water, which is detailed in the impact salinity can play on its morphology (Simon *et al.*, 2022). On the contrary, the major disadvantage which surrounds the use of macroalgae is related to the high expenditure for infrastructure and the energy demand and costs of harvesting (Kostas *et al.*, 2021).

Due to both environmental (Tang *et al.*, 2021) and economic (Steinbruch *et al.*, 2020) benefits associated with the macroalgae biorefinery, it is expected for this industry to grow exponentially going forward. While the benefits and potential for growth in this industry are clear to see due to the increased growth of the sector ($15.01 billion in 2021 to $24.92 billion in 2028 at a CAGR of 7.51%) (Fortune Business Insights, 2021) there are also numerous challenges to overcome. Section 4.5.3 gives a breakdown of the key challenges that must be overcome for the success of the macroalgae biorefinery going forward and its potential for further growth (Figure 4.5).

4.5.2 Key processes

4.5.2.1 Anaerobic digestion

A key concept in the process of achieving the biorefinery concept associated with macroalgae is anaerobic digestion (AD). AD involves a combination of biological processes by which CO_2 and CH_4 are produced by the breakdown of organic matter under anaerobic conditions (Adekunle & Okolie,

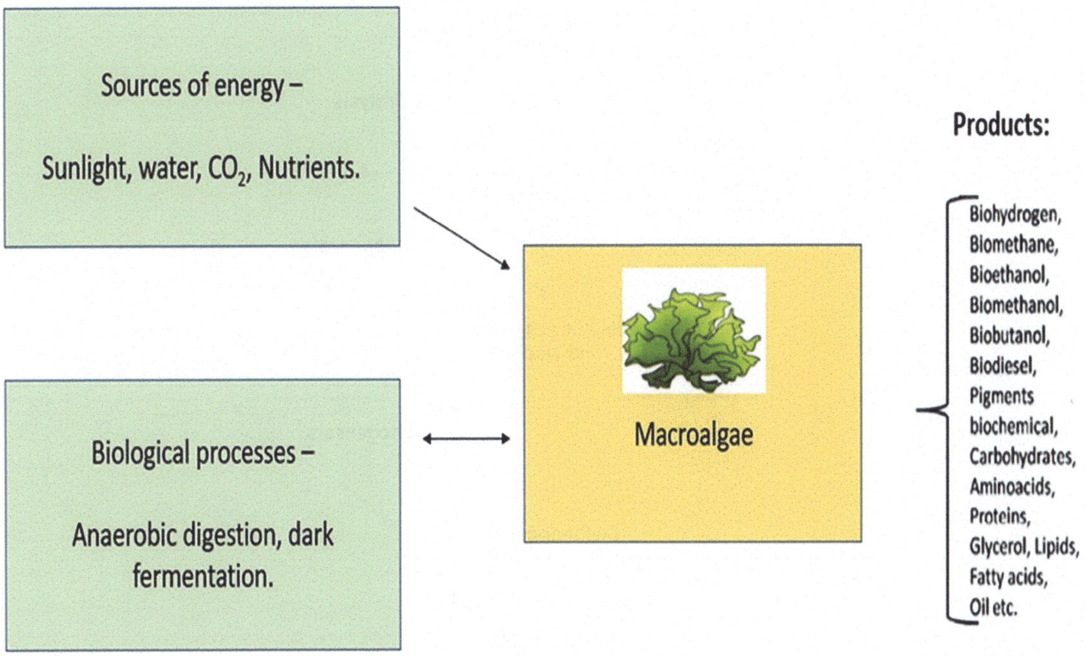

Figure 4.5 Macroalgae products relating to various biorefinery sectors. (Source: Redrawn from Rodionova *et al.*, 2017).

2015). The microbial consortium which acts during AD consists of hydrolytic bacteria, acidogenic bacteria, acetogenic bacteria, and methanogenic archaea. The first step of AD is hydrolysis, during which hydrolytic bacteria break down the substrate, that is, seaweed (macroalgae), into sugars, amino acids, and fatty acids. These compounds are then available to acidogenic bacteria which break down the sugars into VFAs and alcohols during acidogenesis. Following this step, they are converted into acetic acid or H_2 and CO_2 in a process called acetogenesis. The final stage is methanogenesis which involves the production of CH_4 and CO_2 through archaea (Meegoda, *et al.*, 2018).

The typical biogas composition is 60% methane, 38% carbon dioxide, and 2% trace gases (Frank-Whittle *et al.*, 2014). When methanogenesis is blocked, hydrogen gas can be produced in a process known as dark fermentation (Nath & Das, 2004). Dark fermentation ultimately ends with VFAs and hydrogen production by anaerobic fermentative bacteria as highlighted in Figure 4.6. The microorganisms involved include *Escherichia coli, Enterobacter aerogenes, Citrobacter intermedius, Enterobacter cloacae, Ruminococcus albus, Clostridium beijerinckii*, and *Clostridium paraputrificum* (Koutra *et al.*, 2020). Meanwhile, acidogenic fermentation involves maximizing the production of acetate by consuming H_2 to favour the acetogenesis process.

When characterizing AD by its desirable end products it can be broken down into both single-stage and two-stage AD. Two-stage AD offers advantages such as increased energy efficiency, optimal process stability and increased opportunities to control key parameters when compared to one-stage AD (Srisowmeya, *et al.*, 2020). An important aspect of both one and two-stage AD is reactor setup. In stage one, the first three phases of AD are carried out, that is hydrolysis, acidogenesis, and acetogenesis. Hydrolytic bacteria hydrolyse the complex organic polymers into monomers while

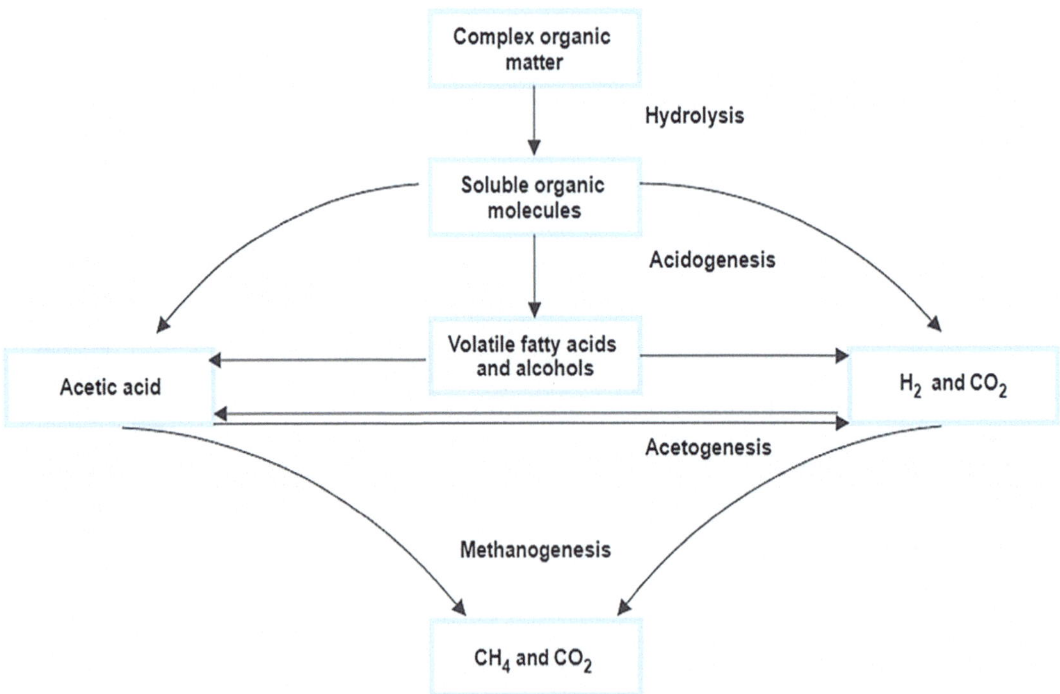

Figure 4.6 Schematic of the processes involved in anaerobic digestion.

acidogens and acetogens convert all the organic acids into acetic acid, H_2 and CO_2. In the second stage of AD, methanogens utilize the products of the first stage to produce CH_4 and CO_2 (Hans *et al.*, 2019). Table 4.3 highlights the stoichiometry involved in each of these four processes.

4.5.2.2 Reactor design

AD and dark fermentation can be carried out in a range of different reactor configurations either as attached or suspended growth systems. Continuously stirred tank reactors (CSTR) (Tabassum *et al.*, 2016)

Table 4.3 Chemical equations involved in anaerobic digestion processes.

Hydrolysis
$(C_6H_{10}O_4)_n + 2H_2O \rightarrow C_6H_{12}O_6 + O_2$
Cellulose → Glucose
$(-RCH(NH_2)COO-)_n + (n-1)H_2O \rightarrow nRCH(NH_2)COOH$
Protein → Amino acids
$(H_2COOC(CH_2)_nCH_3)_n \rightarrow n(H_2COOC(CH_2)_nCH_3)$
Fat → Fatty acids
Acidogenesis
$C_6H_{12}O_6 \rightarrow 2CH_3CH_2OH + 2CO_2$
Glucose → Ethanol
$C_6H_{12}O_6 + 2H_2 \rightarrow 2CH_3CH_2COOH + 2H_2O$
Glucose → Butyrate
$C_6H_{12}O_6 \rightarrow CH_3CH_2CH_2COOH + 2CO_2 + 2H_2$
Glucose → Butyrate
$C_6H_{12}O_6 \rightarrow 3CH_3COOH$
Glucose → Acetate
Acetogenesis
$CH_3CH_2COO^- + 3H_2O \longleftrightarrow CH_3COO^- + H^+ + HCO_3^- + 3H_2$
Propionic → Acetate
$C_6H_{12}O_6 + 2H_2O \longleftrightarrow 2CH_3COOH + 2CO_2 + 4H_2$
Glucose → Acetate
$CH_3CH_2OH + 2H_2O \longleftrightarrow CH_3COO^- + 2H_2 + H^+$
Ethanol → Acetate
Methanogenesis
$CH_3CH_2OH \rightarrow CH_4 + 2CO_2$
Ethanol
$CO_2 + 4H_2 \rightarrow CH_4 + 2H_2O$
$2CH_3CH_2OH + CO_2 \rightarrow CH_4 + 2CH_3COOH$
Ethanol → Acetate

Source: Adapted from Hans and Kumar (2019).

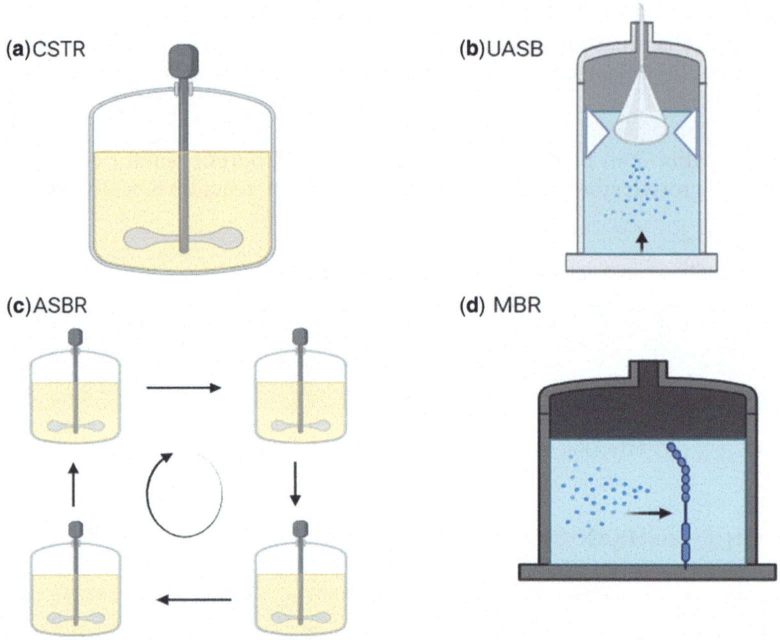

Figure 4.7 Reactor configurations for biogas production. (a) CSTR, (b) UASB, (c) ASBR and (d) MBR.

(Figure 4.7a) and up-flow anaerobic sludge blanket (UASB) (Liu *et al.*, 2013) (Figure 4.7b) reactors are the most performing and used designs. Alternatively, both anaerobic sequencing batch reactor (Figure 4.7c) and membrane bioreactor (MBR) (Figure 4.7d) are also used. CSTRs are frequently used in continuous hydrogen production. In comparison with batch reactors, microbial cultures in a CSTR are evenly suspended in the liquor with lower resistance in mass transfer. The CSTR has continuous input and output of material. The CSTR is well mixed with no dead zones or bypasses in ideal operation with typical total solid content reaching 10% (Thakur *et al.*, 2023).

UASB reactors tend to be used for high strength wastewater rich in COD, the influent wastewater flows from the bottom into the reactor, and it is distributed in an up-flow mode through a blanket of granular sludge. Wastewater flows upward through the blanket and is processed by anaerobic microorganisms. Following this, the treated effluent passes out around the edges of a funnel. UASB reactors offer several benefits such as greater contact surface area, operation at higher OLR, better settleability, enhanced solids retention and efficient solid separation from treated effluent. The sludge blanket is suspended by gravity-settling coupled with the upward flow of the effluent. Dense, spherical, compact biofilms are referred to as granules, with active methanogenic microbial consortia. Nevertheless, the disadvantage of UASB reactors is that the feedstock with high solids content prevents the development of dense granular sludge.

Anaerobic sequencing batch reactors (ASBR) are high-rate liquid digestion systems which rely on the sequential feeding of the reactor followed by mixing and the settling of solids (Figure 4.7c). The operation of this reactor is heavily influenced by factors such as the organic loading rate, temperature, pH, and substrate concentration. An ASBR reactor typically performs more effectively in terms of hydrogen production (hydrogen content $29.2 \pm 8.8\%$) when operated at a hydraulic retention time (HRT) of 24 h (Buitrón & Carvajal, 2010). Alternatively, MBRs (Figure 4.7d) have also been shown to be highly successful in terms of the dark fermentation processes due to their ability to control the biomass concentration (Show *et al.*, 2011), while Kim *et al.* (2006) found hydrogen production to increase with higher glucose concentrations (10–35 g/L). An MBR reactor offers further advantages in terms of improving effluent quality and having a smaller footprint. In choosing this reactor type it is also key to look at the desired HRT as typically membrane reactors are operated at a high volumetric rate thus resulting in much lower HRTs (<24 hrs) while it is also beneficial to keep a lower HRT to avoid membrane fouling (Rahman *et al.*, 2023).

4.5.3 Key challenges of macroalgal biorefineries

A major drawback in terms of the possibilities of utilizing macroalgae in a biorefinery concept is attributed to the scale of cultivation needed to produce enough macroalgae to make a significant impact should natural recurring resources become limited. To meet 1% of the UK's total energy demand it would require an area of cultivation of almost 5440 km^2 which is equal to half of the current global aquaculture production area (Hughes *et al.*, 2013). Needing an area of such size would cause huge problems in terms of the availability of feedstock supply chains as well as species selectivity and suitability. The impact of seasonality on cultivation is also a huge problem due to impacts on the macroalgae's biological composition and thus bioproduct potential (Kostas *et al.*, 2021). Such problems will have noticeable knock-on effects in terms of the scalability and large-scale integration of macroalgal biorefineries. Research to date has focused much on laboratory-scale projects. Further research and development are thus needed in terms of upscaling and overcoming potential challenges. Key challenges to note are in terms of complex licensing regulations (Camarena-Gómez *et al.*, 2022), which vary across the world as well as seasonal issues that can have an impact on the biochemical composition of the macroalgal species produced.

A second key issue that may hinder the progression of macroalgae biorefinery technologies is linked to the extensive costs involved in the removal of water, washing and drying steps that commence post-harvesting in preparation of the biomass. Considerable amounts of fresh water are required to wash the biomass of salts, epiphytes, and sand (Chisti, 2013). This step is also key in preservation of bioreactors as large amounts of salt contained in biomasses are known to cause both corrosive damage in bioprocess infrastructure as well as the bioreactor itself. For this reason, it is pivotal to look at ways of preserving and saving freshwater in terms of the whole process. Research to date has focused on a variety of methods such as closed production facilities and water recycling strategies (Pate *et al.*, 2011). Novel technologies such as using advanced textiles for cultivation of seaweed as trialled off the west coast of Ireland (Taelman *et al.*, 2015) have sought to cut costs associated with macroalgae farming. While early research has indicated that a complete salt-based macroalgal biorefinery concept is viable (Kostas, *et al.*, 2021), further research and development is needed in this area.

To allow for the continued growth and prosperity of the seaweed industry, it is pivotal to create a complete understanding of the optimal integrated bioprocessing pathways for each macroalgae species that is currently being cultivated. With this information, a sustainable and environmentally friendly biorefineries that generate bioproducts and biofuels from macroalgae biomasses can be generated. Going forward, industry and academia need to work together to identify the optimal bioprocessing routes for each species of seaweed utilizing environmental assessment tools for each individual bioprocess to combat the ongoing fossil fuel shortages (Su *et al.*, 2023). Such tools include attributional and consequential life cycle analysis (LCA), exergy and energy-based models. The use of such models allows for quantifiable and clear comparisons of energy yields of different feedstocks and how best to maximize and develop the numerous macroalgal biorefinery pathways.

4.6 CONCLUSION

Macroalgae have potential in terms of a feedstock for biorefinery industries to produce biochemicals and bioenergy while tackling the current issue of depleting petrochemical resources. Due to ongoing research and further scientific breakthroughs, it is expected that the macroalgae biorefinery industry will continue to grow due to the many financial and environmental benefits it possesses. This has been evidenced by the industries increasing compound annual growth rate (CAGR). While its potential for use in the production of biofuels, bioproducts and high-value products is clear, it is also obvious that further research is needed in terms of enhancement and scaling up of biorefinery systems. Problems related to cultivation and biorefinery design must be considered in terms of unlocking the full potential of this industry and the many possible economic and environmental benefits it possesses. With further collaboration between industry and academia, the seaweed biorefinery industry will become more established and continue to contribute to the creation of a low-carbon economy.

REFERENCES

Adekunle K. and Okolie J. (2015). A review of biochemical process of anaerobic digestion. *Advances in Bioscience and Biotechnology*, **6**, 205–212, https://doi.org/10.4236/abb.2015.63020

Allen E., Browne J., Hynes S. and Murphy J. D. (2013). The potential of algae blooms to produce renewable gaseous fuel. *Waste Management*, **33**(11), 2425–2433, https://doi.org/10.1016/j.wasman.2013.06.017

Allen E., Wall D. M., Herrmann C. and Murphy J. D. (2014). Investigation of the optimal percentage of green seaweed that may be co-digested with dairy slurry to produce gaseous biofuel. *Bioresource Technology*, **170**, 436–444, https://doi.org/10.1016/j.biortech.2014.08.005

Allouache A., Majda A., Toudert A. Z., Amrane A. and Ballesteros M. (2021). Cellulosic bioethanol production from *Ulva lactuca* macroalgae. *Cellulose Chemistry and Technology*, **55**(5–6), 629–635, https://doi.org/10.35812/CelluloseChemTechnol.2021.55.51

Amador-Noguez D., Brasg I. A., Feng X. J., Roquet N. and Rabinowitz J. D. (2011). Metabolome remodeling during the acidogenic-solventogenic transition in *Clostridium acetobutylicum*. *Applied and Environmental Microbiology*, **77**(22), 7984–7997, https://doi.org/10.1128/AEM.05374-11

Aslanbay Guler B., Gurlek C., Sahin Y., Oncel S. S. and Imamoglu E. (2023). Renewable bioethanol for a sustainable green future. In: A Sustainable Green Future: Perspectives on Energy, Economy, Industry, Cities and Environment. Springer International Publishing, Cham, pp. 449–480.

Awang G. M., Jones G. A., Ingledew W. M. and Kropinski A. M. B. (1988). The acetone-butanol-ethanol fermentation. *CRC Critical Reviews in Microbiology*, **15**(Suppl. 1), S33–S67, https://doi.org/10.3109/10408418809104464

Battista F., Fino D., Mancini G. and Ruggeri B. (2016). Mixing in digesters used to treat high viscosity substrates: the case of olive oil production wastes. *Journal of Environmental Chemical Engineering*, **4**(1), 915–923, https://doi.org/10.1016/j.jece.2015.12.032

Ben Amor C., Jmel M. A., Chevallier P., Mantovani D. and Smaali I. (2021). Efficient extraction of a high molecular weight ulvan from stranded *Ulva* sp. biomass: application on the active biomembrane synthesis. *Biomass Conversion and Biorefinery*, **13**, 1–11, https://doi.org/10.1007/s13399-021-01426-9

Bikker P., van Krimpen M. M., van Wikselaar P., Houweling-Tan B., Scaccia N., van Hal J. W., Huijgen W. J., Cone J. W. and López-Contreras A. M. (2016). Biorefinery of the green seaweed *Ulva lactuca* to produce animal feed, chemicals, and biofuels. *Journal of Applied Phycology*, **28**, 3511–3525, https://doi.org/10.1007/s10811-016-0842-3

Biris-Dorhoi E. S., Michiu D., Pop C. R., Rotar A. M., Tofana M., Pop O. L., Socaci S. A. and Farcas A. C. (2020). Macroalgae – a sustainable source of chemical compounds with biological activities. *Nutrients*, **12**(10), 3085, https://doi.org/10.3390/nu12103085

Bojorges H., Fabra M. J., López-Rubio A. and Martínez-Abad A. (2022). Alginate industrial waste streams as a promising source of value-added compounds valorization. *Science of the Total Environment*, **838**, 156394, https://doi.org/10.1016/j.scitotenv.2022.156394

Bolzonella D., Battista F., Cavinato C., Gottardo M., Micolucci F., Lyberatos G. and Pavan P. (2018). Recent developments in biohythane production from household food wastes: a review. *Bioresource Technology*, **257**, 311–319, https://doi.org/10.1016/j.biortech.2018.02.092

Bruhn A., Dahl J., Nielsen H. B., Nikolaisen L., Rasmussen M. B., Markager S., Olesen B., Arias C. and Jensen P. D. (2011). Bioenergy potential of *Ulva lactuca*: biomass yield, methane production and combustion. *Bioresource Technology*, **102**(3), 2595–2604, https://doi.org/10.1016/j.biortech.2010.10.010

Buitrón G. and Carvajal C. (2010). Biohydrogen production from tequila vinasses in an anaerobic sequencing batch reactor: effect of initial substrate concentration, temperature, and hydraulic retention time. *Bioresource Technology*, **101**(23), 9071–9077, https://doi.org/10.1016/j.biortech.2010.06.127

Cabanillas-Terán N., Hernández-Arana H. A., Ruiz-Zárate MÁ, Vega-Zepeda A. and Sanchez-Gonzalez A. (2019). *Sargassum* blooms in the Caribbean alter the trophic structure of the sea urchin *Diadema antillarum*. *PeerJ*, **7**, e7589, https://doi.org/10.7717/peerj.7589

Camarena-Gómez M. T., Lähteenmäki-Uutela A. and Spilling K. (2022). Macroalgae production in Northern Europe: business and government perspectives on how to regulate a novel blue bioeconomy. *Aquaculture*, **560**, 738434, https://doi.org/10.1016/j.aquaculture.2022.738434

Cao Y., Liu H., Liu W., Guo J. and Xian M. (2022). Debottlenecking the biological hydrogen production pathway of dark fermentation: insight into the impact of strain improvement. *Microbial Cell Factories*, **21**(1), 166, https://doi.org/10.1186/s12934-022-01893-3

CBI (2023). The European market potential for seaweed, Available at: https://www.cbi.eu/market-information/fish-seafood/seaweed/market-potentia

Cecchi F., Pavan P. and Mata-Alvarez J. (1996). Anaerobic co-digestion of sewage sludge: application to the macroalgae from the Venice lagoon. *Resources, Conservation and Recycling*, **17**(1), 57–66, https://doi.org/10.1016/0921-3449(96)88182-1

Chen Y., Cheng J. J. and Creamer K. S. (2008). Inhibition of anaerobic digestion process: a review. *Bioresource Technology*, **99**(10), 4044–4064, https://doi.org/10.1016/j.biortech.2007.01.057

Chisti Y. (2013). Constraints to commercialization of algal fuels. *Journal of Biotechnology*, **167**(3), 201–214, https://doi.org/10.1016/j.jbiotec.2013.07.020

Costa J. C., Oliveira J. V., Pereira M. A., Alves M. M. and Abreu A. A. (2015). Biohythane production from marine macroalgae *Sargassum* sp. coupling dark fermentation and anaerobic digestion. *Bioresource Technology*, **190**, 251–256, https://doi.org/10.1016/j.biortech.2015.04.052

Dahlgren S., Kanda W. and Anderberg S. (2022). Drivers for and barriers to biogas use in manufacturing, road transport and shipping: a demand-side perspective. *Biofuels*, **13**(2), 177–188, https://doi.org/10.1080/17597269.2019.1657661

de Fournas N. and Wei M. (2022). Techno-economic assessment of renewable methanol from biomass gasification and PEM electrolysis for decarbonization of the maritime sector in California. *Energy Conversion and Management*, **257**, 115440, https://doi.org/10.1016/j.enconman.2022.115440

Deyab M. A. (2016). Corrosion inhibition of aluminum in biodiesel by ethanol extracts of Rosemary leaves. *Journal of the Taiwan Institute of Chemical Engineers*, **58**, 536–541, https://doi.org/10.1016/j.jtice.2015.06.021

Ding L., Cheng J., Lin R., Deng C., Zhou J. and Murphy J. D. (2020). Improving biohydrogen and biomethane co-production via two-stage dark fermentation and anaerobic digestion of the pretreated seaweed *Laminaria digitata*. *Journal of Cleaner Production*, **251**, 119666, https://doi.org/10.1016/j.jclepro.2019.119666

Dogmaz S. and Cavas L. (2023). Biohydrogen production via green silver nanoparticles synthesized through biomass of bloomed *Ulva lactuca*. *Bioresource Technology*, **379**, 129028, https://doi.org/10.1016/j.biortech.2023.129028

Drapcho C. M., Nghim N. P. and Walker T. (2008). Biofuels Engineering Process Technology. McGraw-Hill Education, New York, USA.

Dutta K., Daverey A. and Lin J. G. (2014). Evolution retrospective for alternative fuels: first to fourth generation. *Renewable Energy*, **69**, 114–122, https://doi.org/10.1016/j.renene.2014.02.044

EEA (2021a). EEA. https://www.eea.europa.eu/highlights/eu-achieves-20-20-20

EEA (2021b). EEA. https://www.eea.europa.eu/ims/use-of-renewable-energy-for

EEA (2023). EEA. https://www.eea.europa.eu/en/topics/in-depth/renewable-energy#:~:text=As%20a%20result%2C%20greenhouse%20gas,and%20ordinary%20sights%20across%20Europe

Enquist-Newman M., Faust A. M. E., Bravo D. D., Santos C. N. S., Raisner R. M., Hanel A., Sarvabhowman P., Le C., Regitsky D. D., Cooper S. R. and Peereboom L. (2014). Efficient ethanol production from brown macroalgae sugars by a synthetic yeast platform. *Nature*, **505**(7482), 239–243, https://doi.org/10.1038/nature12771

Farghali M., Yuhendra A. P., Mohamed I. M., Iwasaki M., Tangtaweewipat S., Ihara I., Sakai R. and Umetsu K. (2021). Thermophilic anaerobic digestion of *Sargassum fulvellum* macroalgae: biomass valorization and biogas optimization under different pre-treatment conditions. *Journal of Environmental Chemical Engineering*, **9**(6), 106405.

Forrest K., Mac Kinnon M., Tarroja B. and Samuelsen S. (2020). Estimating the technical feasibility of fuel cell and battery electric vehicles for the medium and heavy-duty sectors in California. *Applied Energy*, **276**, 115439, https://doi.org/10.1016/j.apenergy.2020.115439

Fortune Business Insights (2021). Fortune Business Insights. https://www.fortunebusinessinsights.com/industry-reports/commercial-seaweed-market-100077

Frank-Whittle I. H., Walter A., Ebner C. and Insam H. (2014). Investigation into the effect of high concentration of volatile fatty acids in anaerobic digestion on methanogenic communities. *Waste Management*, **34**, 2080–2089, https://doi.org/10.1016/j.wasman.2014.07.020

Ganesan K., Budtova T., Ratke L., Gurikov P., Baudron V., Preibisch I., Niemeyer P., Smirnova I. and Milow B. (2018). Review on the production of polysaccharide aerogel particles. *Materials*, **11**(11), 2144, https://doi.org/10.3390/ma11112144

Ghadiryanfar M., Rosentrater K. A., Keyhani A. and Omid M. (2016). A review of macroalgae production, with potential applications in biofuels and bioenergy. *Renewable and Sustainable Energy Reviews*, **54**, 473–481, https://doi.org/10.1016/j.rser.2015.10.022

Ghimire A., Kumar G., Sivagurunathan P., Shobana S., Saratale G. D., Kim H. W., Luongo V., Esposito G. and Munoz R. (2017). Bio-hythane production from microalgae biomass: key challenges and potential opportunities for algal bio-refineries. *Bioresource Technology*, **241**, 525–536, https://doi.org/10.1016/j.biortech.2017.05.156

Gray N., McDonagh S., O'Shea R., Smyth B. and Murphy J. D. (2021). Decarbonising ships, planes, and trucks: An analysis of suitable low-carbon fuels for the maritime, aviation and haulage sectors. *Advances in Applied Energy*, **1**, 100008, https://doi.org/10.1016/j.adapen.2021.100008

Guiry M. D. (2023). AlgaeBase. world-wide electronic publication, National University of Ireland, Galway. Available at: http://www.algaebase.org (Accessed July 11, 2023).

Hallenbeck P. C. (2013). Fundamentals of biohydrogen. In: Biohydrogen, A. Pandey, J. S. Chang, P. Hallenbeck and C. Larroche (eds.), **588**, Elsevier, Netherlands, pp. 25–43.

Han D., Zhu T. and Row K. H. (2011). Ultrasonic extraction of phenolic compounds from *Laminaria japonica* Aresch using ionic liquid as extraction solvent. *Bulletin of the Korean Chemical Society*, **32**(7), 2212–2216, https://doi.org/10.5012/bkcs.2011.32.7.2212

Hans M. and Kumar S. (2019). Biohythane production in two-stage anaerobic digestion system. *International Journal of Hydrogen Energy*, **44**(32), 17363–17380, https://doi.org/10.1016/j.ijhydene.2018.10.022

Hans M., Kumar S., Chandel A. K. and Polikarpov I. (2019). A review on bioprocessing of paddy straw to ethanol using simultaneous saccharification and fermentation. *Process Biochemistry*, **85**, 125–134, https://doi.org/10.1016/j.procbio.2019.06.019

Harb T. B. and Chow F. (2022). An overview of beach-cast seaweeds: potential and opportunities for the valorization of underused waste biomass. *Algal Research*, **62**, 102643, https://doi.org/10.1016/j.algal.2022.102643

Hierholtzer A., Chatellard L., Kierans M., Akunna J. C. and Collier P. J. (2013). The impact and mode of action of phenolic compounds extracted from brown seaweed on mixed anaerobic microbial cultures. *Journal of Applied Microbiology*, **114**(4), 964–973, https://doi.org/10.1111/jam.12114

Hinds C., Oxenford H., Cumberbatch J., Fardin F. and Cashman A. (2016). *Sargassum* Management Brief.

Holdt S. L. and Kraan S. (2011). Bioactive compounds in seaweed: functional food applications and legislation. *Journal of Applied Phycology*, **23**, 543–597, https://doi.org/10.1007/s10811-010-9632-5

Hughes A. D., Black K. D., Campbell I., Heymans J. J., Orr K. K., Stanley M. S. and Kelly M. S. (2013). Comments on 'Prospects for the use of macroalgae for fuel in Ireland and UK: an overview of marine management issues'. *Marine Policy*, **38**, 554–556, https://doi.org/10.1016/j.marpol.2012.08.001

IEA (2014). World Energy Outlook 2014. IEA, Paris. https://www.iea.org/reports/world-energy-outlook-2014, License: CC BY 4.0.

Jin C., Yao M., Liu H., Chia-fon F. L. and Ji J. (2011). Progress in the production and application of n-butanol as a biofuel. *Renewable and Sustainable Energy Reviews*, **15**(8), 4080–4106, https://doi.org/10.1016/j.rser.2011.06.001

Keskin T., Abubackar H. N., Arslan K. and Azbar N. (2019). Biohydrogen production from solid wastes. In: Biohydrogen, A. Pandey, S. Venkata Mohan, J.-S. Chang, P. C. Hallenbeck and C. Larroche (eds.), 2nd edn, Elsevier, pp. 321–346.

Khalafalla S. S., Zahid U., Abdul Jameel A. G., Ahmed U., Alenazey F. S. and Lee C. J. (2020). Conceptual design development of coal-to-methanol process with carbon capture and utilization. *Energies*, **13**(23), 6421, https://doi.org/10.3390/en13236421

Kidgell J. T., Magnusson M., de Nys R. and Glasson C. R. (2019). Ulvan: a systematic review of extraction, composition, and function. *Algal Research*, **39**, 101422, https://doi.org/10.1016/j.algal.2019.101422

Kidgell J. T., Glasson C. R., Magnusson M., Vamvounis G., Sims I. M., Carnachan S. M., Hinkley S. F., Lopata A. L., de Nys R. and Taki A. C. (2020). The molecular weight of ulvan affects the in vitro inflammatory response of a murine macrophage. *International Journal of Biological Macromolecules*, **150**, 839–848, https://doi.org/10.1016/j.ijbiomac.2020.02.071

Kim S. H., Han S. K. and Shin H. S. (2006). Effect of substrate concentration on hydrogen production and 16S rDNA-based analysis of the microbial community in a continuous fermenter. *Process Biochemistry*, **41**(1), 199–207, https://doi.org/10.1016/j.procbio.2005.06.013

Kober T., Schiffer H. W., Densing M. and Panos E. (2020). Global energy perspectives to 2060–WEC's World Energy Scenarios 2019. *Energy Strategy Reviews*, **31**, 100523, https://doi.org/10.1016/j.esr.2020.100523

Korzen L., Pulidindi I. N., Israel A., Abelson A. and Gedanken A. (2015). Single step production of bioethanol from the seaweed *Ulva rigida* using sonication. *RSC Advances*, **5**(21), 16223–16229, https://doi.org/10.1039/C4RA14880K

Kostas E. T., Adams J. M., Ruiz H. A., Durán-Jiménez G. and Lye G. J. (2021). Macroalgal biorefinery concepts for the circular bioeconomy: a review on biotechnological developments and future perspectives. *Renewable and Sustainable Energy Reviews*, **151**, 111553, https://doi.org/10.1016/j.rser.2021.111553

Koutra E., Tsafrakidou P., Sakarika M. and Kornaros M. (2020). Microalgal biorefinery. In: Microalgae Cultivation for Biofuels Production, A. Yousuf (ed.), Academic Press, pp. 163–185.

Kumar M. D., Sharmila V. G., Kumar G., Park J. H., Al-Qaradawi S. Y. and Banu J. R. (2022). Surfactant induced microwave disintegration for enhanced biohydrogen production from macroalgae biomass: thermodynamics and energetics. *Bioresource Technology*, **350**, 126904, https://doi.org/10.1016/j.biortech.2022.126904

Kumar N., Chaudhary A., Ahlawat O. P., Naorem A., Upadhyay G., Chhokar R. S., Gill S. C., Khippal A., Tripathi S. C. and Singh G. P. (2023). Crop residue management challenges, opportunities and way forward for sustainable food-energy security in India: a review. *Soil and Tillage Research*, **228**, 105641, https://doi.org/10.1016/j.still.2023.105641

Lahaye M. and Robic A. (2007). Structure and functional properties of ulvan, a polysaccharide from green seaweeds. *Biomacromolecules*, **8**(6), 1765–1774, https://doi.org/10.1021/bm061185q

Lakshmi D. S., Sankaranarayanan S., Gajaria T. K., Li G., Kujawski W., Kujawa J. and Navia R. (2020). A short review on the valorization of green seaweeds and ulvan: feedstock for chemicals and biomaterials. *Biomolecules*, **10**(7), 991, https://doi.org/10.3390/biom10070991

Lay C. H., Kumar G., Mudhoo A., Lin C. Y., Leu H. J., Shobana S. and Nguyen M. L. T. (2020). Recent trends and prospects in biohythane research: an overview. *International Journal of Hydrogen Energy*, **45**(10), 5864–5873, https://doi.org/10.1016/j.ijhydene.2019.07.209

Lehahn Y., Ingle K. N. and Golberg A. (2016). Global potential of offshore and shallow waters macroalgal biorefineries to provide for food, chemicals, and energy: feasibility and sustainability. *Algal Research*, **17**, 150–160, https://doi.org/10.1016/j.algal.2016.03.031

Li Y., Tang W., Chen Y., Liu J. and Chia-fon F. L. (2019). Potential of acetone-butanol-ethanol (ABE) as a biofuel. *Fuel*, **242**, 673–686, https://doi.org/10.1016/j.fuel.2019.01.063

Liu T., Li Z., Li W., Shi C. and Wang Y. (2013). Preparation and characterization of biomass carbon-based solid acid catalyst for the esterification of oleic acid with methanol. *Bioresource Technology*, **133**, 618–621, https://doi.org/10.1016/j.biortech.2013.01.163

Liu C., Lin F., Kong X., Fan Y., Xu W., Lei M. and Xiao R. (2022). Lignin-first biorefinery of corn stalk via zirconium (IV) chloride/sodium hydroxide-catalyzed aerobic oxidation to produce phenolic carbonyls. *Bioresource Technology*, **354**, 127183, https://doi.org/10.1016/j.biortech.2022.127183

Liu Y., Cao L., Cheung W. W. and Sumaila U. R. (2023). Global estimates of suitable areas for marine algae farming. *Environmental Research Letters*, **18**(6), 064028, https://doi.org/10.1088/1748-9326/acd398

Llano T., Arce C., Gallart L. E., Perales A. and Coz A. (2023). Techno-economic analysis of macroalgae biorefineries: a comparison between ethanol and butanol facilities. *Fermentation*, **9**(4), 340, https://doi.org/10.3390/fermentation9040340

Loret E. A., Lluesma M., Cassien A., Brinvillier D., Villard P. H. and Loret E. (2020). A marine virus controls the green algae *Ulva lactuca* proliferation in the bay of Marseille.

Madejón E., Panettieri M., Madejón P. and Perez-de-Mora A. (2022). Composting as sustainable managing option for seaweed blooms on recreational beaches. *Waste and Biomass Valorization*, **13**(2), 863–875, https://doi.org/10.1007/s12649-021-01548-1

Manikandan N. A. and Lens P. N. (2022a). Biorefining of green macroalgal (*Ulva* sp.) biomass and its application in the adsorptive recovery of rare earth elements (REEs). *Separation and Purification Technology*, **303**, 122200, https://doi.org/10.1016/j.seppur.2022.122200

Manikandan N. A. and Lens P. N. (2022b). Green extraction and esterification of marine polysaccharide (ulvan) from green macroalgae *Ulva* sp. using citric acid for hydrogel preparation. *Journal of Cleaner Production*, **366**, 132952, https://doi.org/10.1016/j.jclepro.2022.132952

Manikandan N. A. and Lens P. N. (2023). Sustainable biorefining and bioprocessing of green seaweed (*Ulva* sp.) for the production of edible (ulvan) and non-edible (polyhydroxyalkanoate) biopolymeric films. *Microbial Cell Factories*, **22**(1), 140, https://doi.org/10.1186/s12934-023-02154-7

McKennedy J. and Sherlock O. (2015). Anaerobic digestion of marine macroalgae: a review. *Renewable and Sustainable Energy Reviews*, **52**, 1781–1790, https://doi.org/10.1016/j.rser.2015.07.101

Meegoda J. N., Li B., Patel K. and Wang L. B. (2018). A review of the processes, parameters, and optimization of anaerobic digestion. *International Journal of Environmental Research and Public Health*, **15**(10), 2224, https://doi.org/10.3390/ijerph15102224

Mhatre A., Gore S., Mhatre A., Trivedi N., Sharma M., Pandit R., Anil A. and Lali A. (2019). Effect of multiple product extractions on bio-methane potential of marine macrophytic green alga *Ulva lactuca*. *Renewable Energy*, **132**, 742–751, https://doi.org/10.1016/j.renene.2018.08.012

Mordor intelligence (2022). Mordor intelligence, Available at: https://www.mordorintelligence.com/industry-reports/commercial-seaweed-market

Morelli A. and Chiellini F. (2010). Ulvan as a new type of biomaterial from renewable resources: functionalization and hydrogel preparation. *Macromolecular Chemistry and Physics*, **211**, 821–832, https://doi.org/10.1002/macp.200900562

Murphy J. D., Drosg B., Allen E., Jerney J., Xia A. and Herrmann C. (2015). A Perspective on Algal Biogas. IEA Bioenergy, Paris, France, pp. 1–40.

Nath K. and Das D. (2004). Improvement of fermentative hydrogen production: various approaches. *Applied Microbiology and Biotechnology*, **65**, 520–, https://doi.org/10.1007/s00253-004-1644-0

Nigam P. S. and Singh A. (2011). Production of liquid biofuels from renewable resources. *Progress in Energy and Combustion Science*, **37**(1), 52–68, https://doi.org/10.1016/j.pecs.2010.01.003

Oliva A., Papirio S., Esposito G. and Lens P. N. L. (2022). Pretreatment of lignocellulosic materials to enhance their methane potential. In: Renewable Energy Technologies for Energy Efficient Sustainable Development, A. Sinharoy, P. N. L. Lens (eds.), Applied Environmental Science and Engineering for a Sustainable Future, Springer, Cham, 85–120.

Olsson J., Toth G. B. and Albers E. (2020). Biochemical composition of red, green, and brown seaweeds on the Swedish west coast. *Journal of Applied Phycology*, **32**(5), 3305–3317, https://doi.org/10.1007/s10811-020-02145-w

Ortiz J., Romero N., Robert P., Araya J., Lopez-Hernández J., Bozzo C., Navarrete E., Osorio A. and Rios A. (2006). Dietary fiber, amino acid, fatty acid and tocopherol contents of the edible seaweeds Ulva lactuca and Durvillaea antarctica. *Food chemistry*, **99**(1), 98–104.

Oxenford H. A., Cox S. A., van Tussenbroek B. I. and Desrochers A. (2021). Challenges of turning the *Sargassum* crisis into gold: current constraints and implications for the Caribbean. *Phycology*, **1**(1), 27–48, https://doi.org/10.3390/phycology1010003

Papadis E. and Tsatsaronis G. (2020). Challenges in the decarbonization of the energy sector. *Energy*, **205**, 118025, https://doi.org/10.1016/j.energy.2020.118025

Patakova P., Linhova M., Rychtera M., Paulova L. and Melzoch K. (2013). Novel and neglected issues of acetone–butanol–ethanol (ABE) fermentation by *Clostridia*: *Clostridium* metabolic diversity, tools for process mapping and continuous fermentation systems. *Biotechnology Advances*, **31**(1), 58–67, https://doi.org/10.1016/j.biotechadv.2012.01.010

Pate R., Klise G. and Wu B. (2011). Resource demand implications for US algae biofuels production scale-up. *Applied Energy*, **88**(10), 3377–3388, https://doi.org/10.1016/j.apenergy.2011.04.023

Pereira L. (2021). Macroalgae. *Encyclopedia*, **1**(1), 177–188, https://doi.org/10.3390/encyclopedia1010017

Potts T., Du J., Paul M., May P., Beitle R. and Hestekin J. (2012). The production of butanol from Jamaica bay macroalgae. *Environmental Progress & Sustainable Energy*, **31**(1), 29–36.

Potts L. D., Perez Calderon L. J., Gontikaki E., Keith L., Gubry-Rangin C., Anderson J. A. and Witte U. (2018). Effect of spatial origin and hydrocarbon composition on bacterial consortia community structure and hydrocarbon biodegradation rates. *FEMS Microbiology Ecology*, **94**(9), fiy127, https://doi.org/10.1093/femsec/fiy127

Pouillart P. R. (1998). Role of butyric acid and its derivatives in the treatment of colorectal cancer and hemoglobinopathies. *Life Sciences*, **63**(20), 1739–1760, https://doi.org/10.1016/S0024-3205(98)00279-3

Qu X., Zeng H., Gao Y., Mo T. and Li Y. (2022). Bio-hydrogen production by dark anaerobic fermentation of organic wastewater. *Frontiers in Chemistry*, **10**, 978907, https://doi.org/10.3389/fchem.2022.978907

Rahman T. U., Roy H., Islam M. R., Tahmid M., Fariha A., Mazumder A., Tasnim N., Pervez M. N., Cai Y., Naddeo V. and Islam M. S. (2023). The advancement in membrane bioreactor (MBR) technology toward sustainable industrial wastewater management. *Membranes*, **13**(2), 181, https://doi.org/10.3390/membranes13020181

Rakopoulos D. C., Rakopoulos C. D., Giakoumis E. G., Dimaratos A. M. and Kyritsis D. C. (2010). Effects of butanol–diesel fuel blends on the performance and emissions of a high-speed DI diesel engine. *Energy Conversion and Management*, **51**(10), 1989–1997, https://doi.org/10.1016/j.enconman.2010.02.032

Rasyid A. (2017). Evaluation of nutritional composition of the dried seaweed *Ulva lactuca* from Pameungpeuk waters, Indonesia. *Tropical Life Sciences Research*, **28**(2), 119–125, https://doi.org/10.21315/tlsr2017.28.2.9

Reith J. H., Wijffels R. H. and Barten H. (2003). Bio-methane & Bio-Hydrogen, Status and Perspectives of Biological Methane and Hydrogen Production. Dutch Biological Hydrogen Foundation, The Netherlands, pp. 118–125.

Resiere D., Valentino R., Nevière R., Banydeen R., Gueye P., Florentin J., Cabié A., Lebrun T., Mégarbane B., Guerrier G. and Mehdaoui H. (2018). *Sargassum* seaweed on Caribbean islands: an international public health concern. *The Lancet*, **392**(10165), 2691, https://doi.org/10.1016/S0140-6736(18)32777-6

Richter H., Molitor B., Wei H., Chen W., Aristilde L. and Angenent L. T. (2016). Ethanol production in syngas-fermenting *Clostridium ljungdahlii* is controlled by thermodynamics rather than by enzyme expression. *Energy & Environmental Science*, **9**(7), 2392–2399, https://doi.org/10.1039/C6EE01108J

Rinaldi R. and Visconti C. G. (2023). Flexible operations of a multi-tubular reactor for methanol synthesis from biogas exploiting green hydrogen. *Chemical Engineering Science*, **272**, 118611, https://doi.org/10.1016/j.ces.2023.118611

Rodionova M. V., Poudyal R. S., Tiwari I., Voloshin R. A., Zharmukhamedov S. K., Nam H. G., Zayadan B. K., Bruce B. D., Hou H. J. and Allakhverdiev S. I. (2017). Biofuel production: challenges and opportunities. *International Journal of Hydrogen Energy*, **42**(12), 8450–8461, https://doi.org/10.1016/j.ijhydene.2016.11.125

Rodríguez-Martínez R. E., Torres-Conde E. G. and Jordán-Dahlgren E. (2023). Pelagic *Sargassum* cleanup cost in Mexico. *Ocean & Coastal Management*, **237**, 106542, https://doi.org/10.1016/j.ocecoaman.2023.106542

Ryther J. H., Debusk T. A. and Blakeslee M. (1984). Cultivation and conversion of marine macroalgae. [*Gracilaria* and *Ulva*] (No. SERI/STR-231-2360). Harbor Branch Foundation, Fort Pierce, FL (USA).

Sanap P. P., Diwan A. G. and Mahajan Y. S. (2023). Ethanol blending in petrol: a techno-commercial overview. *Materials Today: Proceedings*, https://doi.org/10.1016/j.matpr.2023.04.055

Saqib A., Tabbssum M. R., Rashid U., Ibrahim M., Gill S. S. and Mehmood M. A. (2013). Marine macroalgae *Ulva*: a potential feed-stock for bioethanol and biogas production. *Asian Journal of Agriculture and Biology*, **1**(3), 155–163.

Saraf N. and Shastri Y. (2023). System dynamics-based assessment of novel transport options adoption in India. *Clean Technologies and Environmental Policy*, **25**(3), 799–823, https://doi.org/10.1007/s10098-022-02398-8

Shi X., Jung K. W., Kim D. H., Ahn Y. T. and Shin H. S. (2011). Direct fermentation of *Laminaria japonica* for biohydrogen production by anaerobic mixed cultures. *International Journal of Hydrogen Energy*, **36**(10), 5857–5864, https://doi.org/10.1016/j.ijhydene.2011.01.125

Shi J., Zhu Y., Feng Y., Yang J. and Xia C. (2023). A prompt decarbonization pathway for shipping: green hydrogen, ammonia, and methanol production and utilization in marine engines. *Atmosphere*, **14**(3), 584, https://doi.org/10.3390/atmos14030584

Show K. Y., Lee D. J. and Chang J. S. (2011). Bioreactor and process design for biohydrogen production. *Bioresource Technology*, **102**(18), 8524–8533, https://doi.org/10.1016/j.biortech.2011.04.055

Simon C., McHale M. and Sulpice R. (2022). Applications of *Ulva* biomass and strategies to improve its yield and composition: a perspective for *Ulva* aquaculture. *Biology*, **11**(11), 1593, https://doi.org/10.3390/biology11111593

Smith V. H., Sturm B. S., Denoyelles F. J. and Billings S. A. (2010). The ecology of algal biodiesel production. *Trends in Ecology & Evolution*, **25**(5), 301–309, https://doi.org/10.1016/j.tree.2009.11.007

Snehya A. V., Sundaramahalingam M. A., Rajeshbanu J., Anandan S. and Sivashanmugam P. (2021). Studies on evaluation of surfactant coupled sonication pretreatment on *Ulva fasciata* (marine macroalgae) for enhanced biohydrogen production. *Ultrasonics Sonochemistry*, **81**, 105853, https://doi.org/10.1016/j.ultsonch.2021.105853

Snehya A. V., Sundaramahalingam M. A., Rajeshbanu J. and Sivashanmugam P. (2022). Studies on evaluation of pretreatment efficiency of surfactant mediated ultrasonic pretreatment of *Sargassum tennerimum* (marine macroalgae) for achieving profitable biohydrogen production. *International Journal of Hydrogen Energy*, **47**(39), 17184–17193, https://doi.org/10.1016/j.ijhydene.2022.03.216

Soares J. F., Confortin T. C., Todero I., Mayer F. D. and Mazutti M. A. (2020). Dark fermentative biohydrogen production from lignocellulosic biomass: technological challenges and future prospects. *Renewable and Sustainable Energy Reviews*, **117**, 109484, https://doi.org/10.1016/j.rser.2019.109484

Srisowmeya G., Chakravarthy M. and Devi G. N. (2020). Critical considerations in two-stage anaerobic digestion of food waste – a review. *Renewable and Sustainable Energy Reviews*, **119**, 109587, https://doi.org/10.1016/j.rser.2019.109587

Steinbruch E., Drabik D., Epstein M., Ghosh S., Prabhu M. S., Gozin M., Kribus A. and Golberg A. (2020). Hydrothermal processing of a green seaweed *Ulva* sp. for the production of monosaccharides, polyhydroxyalkanoates, and hydrochar. *Bioresource Technology*, **318**, 124263, https://doi.org/10.1016/j.biortech.2020.124263

Su C. W., Pang L. D., Qin M., Lobonţ O. R. and Umar M. (2023). The spillover effects among fossil fuel, renewables and carbon markets: evidence under the dual dilemma of climate change and energy crises. *Energy*, **274**, 127304, https://doi.org/10.1016/j.energy.2023.127304

Sunwoo I. Y., Kwon J. E., Nguyen T. H., Ra C. H., Jeong G. T. and Kim S. K. (2017). Bioethanol production using waste seaweed obtained from Gwangalli beach, Busan, Korea by co-culture of yeasts with adaptive evolution. *Applied Biochemistry and Biotechnology*, **183**, 966–979, https://doi.org/10.1007/s12010-017-2476-6

Tabassum M. R., Xia A. and Murphy J. D. (2016). Seasonal variation of chemical composition and biomethane production from the brown seaweed *Ascophyllum nodosum*. *Bioresource Technology*, **216**, 219–226, https://doi.org/10.1016/j.biortech.2016.05.071

Tabassum M. R., Xia A. and Murphy J. D. (2017a). Comparison of pre-treatments to reduce salinity and enhance biomethane yields of *Laminaria digitata* harvested in different seasons. *Energy*, **140**, 546–551, https://doi.org/10.1016/j.energy.2017.08.070

Tabassum M. R., Xia A. and Murphy J. D. (2017b). Potential of seaweed as a feedstock for renewable gaseous fuel production in Ireland. *Renewable and Sustainable Energy Reviews*, **68**, 136–146, https://doi.org/10.1016/j.rser.2016.09.111

Taelman S. E., Champenois J., Edwards M. D., De Meester S. and Dewulf J. (2015). Comparative environmental life cycle assessment of two seaweed cultivation systems in North West Europe with a focus on quantifying sea surface occupation. *Algal Research*, **11**, 173–183, https://doi.org/10.1016/j.algal.2015.06.018

Tang T., Effiong K., Hu J., Li C. and Xiao X. (2021). Chemical prevention and control of the green tide and fouling organism *Ulva*: key chemicals, mechanisms, and applications. *Frontiers in Marine Science*, **8**, 618950, https://doi.org/10.3389/fmars.2021.618950

Thakur H., Verma N. K., Dhar A. and Powar S. (2023). Investigation of continuous stirred tank reactors for improving the mixing in anaerobic digestion: A numerical study. *Results in Engineering*, **19**, 101317, https://doi.org/10.1016/j.rineng.2023.101317

Thompson T. M., Young B. R. and Baroutian S. (2020). Pelagic *Sargassum* for energy and fertiliser production in the Caribbean: a case study on Barbados. *Renewable and Sustainable Energy Reviews*, **118**, 109564, https://doi.org/10.1016/j.rser.2019.109564

Thorsteinsson M., Weisbjerg M. R., Lund P., Battelli M., Chassé É, Bruhn A. and Nielsen M. O. (2023). Effects of seasonal and interspecies differences in macroalgae procured from temperate seas on the Northern hemisphere on in vitro methane mitigating properties and rumen degradability. *Algal Research*, **73**, 103139, https://doi.org/10.1016/j.algal.2023.103139

Titlyanov E. A. and Titlyanova T. V. (2010). Seaweed cultivation: methods and problems. *Russian Journal of Marine Biology*, **36**, 227–242.

Wang X., Zhang Y., Xia C., Alqahtani A., Sharma A. and Pugazhendhi A. (2023). A review on optimistic biorefinery products: biofuel and bioproducts from algae biomass. *Fuel*, **338**, 127378, https://doi.org/10.1016/j.fuel.2022.127378

Wekre M. E., Hellesen Brunvoll S. and Jordheim M. (2022). Advancing quantification methods for polyphenols in brown seaweeds – applying a selective qNMR method compared with the TPC assay. *Phytochemical Analysis*, **33**(7), 1099–1110, https://doi.org/10.1002/pca.3162

Wekre M. E., Holmelid B., Underhaug J., Pedersen B., Kopplin G. and Jordheim M. (2023). Characterization of high value products in the side-stream of *Laminaria hyperborea* alginate production – targeting the phenolic content. *Algal Research*, **72**, 103109, https://doi.org/10.1016/j.algal.2023.103109

Xia A., Cheng J. and Murphy J. D. (2016). Innovation in biological production and upgrading of methane and hydrogen for use as gaseous transport biofuel. *Biotechnology Advances*, **34**(5), 451–472, https://doi.org/10.1016/j.biotechadv.2015.12.009

Yaich H., Amira A. B., Abbes F., Bouaziz M., Besbes S., Richel A., Blecker C., Attia H. and Garna H. (2017). Effect of extraction procedures on structural, thermal and antioxidant properties of ulvan from *Ulva lactuca* collected in Monastir coast. *International Journal of Biological Macromolecules*, **105**, 1430–1439, https://doi.org/10.1016/j.ijbiomac.2017.07.141

Yin Y. and Wang J. (2018). Pretreatment of macroalgal *Laminaria japonica* by combined microwave-acid method for biohydrogen production. *Bioresource Technology*, **268**, 52–59, https://doi.org/10.1016/j.biortech.2018.07.126

Yuhendra A. P., Farghali M., Mohamed I. M., Iwasaki M., Tangtaweewipat S., Ihara I., Sakai R. and Umetsu K. (2021). Potential of biogas production from the anaerobic digestion of *Sargassum fulvellum* macroalgae: influences of mechanical, chemical, and biological pretreatments. *Biochemical Engineering Journal*, **175**, 108140, https://doi.org/10.1016/j.bej.2021.108140

Yusri I. M., Mamat R., Akasyah M. K., Jamlos M. F. and Yusop A. F. (2019). Evaluation of engine combustion and exhaust emissions characteristics using diesel/butanol blended fuel. *Applied Thermal Engineering*, **156**, 209–219, https://doi.org/10.1016/j.applthermaleng.2019.02.028

Zhang L., Liao W., Huang Y., Wen Y., Chu Y. and Zhao C. (2022). Global seaweed farming and processing in the past 20 years. *Food Production. Processing and Nutrition*, **4**(1), 23, https://doi.org/10.1186/s43014-022-00103-2

Part 2

Algae-Based Wastewater Treatment

doi: 10.2166/9781789063547_0077

Chapter 5

Wastewater treatment by microalgae-based processes

F. G. Acien[1], C. Gómez[1], A. Morillas[1], A. Zouhayr[2], A. Sanchez[1], R. Nordio[1], E. Rodriguez[3], J. L. Guzman[3] and J. M. Fernández-Sevilla[1]

[1]Department of Chemical Engineering-CIESOL, University of Almeria, Cañada San Urbano 04007 Almeria, Spain
[2]Aqualia, Av. del Camino de Santiago, 40. 28050 Madrid, Spain
[3]Department of Informatics-CIESOL, University of Almeria, Cañada San Urbano 04007 Almeria, Spain

ABSTRACT

This chapter summarizes the status, major challenges and potential contribution of microalgae-related wastewater treatment processes. Although the use of microalgae for wastewater treatment was proposed in the last century, technology was not sufficiently efficient and robust to be applied at a commercial scale. Only recent advances in the knowledge of biological systems, the engineering of reactors and the harvesting and processing of the produced biomass allow the development of the first industrial demonstrations. Facilities of several hectares are already in operation demonstrating the feasibility of this technology. However, challenges remain for the further improvement and enlargement of these systems. They are related to (1) the improvement of knowledge and management of biological systems, (2) the development of adequate strategies for the allocation and implementation of large-scale facilities, (3) the definition of optimal operation conditions including the development of non-assisted systems capable of operating under variable environmental conditions and (4) the development of adequate routes for the valorization of biomass. Much effort is being devoted to solving these challenges and thus making this technology reliable for industrial applications. Once it is achieved, the use of microalgae will be incorporated into the portfolio of available technologies for wastewater treatment. In this respect, no single technology is capable of solving all the scenarios related to wastewater treatment, but microalgae-related processes represent a semi-intensive technology capable of contributing to efficiently treating wastewater while recovering nutrient-energy-water scenarios related to temperate climates with no severe land restrictions. Moreover, the use of microalgae represents a change of paradigm in the field of wastewater treatment because by using this type of microorganisms it is possible to produce valuable biomass at a higher price than the wastewater treatment cost. The potential of microalgae-related wastewater treatment processes is thus highly relevant, and valuable to achieve the sustainable development goals defined by the United Nations.

Keywords: wastewater treatment, microalgae, raceway, nutrients recovery, biomass production, valorization, modelling, advanced control.

5.1 INTRODUCTION

Wastewater treatment is a continuous process due to the increase in population and the enhancement of lifestyle. Additionally, the current scenario of global warming imposes the necessity to reduce the impact of existing processes and to recover resources from waste. Specifically, the sustainable development goals (SDGs) of the United Nations identify 'Clean water and sanitation' as a major objective in addition to others such as 'Climate action', 'Life below water' and 'Life on land' (United Nations, 2015). Thus, to mitigate water and nutrient scarcity the recovery of water, energy and nutrients contained in wastewater is mandatory, and this fact drives the development of new processes alternative to conventional processes based on activated sludge (Muga & Mihelcic, 2008). Moreover, energy saving or energy recovery is also a necessity to mitigate global warming related to the emission of greenhouse gases involved in energy production systems. In this scenario, the use of microalgae emerges as an interesting alternative.

Microalgae naturally occur in conventional wastewater treatment processes, but usually, the presence of this type of microorganisms is disregarded or prevented. However, they can be an interesting partaker for wastewater treatment due to their capacity to produce oxygen (O_2) and fix compounds such as carbon, nitrogen and phosphorous, into valuable biomass. This makes microalgae-based wastewater treatment one of the most promising alternatives to conventional methods (Cano *et al.*, 2022; Lundquist *et al.*, 2010). Thus, wastewater treatment processes based on microalgae were already proposed in the last century but until now only a few examples of large-scale processes exist (Arbib *et al.*, 2022; Craggs *et al.*, 2013; Mehrabadi *et al.*, 2017). The reason for that was the lack of the necessary knowledge of the process and the low capacity of the technology that was used. Recent advances in both fields allow for overpassing these barriers, and the first demonstrators of the technology are already in operation. However, the knowledge of processes and technologies currently utilized must be still improved. Moreover, the gap between the efficiency of the current processes and the theoretical values remains high. Thus, large opportunities exist to improve technology and make it more robust and reliable for its commercial development.

One of the major aspects to be considered when developing wastewater treatment processes based on microalgae is the existence of microalgae and bacteria consortia. This fact is highly relevant because it imposes the necessity to design bioreactors and the overall production system as a function of culture conditions required for each of them, which are frequently different for both microalgae and bacteria (Umamaheswari & Shanthakumar, 2016). In the case of bacteria, the scenario is similar to an activated sludge-based process, which means that organic matter and O_2 concentrations determine the growth of heterotrophic bacteria, whereas other parameters such as nitrogen and total inorganic carbon concentration influence the behaviour of other microorganisms such as nitrifying and denitrifying bacteria. Concerning O_2, to ensure aerobic conditions the dissolved O_2 concentration must be higher than 2 mg/L, under these conditions it can still operate but the maximal performance is achieved when operating at 10 mg/L. The behaviour of bacteria is independent of availability of light and they prefer moderate pH and temperature values; a slightly high temperature and low pH are recommendable. In the case of microalgae, the growth is fully linked to the availability of light and independent of the presence of organic matter, the major nutrients being inorganic carbon, nitrogen and phosphorous. A large capacity of microalgae cells to produce O_2 is remarkable, although they have a low tolerance to high dissolved O_2 concentrations (higher than 20 mg/L) that induces photorespiration phenomena. Regarding pH and temperature, most of the microalgae prefer slightly alkaline pH and moderate temperature. In practice, the major difference between microalgae and bacteria is the necessity of light for microalgae growth. This fact imposes the necessity to use outdoor photobioreactors capturing natural sunlight as the driver of the process. The influence of culture conditions on the behaviour of major types of microorganisms present in microalgae-related wastewater treatment processes has been studied (López Muñoz & Bernard, 2021; Sánchez-Zurano *et al.*, 2022; Solimeno & García, 2019).

The necessity of providing light to microalgae-related wastewater treatment implies the use of specifically designed photobioreactors instead of conventional bioreactors. In this sense, although in some cases the use of artificially illuminated reactors has been proposed, the energy consumption and the required investment costs make it unfeasible for large-scale systems. Thus, the maximum efficiency of the photosynthesis process is 10% of photosynthetically active radiation. To produce 1 kg of microalgae biomass (on average with an energy content of 20 MJ/kg), up to 200 MJ of light are required which is equivalent to 5.6 kWh (Acién *et al.*, 2016). Considering a 100% conversion of electricity from light and an electricity cost of 0.1 €/kWh, it means that a minimum of 0.56 €/kg is needed, but this value increases to 5.6 €/kg when reducing the efficiency of the photosynthesis process to 1%. According to the nutrient content of wastewater, up to 1 kg of biomass can be produced per m^3 of wastewater treated, which means that only the cost of artificial illumination is higher than the cost of wastewater treatment using conventional technologies (0.2 €/m^3). Focusing on outdoor reactors the use of open raceways is the most suitable alternative because of the low cost (below 20 €/m^2 installation cost just considering the construction of raceway ponds, not ancillaries) and energy consumption (below 5 W/m^2) of this type of system (Acién *et al.*, 2016). For a base case of 1 ha, the systems already in operation demonstrate to be feasible for treating up to 5,000 m^3/day and producing 100 ton/year of dry matter, fixing up to 10 ton N/year and 2 ton P/year, while consuming less energy compared to conventional wastewater treatment systems, below 0.2 kWh/m^3 versus up to 0.8 kWh/m^3 for activated sludge systems (Acién Fernández *et al.*, 2017). These figures make the development of microalgae-based wastewater treatment processes very attractive for small- and medium-sized cities.

In this chapter, the major aspects of technologies that are currently utilized are summarized to provide an overview of the current status of the art, and based on that the major challenges to be faced are analysed. Some alternatives to improve the reliability of microalgae-related wastewater treatment processes are then provided. Finally, the relevance of improving and expanding the use of this type of technology is discussed.

5.2 CURRENT STATUS OF MICROALGAE-RELATED WASTEWATER TREATMENT PROCESSES

Although microalgae-related wastewater treatment processes were developed by Oswald in the last century (Oswald & Golueke, 1960), their development has been quite limited, with the capacity of previous processes remaining lower than required for commercial development. Thus, the size of more relevant demonstration facilities was scaled up to a few thousand square metres. However, the basis of these previous processes allows understanding the principles of the process and identifying the major barriers to solve for industrial development (Craggs *et al.*, 2012; Olguín, 2012; Park *et al.*, 2013).

5.2.1 Biology of microalgae–bacteria consortia

When considering the treatment of wastewater using microalgae it is necessary to understand that always a consortium of microalgae and bacteria exists. No specific microalgae strains are utilized; equally as in conventional processes based on activated sludge naturally occurring bacteria are managed. The challenge is to know the most relevant microorganisms involved in the process and their optimal culture conditions to maintain these conditions in the bioreactor, and then favour the development of positive microorganisms versus other competitors. In the case of microalgae–bacteria consortia, the most relevant microorganisms include microalgae and heterotrophic bacteria, in addition to others such as nitrifying or denitrifying bacteria. Microalgae are the microorganisms responsible for O_2 production and inorganic carbon fixation through the photosynthesis process, whereas heterotrophic bacteria are responsible for organic matter removal producing carbon dioxide (CO_2) and consuming O_2. The nexus between both microalgae and heterotrophic bacteria is the O_2; then the O_2 produced by microalgae and consumed by heterotrophic bacteria is ideally equal.

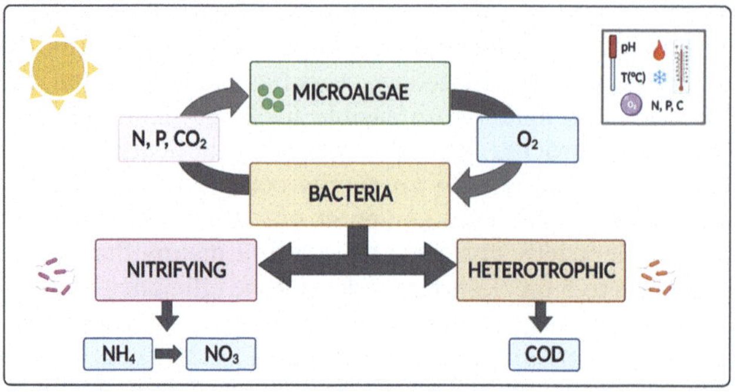

Figure 5.1 Scheme of major phenomena taking place in microalgae–bacteria consortia developing in microalgae-related wastewater treatment processes.

Knowing the phenomena and interactions taking place when managing microalgae–bacteria consortia is critical; a basic scheme of these phenomena is shown in Figure 5.1. To ensure the degradation of biodegradable organic matter a minimum population of heterotrophic bacteria is required according to the load of organic matter to be removed, and then the amount of O_2 required for this process can be calculated. This helps to define the minimum population of microalgae required to produce the necessary O_2 for heterotrophic bacteria by knowing the O_2 production capacity of the microalgae cells. The latter is a function of availability of light, and then the relevance of light for the bioreactor is determined. In addition to this basic phenomenon, other phenomena such as nitrogen and phosphorous solubilization, their release by bacteria and their consumption by both microalgae and bacteria must be considered. Moreover, the presence of other types of microorganisms such as nitrifying and denitrifying bacteria, or phosphorous-related bacteria, could be also important. Finally, the effect of culture conditions such as nutrient concentration, temperature and dissolved O_2 concentration on the performance of all these microorganisms must be taken into account.

Fortunately, there is a large set of knowledge and models already developed to simulate the behaviour of this type of biological systems. Different biological models such as BIO-ALGAE, ABACO and ALBA have been reported, all of them based on the same phenomena described but with some assumptions or differences about the chemistry of the water or phenomena considered (Casagli *et al.*, 2021b; Sánchez-Zurano *et al.*, 2021d; Solimeno *et al.*, 2019). Unfortunately, the diversity of models is large and still a unified model does not exist; however, similar to the activated sludge model it is expected to achieve a unified microalgae sludge model very soon. Related to this objective, recently the use of photo-respirometry methods to evaluate the performance of microalgae–bacteria consortia has been established (Sánchez-Zurano *et al.*, 2020; Sforza *et al.*, 2019). This tool allows a fast and reliable evaluation of the status of the biological system thus helping in the decision-making process for the operation of industrial facilities. Some of these models are already implemented in easy-to-use tools such as simulators, facilitating the analysis of scenarios and comparison of alternatives (Sánchez-Zurano *et al.*, 2021a).

Availability of light is a major parameter in the management of microalgae–bacteria consortia prevailing in wastewater treatment processes. This parameter is resumed in the average irradiance inside the culture, which is a function of solar radiation on the reactor surface and the attenuation of the light by the biomass, which is a function of biomass concentration, the attenuation properties of the biomass and the water depth (Grima *et al.*, 1994). The average irradiance is a key factor in determining the performance of the biological system and thus both the wastewater treatment and biomass production capacities. In this sense, two main scenarios are usually considered. If the objective is to maximize the capacity of wastewater treatment, microalgae are used mainly as O_2

producers, and then high water depths are used (30–40 cm) because enough average irradiance will exist to produce the O_2 required by bacteria to degrade the organic matter (Sutherland *et al.*, 2014). However, in this scenario, the percentage of bacteria in the biomass is high (up to 60%) and then the quality of the produced biomass reduces. Regarding nitrogen removal, the assimilation into biomass only accounted for 57% of the inlet nitrogen under the best conditions because nitrification and volatilization reduced the availability of this element. On the contrary, if the objective is to maximize the recovery of resources from wastewater, the production of microalgae biomass must be prioritized. For that, low water depths must be utilized to enlarge the average irradiance inside the reactor and then the increased growth of microalgae biomass (Morillas-España *et al.*, 2021). In this scenario, the amount of O_2 produced by the microalgae is more than that required by bacteria. Moreover, excess dissolved O_2 concentrations can prevail and the desorption of O_2 by aeration using adequate mass transfer units is necessary. In addition, if an inadequate C/N ratio is present in the wastewater additional inorganic carbon must be provided to allow the microalgae cells to fix the nitrogen and phosphorous present in the wastewater. Under these conditions, the wastewater is used as a nutrient source and the flow of wastewater treated is reduced compared to the first strategy. The operation at short hydraulic retention times presented a more interesting performance with higher biomass productivity (de Godos *et al.*, 2016). The priority and operation mode is very relevant because they determine the quality of the produced biomass in terms of microalgae and bacteria composition. This is very relevant for further applications of biomass and to ensure the accomplishment of regulation in terms of wastewater treatment (Nordio *et al.*, 2023).

5.2.2 Engineering of photobioreactors

Microalgae-related wastewater treatment is performed in raceway reactors, as shown in Figure 5.2. The reasons for that include the low cost and energy consumption of this technology and its well-established technology (Lundquist *et al.*, 2010). However, the design and operation of this type of reactor are far from optimal values, and large improvements are possible. Raceways consist of horizontal surfaces on which a liner is installed, the perimeter of the reactor being defined using concrete blocks or sand barriers. The reactor usually consists of two channels along which the culture is continuously recirculated using a paddlewheel. The water depth is in the range of 20–40 cm,

Figure 5.2 Image of raceway reactors utilized for wastewater treatment at Agramon (Spain) designed and operated by Aqualia.

and the paddlewheel is designed specifically based on these data to ensure adequate efficiency. In general, a length-to-width ratio of 10 : 20 is preferred. In summary, the design and construction of raceway reactors is a hydrodynamic problem, the objective being to minimize the pressure drop along the reactor to minimize the energy consumption into the paddlewheel and to enlarge the size of the reactor. In this sense, the overall size of the raceway ponds is limited because the only input of energy is provided by the paddlewheel and the capacity of this system to provide energy to the liquid requires a maximum water drop of 15 cm between the inlet and outlet of the paddlewheel. For reactors operating at high water depths of 30–40 cm the overall surface can be up to 10,000 m^2, whereas for reactors operating at low water depths of 10–20 cm the overall surface can be up to 5,000 m^2 (Craggs *et al.*, 2012). Traditionally, the design of raceway reactors was performed based on the Manning equation. This equation is accurate mainly for channels, but not for bends and other accessories, then the use of the Bernoulli equation proves a more general and accurate design (Mendoza *et al.*, 2013a).

A critical part of raceway reactors is the bends connecting the channels. This part represents a relevant pressure drop in these systems and thus the number of bends must be minimized, according to the dimensions of the available land. The design of the bends must be carefully optimized, and the installation of 'islands' or baffles facilitating the circulation of the culture is necessary (Sompech *et al.*, 2012). The use of baffles is simple and 50% cheaper, thus it is the recommendable solution (Mendoza *et al.*, 2013a). Concerning the paddlewheel it must be designed according to the water depth. The recommendation is to install systems with a total diameter equal to four times the water depth and include a range of 10–12 paddles (Weissman & Goebel, 1987). The minimization of hydraulic losses in the system allows for the optimization of the performance of the paddlewheel, the challenge being to avoid split velocity between the rotation of the paddlewheel and the liquid velocity. It means that ideally the rotation velocity must be adequate to provide a tangential velocity equal to the liquid velocity. As the recommended liquid velocity is 0.2 m/s it means that rotation velocity must be in the range of 2–5 rpm for water depths ranging from 0.4 to 0.2 m. Commercial systems normally operate at higher rotational velocities, in the range of 6–10 rpm. The design and operation of the paddlewheel largely determine the energy consumption of the system, and inadequate design increases the energy consumption (Mendoza *et al.*, 2013a). The adequate design allows maintaining the energy consumption of this type of reactor in the range of 1–10 W/m^2. This energy consumption can be reduced by using turbines instead of paddlewheels, although the investment cost increases. The low-energy algae reactor developed by Aqualia is based on this concept and allows for a reduction in energy consumption by up to 25% of the initial value (Arbib *et al.*, 2022). Recent advances are being developed thanks to the use of computational fluid dynamic tools such as ANSYS Fluent for the optimal design and operation of raceway reactors (Hreiz *et al.*, 2014; Inostroza *et al.*, 2021).

Whatever the impulsion system, two problems of raceway reactors are (1) the inadequate light regime to which the cells are exposed in this type of reactor and (2) the low mass transfer capacity of these systems. Concerning the light regime, both experimental and computer-assisted analyses of flow patterns in raceway reactors demonstrated that it is laminar, the vertical movement of cells being very scarce (Barceló-Villalobos *et al.*, 2019). Vertical velocity has been estimated to be 0.02 m/s, which imposes average frequencies of light exposition in the order of 10^{-2} Hz, a hundred times lower than that required for integration of light, of 1 Hz. To solve this problem it would be possible to increase the liquid velocity into the channel. However, this will require a large increase in energy consumption which makes it not feasible. Moreover, the increase in liquid velocity only allows for an increase in the frequency of light exposition on a limited amount, up to a maximum of 10^{-1} Hz. Another alternative proposed has been the use of airfoils favouring vertical mixing (Figure 5.3). These airfoils must be carefully designed and installed, a large number of them being required because their effect is limited to short distances (Inostroza *et al.*, 2023). Although promising results have been provided, still no industrial demonstrator of this technology has been developed. Concerning mass transfer, raceway reactors are designed to minimize energy consumption while allowing the exposure of large volumes

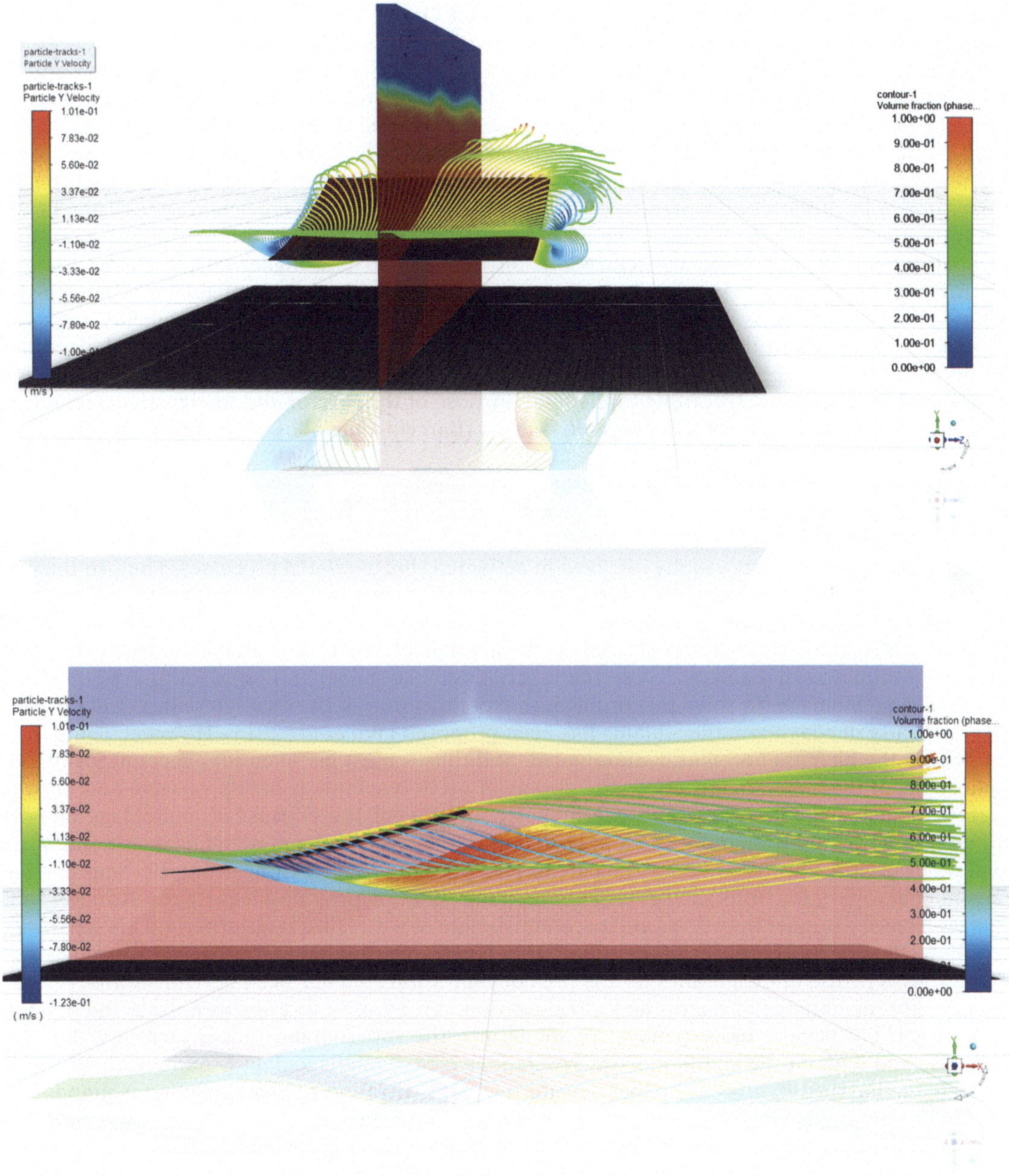

Figure 5.3 Example of baffles that can be used to improve the vertical mixing in raceway reactors to maximize the light utilization efficiency by the microalgae cells.

of culture to the sunlight, the exchange of gases such as CO_2 and O_2 with the atmosphere taking place only through the reactor surface. This exchange is very low due to the scarce mixing between the liquid and air in the channel, thus limiting biomass productivity both by carbon limitation and O_2 oversaturation (Mendoza *et al.*, 2013b). Regarding O_2, the dissolved O_2 concentration in a raceway pond long-term operated in south Spain at a large scale exhibited pronounced daily variations (Arbib *et al.*, 2017). The differences between the representative days of each season were also noticeable. Maximum values were recorded during June with concentrations up to 32 mg/L at midday, whereas at the same hour of the day during winter considerable lower values were achieved, ≈12 mg/L. The concentration of dissolved O_2 during the no photosynthetically active period (night) also presented seasonal differences with lower concentrations during summer and higher during winter (ranging from 1.1 ± 0.7 to 5.2 ± 1.2 mg O_2/L), and in many cases, the dissolved O_2 attained 0 mg/L during the night. To improve the mass transfer capacity the installation of sumps is recommendable. The sump allows for an increase in the mass transfer capacity and adjusts it to the necessities of the systems, both in terms of CO_2 supply and O_2 removal. Recent advances in this field allow adequate design and operation of these systems according to the final purpose of the reactor and its size, thus facilitating the operation and management of large-scale systems (Barceló-Villalobos *et al.*, 2018).

5.2.3 Harvesting and processing of the biomass

One of the major problems related to wastewater treatment involving microalgae is the recovery of biomass for the clarification of water. Microalgae biomass has different and highly variable properties in comparison with bacteria usually involved in activated sludge-related processes. Thus, in the case of microalgae-related processes, the biomass concentration is lower, in the range of 0.35–1.5 g/L, whereas the amount of the solids is also smaller, being in the range of 5–20 μm. In processes involving activated sludge, the concentration of solids is in the range of 1.5–3.0 g/L and the biomass is in flocs of several millimetres in diameter. These characteristics make the separation of solids difficult due to the microalgae biomass. To solve this problem the modification of the characteristics of microalgae cells by the use of coagulants/flocculants has been proposed (Wu *et al.*, 2015). However, using these compounds the biomass becomes contaminated, and thus it must be taken into account for further applications of the biomass. Moreover, the dosage of these reactants is highly variable and must be carefully adjusted for each case and continuously, to prevent the exhaust of microalgae biomass and especially the not accomplishment of regulation in terms of content of solids at the end of the wastewater treatment process. Once the biomass is flocculated, the settling properties improve and conventional separation units such as sedimentation or flotation provide adequate results in terms of biomass recovery (higher than 90%) and accomplishment of water discharge criteria. This separation step allows the pre-concentration of the culture to achieve a maximal solid content up to 40 g/L (Arbib *et al.*, 2022). The use of membranes for the pre-concentration step has been developed. In this case, non-pressure membranes of micro- and ultrafiltration can be used to pre-concentrate the biomass up to 10 g/L but ensure 100% removal of solids into the supernatant, independently of biomass properties and without the addition of any reactant as coagulant/flocculant (Zhang *et al.*, 2010). To complete the harvesting process further dewatering using filtration or centrifugation is mandatory. These unit operations consume more energy and are more expensive than the previous ones, thus their direct use to separate the biomass from the supernatant is not recommendable. Figure 5.4 shows a comparison of different harvesting strategies in terms of (1) biomass concentration at the beginning and end of the process and (2) operation cost/energy consumption. Data clearly show that the two-step process including a pre-concentration plus dewatering step is the best alternative. Although the use of dissolved air flotation and sedimentation is possible, it implies the dosage of flocculants, thus making the process difficult and contaminating the biomass. However, the use of membranes allows for achieving similar results in terms of biomass concentration, while avoiding cost and energy consumption by the use of flocculants. The challenge is to reduce the water content as much as possible to reduce the cost of downstream processes, in some cases including the drying of the

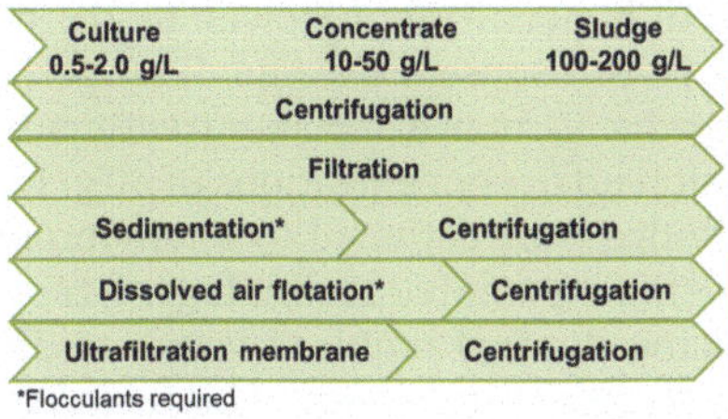

Biomass recovery (%)	Cost (€/kg)	Energy (kWh/m³)	Industrial (Y/N)
95%	0.30	1.00	Yes
95%	0.80	2.60	No
78%	0.05	0.13	No
90%	0.07	0.15	Yes
95%	0.08	0.12	Yes

Figure 5.4 Scheme of different harvesting processes and comparison of their performance at a demonstrative scale (greater than 1 m^3/h). Data from SABANA EU founded project.

biomass being required. However, to save energy and cost the reduction of the water content must be adjusted to that required for the valorization step. It is thus critical to define the more suitable ones at this stage (Arbib *et al.*, 2022).

Although microalgae biomass has been proposed as potentially useful for a large set of applications, when produced in wastewater some of them are strictly forbidden by the regulation, whereas others are also not recommendable according to the prevention criteria. For example, the regulation completely forbids the use of biomass produced from wastewater for direct human use. This is not the case when using biomass for feed although in this case, the feed industry doesn't use this type of material according to the prevention criteria. The remaining applications are related to materials, agriculture or bioenergy (Acién *et al.*, 2016; Arbib *et al.*, 2022). In the case of materials, no relevant examples of the use of microalgae biomass as a source of polymers or composites exist to date, but technically it would be possible (Arias *et al.*, 2020). Concerning agriculture, microalgae contain relevant contents of nutrients such as nitrogen, phosphorous, iron and magnesium, among others. However, more than the inorganic content the organic molecules present in microalgae biomass have been reported as highly valuable for agriculture. In this respect, the amino acid profile of microalgae contains essential amino acids that have been reported to act as plant growth promoters in different crops (Kapoore *et al.*, 2021). Moreover, microalgae biomass is very rich in phytohormones such as auxin, cytokinins and gibberellins improving the root development of plants, and the growth or the development of fruits in crops (Stirk *et al.*, 2002). Microalgae biomass has also been proposed as a source of biopesticides, thus replacing other chemicals less sustainable and toxic compounds already used to prevent diseases in crops (Costa *et al.*, 2019). Microalgae biomass can be used as a raw material to obtain bioenergy, more specifically biofuels. Although the burning of microalgae biomass to produce heat or electricity is possible, the necessity to dry the biomass for this use strongly reduces the efficiency of the process in terms of energy and cost. The production of biofuels is based on the chemical composition of the microalgae biomass: carbohydrates can be converted into bioethanol, lipids can be converted into biodiesel or the whole biomass can be converted into bio-oil by thermochemical treatment, or into biogas/biomethane through anaerobic digestion (Murthy, 2011). This latter process is the most simple and feasible in wastewater treatment plants. This process allows the production of biomethane from microalgae biomass produced in wastewater treatment processes, yielding up to 0.2 kg CH_4/kg volatile solids (VS) equivalent to 0.6 kWh/m^3 of treated wastewater (Arbib *et al.*, 2022).

A techno-economic analysis must be performed to analyse the different alternatives to the application of biomass and to define the most recommendable approach. In this sense, the development of processes capable of proceeding with wet biomass is highly recommendable to avoid the cost and energy consumption of drying processes such as spray-dryers or freeze-dryers. Some applications require the inclusion of cell disruption steps that are usually performed by high-pressure homogenization although the use of ultrasound or milling has been reported as well (Halim *et al.*, 2012). An in-depth analysis of alternatives for the valorization of microalgae biomass produced in wastewater was performed allowing identifying the anaerobic digestion of the biomass to produce biogas or the combination of the production of biofertilizers followed by the production of biogas as the most recommendable (Acién *et al.*, 2016). In this sense, FCC Aqualia is performing the production of biomethane from microalgae biomass produced in wastewater treatment plants in Spain. The biomass is harvested using dissolved air flotation up to concentrations of 40 g/L with biomass recovery efficiencies higher than 95%. The biomass sludge is directly fed to an anaerobic digester to produce biogas, which is cleaned-up to biomethane using a patented technology. According to this company, 1 ha of raceway reactor is capable of producing the fuel required for up to 35 cars annually (Arbib *et al.*, 2022). This example opens the door for the real production of biofuels from microalgae biomass when integrating algal biomass production with wastewater treatment.

5.3 MAJOR CHALLENGES OF MICROALGAE-RELATED WASTEWATER TREATMENT PROCESSES

To facilitate the development of microalgae-related processes for the treatment of wastewater it is necessary to improve the knowledge and the technologies currently used: to improve their efficiency but especially to improve the robustness of processes. The combination of these advances and the implantation of the first commercial units will target the expansion of this technology and thus the enlargement of its contribution to this field.

5.3.1 Improvement of biological systems

To improve the efficiency of microalgae-related wastewater treatment processes it is mandatory to improve the knowledge and management of microalgae–bacteria consortia. In this sense, large efforts are being devoted to know in detail the microbiological composition of this type of consortia, using huge and valuable information being acquired by omics tools in this respect (Casagli *et al.*, 2021a; Clagnan *et al.*, 2022; Robles *et al.*, 2020; Sánchez-Zurano *et al.*, 2021b). Recent advances include the identification of microalgae strains prevailing in this type of biological systems. Data show fast-growing strains such as *Chlorella*, *Scenedesmus* or *Tetradesmus* as prevailing strains, although others such as *Ochromonas*, *Picochlorum*, *Oocystis*, *Dictyosphaerium*, *Poteriospumella* and *Micractinium* can be found as well (Clagnan *et al.*, 2022). These strains are well known and they tolerate large variations of culture conditions such as temperature and pH. Moreover, they are strains with high efficiency in terms of light utilization, which explains their higher adaptability to stringent culture conditions. However, changes in microalgae populations are observed due to changes in environmental and operational conditions, still no distinct patterns are being defined. Concerning bacteria, highly variable results have also been obtained according to the environmental and operational conditions. In general, a high proportion of bacteria have been characterized by genera that harbour pathogenic species, for example *Chryseobacterium*, *Aeromonas*, *Brevundimonas*, *Roseomonas* and *Elizabethkingia*. However, as expected, multiple genera are also involved in biodegradation and bioremediation activities, for example *Arenimonas*, *Phenylobacterium*, *Porphyrobacter*, *Gemmatimonas*, *Leptothrix* and *Polymorphobacter* (Clagnan *et al.*, 2022). The presence of nitrifying or denitrifying bacteria is highly variable according to the culture conditions imposed, especially how suitable they are for the growth of microalgae cells. The main ammonia-oxidizing bacteria and nitrite-oxidizing bacteria (NOB) families found in wastewater treatment-related processes belong to the family *Chromatiaceae*, which includes the genus *Nitrosococcus*. The family *Nitrosomonadacea*, which includes the genera *Nitrosomonas*, *Nitrosospira* and *Nitrosovibrio* was also found (Sánchez-Zurano *et al.*, 2021b). Two common NOBs were detected in these systems *Nitrospiraceae* and *Bradyrhizobiaceae*, within these families, *Nitrospira* and *Nitrobacter* were the main genera detected (Sánchez-Zurano *et al.*, 2021b). Finally, in some cases, microalgal grazers and parasitoids are also found, some of them including *Adineta*, *Brachionus* and *Amoeboaphelidium*. *Adineta* and *Brachionus* are known microalgal grazers, whereas *Amoeboaphelidium* is an algal parasitoid. The presence of these predators could negatively affect biomass yield and lead to the collapse of the system (Clagnan *et al.*, 2022).

The necessity for a better understanding of biological systems managed in microalgae-related wastewater treatment processes imposed the need of (1) developing fast methods for the characterization of the cultures and (2) strategies to optimize their performance. In the first aspect, the development of the photorespirometric method is a valuable tool for the fast monitoring of the performance of the biological system. A scheme showing the steps involved in this methodology is presented in Figure 5.5. This methodology combines light/dark cycles and the addition of specific nutrients such as acetate or ammonium to differentiate the metabolism of microalgae, heterotrophic and nitrifying bacteria (Rossi *et al.*, 2018; Sánchez-Zurano *et al.*, 2022). Largely different results are obtained when applying this methodology to microalgae–bacteria consortia developed using different wastewater types such as urban wastewater, manure or digestate, but also for the same type of wastewater as a function of operating conditions such as availability of light, residence time or organic load. In the second aspect, it is necessary to develop strategies to face the variations in wastewater composition and environmental conditions found in real systems. In this sense, seasonal and daily variations of the culture conditions are inherent to the use of outdoor systems and they must be taken into account. The variations of wastewater composition, usually in organic matter chemical oxygen demand (COD), ammonium concentration and turbidity are more relevant (Figure 5.6). The increase in organic matter favours the development of bacteria, and the increase in the O_2 demand favours the opposite the development of microalgae and the production of O_2. Thus, the COD load influences the O_2 balance and finally the microalgae to bacteria ratio. Control of the concentration of dissolved O_2 is a key factor to fight these variations. Regarding ammonium, a similar trend is

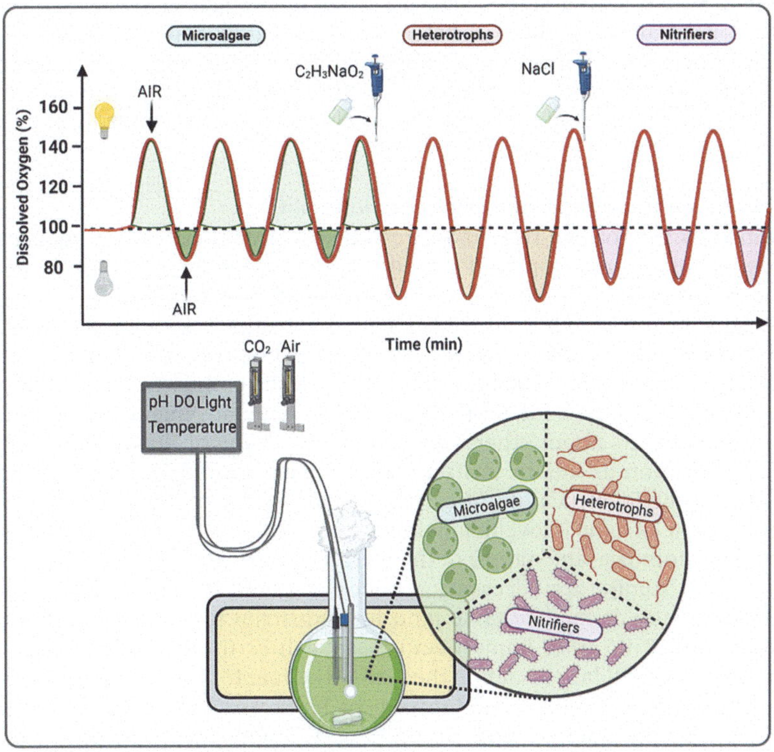

Figure 5.5 Scheme of the photo-respirometric method to evaluate the performance of microalgae–bacteria consortia prevailing in wastewater treatment processes.

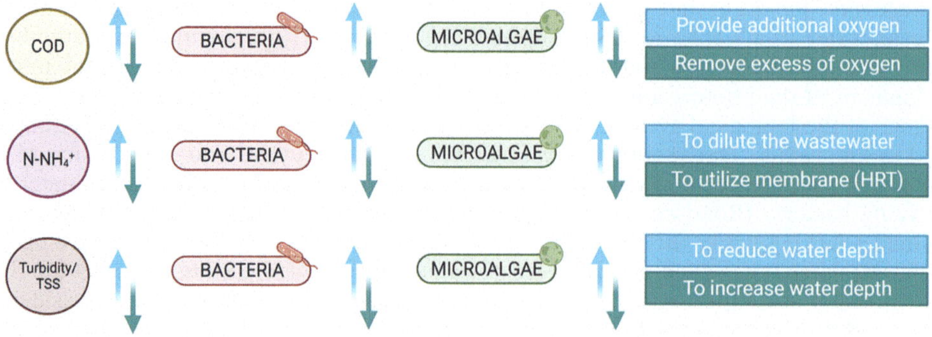

Figure 5.6 Influence of characteristics of wastewater on the performance of microalgae–bacteria consortia and some strategies to mitigate it.

observed, thus the increase in the ammonium concentration favours especially the development of nitrifying bacteria and in some cases damages microalgae cells (Collos & Harrison, 2014). To control these phenomena the load of ammonium must be controlled. Finally, turbidity is a major factor influencing light penetration and microalgae performance. Because no additional filtration systems can be implemented to avoid cost increases the main strategy is to modify the water depth to facilitate the light penetration and the performance of microalgae cells.

Although these general rules allow a better understanding and management of wastewater microalgae-related processes, still much more knowledge about the biology of these systems is required. Reliable models and simulators of the behaviour of these systems are required. Only an in-depth knowledge of biological systems and the development of methods for monitoring and regulating it will allow the development of robust industrial processes.

5.3.2 Allocation and implementation of large-scale facilities

The implementation of microalgae-related wastewater treatment processes imposes the necessity of large surfaces including under optimal conditions because solar radiation is the driver of the process. Previously, up to 10 m^2/pe (population equivalent) was required, but recently this figure has been reduced to 2 m^2/pe (Arbib *et al.*, 2022). That means that a minimum of 1 ha is required for every 5,000 inhabitants. This requirement for large surfaces imposes a challenge in the identification of adequate locations for this type of technology. Factors to be considered in this respect include (1) the topography and nature of the land, (2) the necessity of land movement and landfill management, (3) the land use and conflicts about other uses and (4) the shape of the available plot. Topography and land nature in addition to the necessity of land movement and landfill management imposes serious limitations on the identification of suitable locations for this type of facility. Flat surfaces requiring the minimum of land movement are required otherwise the installation cost can be increased up to 35%. Moreover, the nature of the terrain is critical to support the installation of the reactor, and geological studies are necessary. Concerning land use, this aspect is highly relevant because regulation already defined suitable land for industrial processes, but more relevant than this could be conflicted with other human activities, such as tourism or agriculture. Finally, the shape of the available plot will largely define the size of the reactors and their final arrangement. For example, for a population of 50,000 inhabitants up to 10 ha is required, but due to the restrictions on the geometry of raceway reactors different units of different sizes must be accommodated on it (Figure 5.7). Different scenarios must be studied from a techno-economic point of view as the final step to define the final distribution of the reactors. For large-scale projects the shape of the available land is also relevant; therefore, different design raceway ponds will be needed, from long length

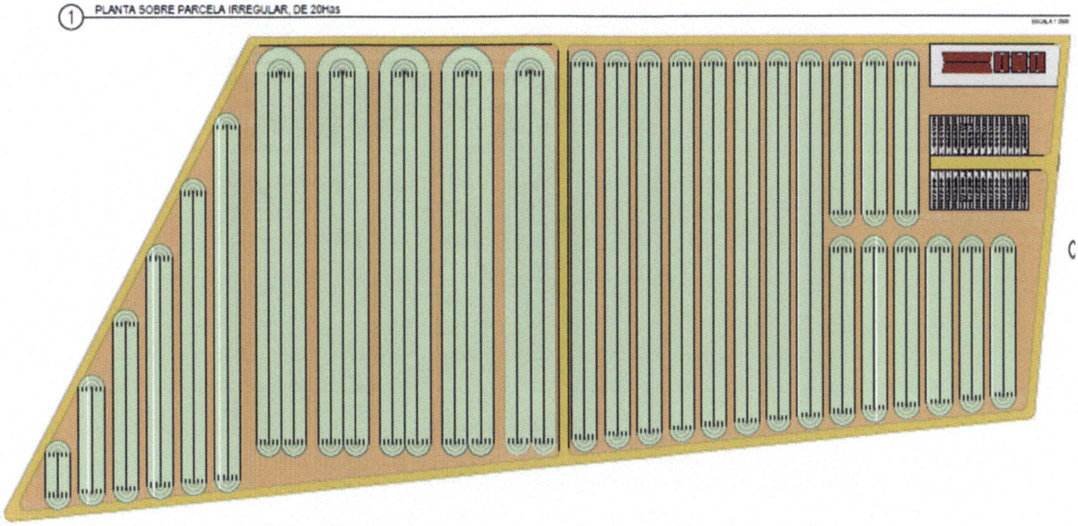

Figure 5.7 Example of distribution of raceway reactors for the installation of a 20 ha facility for wastewater treatment of a city with 100,000 pe, design provided by Aqualia.

two-channel ponds where there is no space limitation, to compact multichannel ones (six channels or more). Figure 5.8 shows real industrial facilities constructed by Aqualia in Merida and Agramon in the framework of the SABANA project.

Construction strategy is another relevant topic. The conventional method consists of the use of reinforced concrete over compacted land and finally covered by the liner (Figure 5.9). This technology is expensive but it is durable and can be shaped in complex ways. As an alternative, the low-cost method is based on land movement, digging the channels, compacting and final lining. Earthen raceways with plastic liners cost little and are easy to build, thus this strategy is less expensive. However, because of the slope required to maintain the wall stability, this design needs more space in comparison with the conventional concrete raceway ponds (Arbib *et al.*, 2022). A third option has been recently developed in which the raceway reactor is made of semi-rigid polyethylene reinforced with a metal. This strategy allows for minimizing the cost of land movement, and avoids the use of concrete while maximizing the use of land. Still, this technology has been only validated at scales up to 1,000 m^2 but it would represent a comfortable option, especially for pilot and demonstration units (Sánchez-Zurano *et al.*, 2021c).

Finally, the design of the reactor and the entire process must be fitted to each case. Similar to other conventional technologies such as the use of activated sludge, according to the capacity and boundary conditions of the process the specific design of the reactor and harvesting/downstream steps must be specifically designed. Concerning the reactor, the water depth is a relevant decision, affecting not only the design and area of the reactor but also to the necessity to incorporate mechanisms for O_2 desorption of inorganic carbon supply. In this sense, the optimal design of sumps allowing improving the mass transfer capacity in raceway reactors has been recently reported (Barceló-Villalobos *et al.*, 2018). Adequate sumps allow for minimizing the energy consumption for O_2 desorption at the same time than maximizing the efficiency of carbon capture when providing CO_2-rich gases as a source of inorganic carbon (Barceló-Villalobos *et al.*, 2022). Moreover, the use of advanced control algorithms allows the optimization of both processes and then the system approaches its theoretical optimal performance (Rodríguez-Torres *et al.*, 2021).

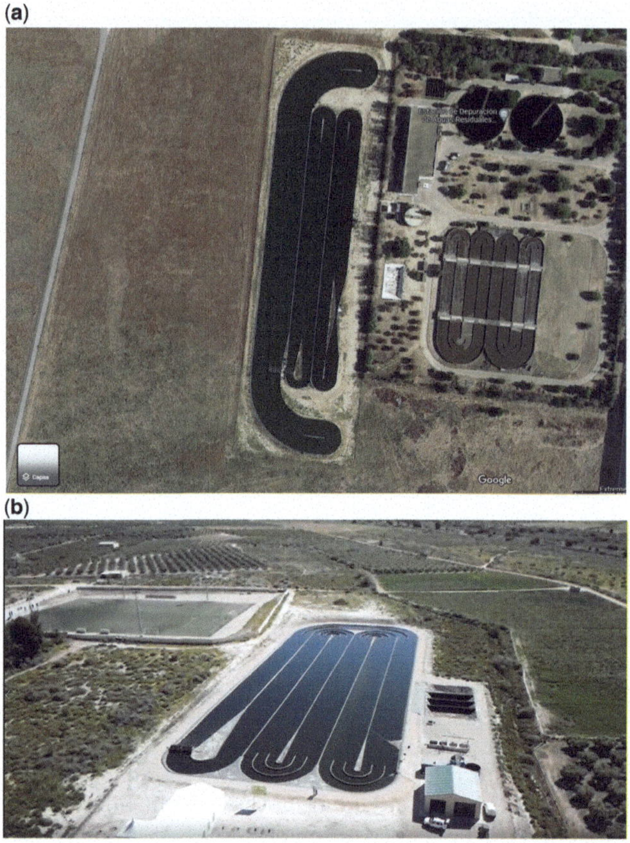

Figure 5.8 (a) Merida plant composed of one raceway of 1 ha and two raceways of 0.5 ha based on two-channel design and (b) Agramon plant composed of a 1 ha compact design composed of six channels (Aqualia).

5.3.3 Optimal operation of processes

Microalgae-related wastewater treatment processes are performed outdoors using natural sunlight as an energy driver. Moreover, large surfaces are required on which the cultures are exposed to both daily and seasonally changing environmental conditions, mainly solar radiation and temperature. Besides, wastewater treatment typically faces problems related to disturbances of both flux and quality of wastewater. Thus, defining the optimal conditions of microalgae-related processes for wastewater treatment can be a difficult task, requiring integrating largely different changing variables. To solve this problem different approaches are possible. The most conventional is the development of mathematical models considering the phenomena taking place in the processes, both biological and physical/chemical. This approach provides an adequate description of the most relevant phenomena, allowing the simulation of different scenarios and to take decisions based on the results from simulations (Hoyo *et al.*, 2022; Sánchez-Zurano *et al.*, 2021a). Moreover, the use of weather forecasts allows preventing failures of the systems and to adapt the operational conditions to the most adequate ones (De-Luca *et al.*, 2019; Rodríguez-Miranda *et al.*, 2022). This approach faces the problem of changes in the composition of the biological system or its adaptation to changing environmental conditions. Thus, a continuous evaluation of the performance of the biological system and recalibration of parameters of the biological model are required.

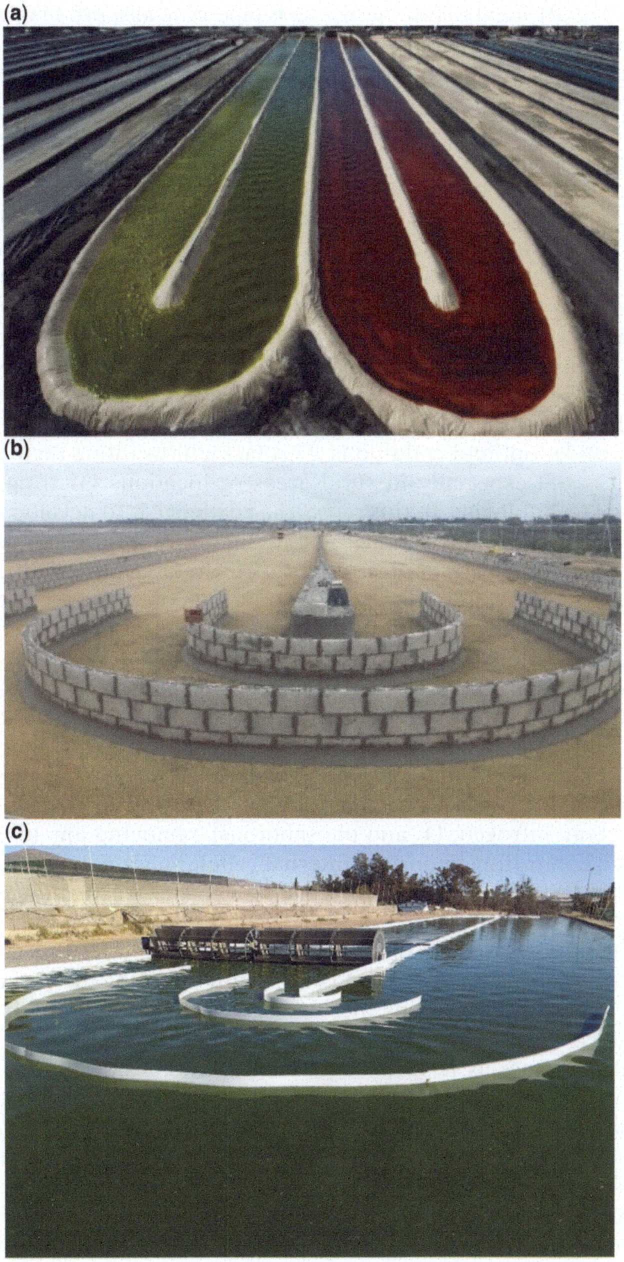

Figure 5.9 Image of raceway reactors. Different construction strategies of raceway reactors for wastewater treatment: (a) use of reinforced concrete, (b) use of dining channels and (c) polyethylene reinforced with a metal.

Alternatively, the use of artificial neural networks is emerging as a promising initiative. Compared to first-principles models, models based on artificial neural networks are faster to run and simpler to re-calibrate on account of their smaller number of parameters and more straightforward formulation. These models allow the simulation of the behaviour of each specific system based on previous data, without an in-depth knowledge of the phenomena taking place. They can infer patterns in the data beyond human comprehension, being especially useful in image or text processing tasks, speech recognition and recommendation management. By providing adequate and enough data, and using adequate algorithms, the neural networks are capable of properly predict the behaviour of the system (Otálora *et al.*, 2021, 2023). Equal to models based on first principles, the models based on neural networks can be fed with weather forecasts to anticipate changes in environmental conditions and to modify the operational parameters to optimize the performance of the systems.

Whatever the operation strategy, based on first-principles models or neural networks, microalgae-related processes are semi-intensive processes much more simple than intensive technologies and require much less supervision and maintenance. The integration of this type of strategy will facilitate the development of fully non-assisted processes, capable of working with the intervention of operators, and always performing under optimal conditions. This fact will facilitate their implantation in small- and medium-sized cities, and especially in rural areas or locations far from large infrastructures required by conventional technologies. In this sense, the challenge is to develop modelling and control frameworks to improve the efficiency, productivity, design and optimization of microalgae-related wastewater treatment processes (Figure 5.10). With this technology, three different objectives could be addressed: (a) maximize biomass production/quality, (b) maximize wastewater treatment capacity or (c) a tradeoff between biomass production and wastewater treatment. According to the selected objective, the process specifications will be imposed by selecting the microalgae strain, biomass quality and required production costs; that is the control requirements for the control optimization problem. Thus, adequate modelling and control approaches are technology solutions that can contribute to better reproducible conditions with competitive market costs by analysing/simulating environmental conditions (solar radiation and ambient temperature), compensating for the permanent non-stationary behaviour of the processes, the presence of disturbances, taking advantage of nutrients provided by wastewater (mainly carbon, nitrogen, O_2 and phosphorous), removing any toxic metabolic products (e.g. CO_2 mitigation) and controlling important internal cellular parameters (e.g. temperature, pH and dissolved O_2) to optimize the biomass production and wastewater treatment (Guzmán *et al.*, 2021).

5.3.4 Develop valuable applications of microalgae biomass

As previously explained, the final use of biomass is highly relevant to the economic reliability of the process. Of all the possible applications those related to agriculture and bioenergy are the most suitable. In the case of agriculture, microalgae biomass can be used as biofertilizers or as biostimulants. The value of microalgae biomass as fertilizer is limited; thus, considering the N and P contents of the biomass and the price of fertilizers currently on the market the maximum price of microalgae biomass for this application is 100 €/ton. This value can be higher if considering microalgae biomass as organic fertilizer authorized for the production of organic foods, thus increasing up to 300 €/ton. However, the value of microalgae biomass as a biostimulant is much higher. Microalgae biomass, including when produced in wastewater, acts as a plant growth promoter favouring the development of plants and fruits, thus improving the production of foods by agriculture in the range of 10–20%, at the same time reducing the regular fertilizers requirement up to 10%, in some cases also reducing the prevalence of diseases and phytopathogens up to 20% (Fernández *et al.*, 2021; Stirk *et al.*, 2013). All these benefits make microalgae biomass a valuable tool for the improvement of food production by conventional agriculture. Companies already selling these products buy dry biomass of *Spirulina* at prices of 5–10 €/kg (dry biomass), which is a minimum of 5,000 €/ton (dry basis). That means that a conservative value for microalgae biomass produced from wastewater can be 1,000 €/ton, much higher than the one for its use as a regular source of N/P or fertilizer.

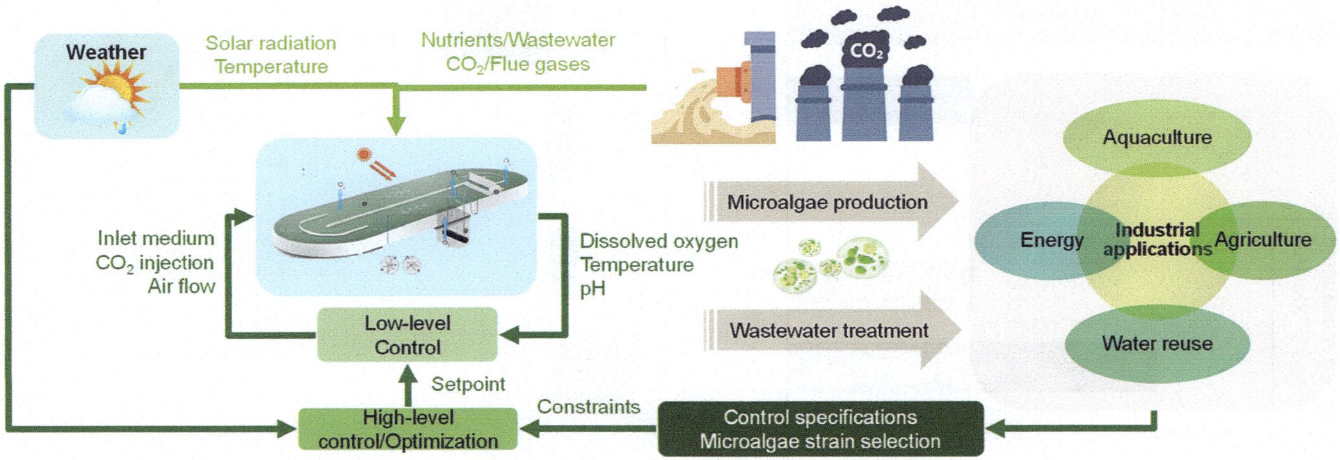

Figure 5.10 Concept of tools for continuous optimization of operating conditions in microalgae-related wastewater treatment processes.

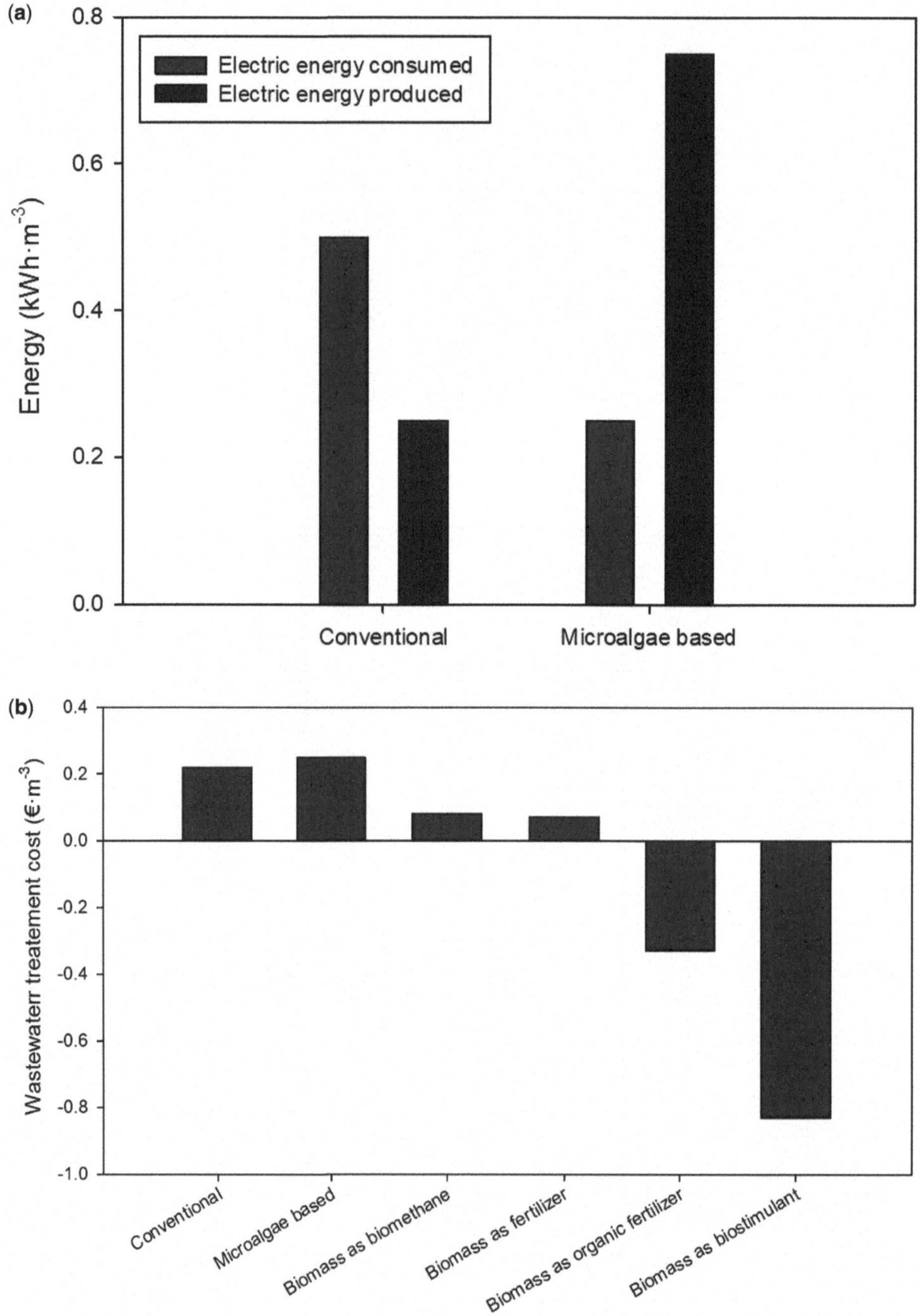

Figure 5.11 Comparison of (a) energy consumption and (b) wastewater treatment cost in different scenarios.

A much more simple application is the production of bioenergy as biomethane. Anaerobic digestion and production of biogas is a well-established technology, already existing in most wastewater treatment plants. The advantage of feeding microalgae to anaerobic reactors is the high methane potential of the biomass in the range of 150–300 L/kg VS depending on the characteristics of the biomass (Posadas *et al.*, 2015; Uggetti *et al.*, 2017). In this sense, FCC Aqualia already demonstrated that energy produced from the anaerobic digestion of microalgae biomass produced in a 1 ha reactor is much higher than the initial energy content of the wastewater, due to the fixation of solar energy by the microalgae cells into the reactor. This energy is released as biomethane during anaerobic digestion, and then the overall process results in a positive energy balance (Figure 5.11). Thus, in conventional wastewater treatment, the energy consumption corresponds to 0.5 kWh/m^3 and the produced biomass when transformed into biogas allows recovering a maximum of 0.25 kWh/m^3. When using microalgae-based processes the energy consumption reduces to half, up to 0.25 kWh/m^3, whereas the amount of biomass increases up to 100 tons/ha/year, equivalent to energy production through anaerobic digestion of 0.75 kWh/m^3. This is a change of paradigm in the wastewater treatment industry because instead of consuming energy, they become energy producers. Moreover, in terms of cost the change of paradigm is more relevant. Thus, the wastewater treatment cost reduces from 0.22 to 0.17 €/m^3 when using microalgae-based processes instead of conventional technologies, mainly due to the reduction of the energy consumption (Figure 5.11). However, if considering the value of microalgae biomass for different applications the net balance can be negative, meaning the value of the biomass compensates the wastewater treatment cost. Thus, if considering the use of biomass for the production of biomethane (0.3 €/kg, 30% conversion) or regular fertilizers (0.1 €/kg, 100% conversion), the wastewater treatment cost reduces to 0.08 and 0.07 €/m^3 respectively, whereas it becomes negative with values up to −0.33 and −0.83 €/m^3 when considering the production of organic fertilizers (0.5 €/kg, 100% conversion) and biostimulants (1.0 €/kg, 100% conversion). This fact opens an avenue for the development of energy and economic positive processes related to wastewater treatment (Arashiro *et al.*, 2018).

5.4 RELEVANCE OF DEVELOPING MICROALGAE-RELATED WASTEWATER TREATMENT PROCESSES

The development of microalgae-related wastewater treatment processes is a non-return way and finally, these processes will be implemented at industrial scale. It will not be a solution suitable for whatever location or scenario, but it will be suitable for certain scenarios. In this section, the relevance of applying microalgae processes for wastewater treatment is analysed.

5.4.1 Improvement of sustainability of wastewater treatment

Microalgae-related processes are one of the most suitable alternatives for wastewater treatment. Data included in this chapter already demonstrate this fact in terms of energy saving (50% of conventional processes), nutrient recovery (up to 90% of those contained in wastewater, equivalent to 10 ton N/ha/year, 2 ton P/ha/year) and production of valuable biomass for agriculture or bioenergy. The European Commission is driving policies to enlarge the sustainability of processes such as 'Green Deal' and 'FarmToFork' programmes in addition to others such as Blue Bioeconomy among others. The United Nations also recommends the development of policies to enlarge the sustainability of human actions, especially related to water management, recovery of nutrients and saving energy. In this sense, microalgae-related wastewater treatment processes demonstrate to be reliable technologies allowing to remove pollutants from wastewater minimizing the release of reusable water, and at the same time recovering nutrients contained in the wastewater as valuable biomass and saving energy or including producing energy as part of the process. The life-cycle assessment of this type of process demonstrates the sustainability of the technology, although the final results are different according to the boundary conditions considered (Arashiro *et al.*, 2018; Colzi Lopes *et al.*, 2018; Garfí *et al.*, 2017).

A favourable life-cycle performance was generally found for microalgae-based systems when displacing conventional energy products (Colzi Lopes *et al.*, 2018). Specifically, the potential environmental impact of the conventional wastewater treatment plant was five times higher than that generated by microalgae-based systems. Even when comparing with other technologies such as constructed wetlands, microalgae-based processes showed to be the less-expensive alternative (Garfí *et al.*, 2017). On the whole, implementing microalgae-based instead of activated sludge systems increases the sustainability and cost-effectiveness of wastewater treatment in small communities, especially if implemented in warm climate regions and coupled with biofertilizer production (Arashiro *et al.*, 2018).

5.4.2 Distributed wastewater treatment

The technology of wastewater treatment is a mature technology capable of offering suitable alternatives for different scenarios, especially for large cities in which the cost of the required infrastructure is well assumed. Moreover, the cost of conventional technologies for wastewater treatment is currently quite reduced due to the continuous improvement of technologies, saving of energy and improvement of control strategies. Thus, both aerobic and anaerobic processes are operating close to optimal conditions under adequate control and supervision. However, these technologies are not cost-effective for medium-small cities, due to excessive cost of infrastructure or inadequate control and supervision. These cities represent the larger fraction of locations that don't accomplish the European Union (EU) regulations. The latest figures for wastewater treatment in Europe show improvements in collection and treatment, even if big differences remain between member states.

It is worth noting that the EU has been working to improve wastewater treatment across member countries through various directives and regulations. The Urban Waste Water Treatment Directive, for instance, sets standards for the collection and treatment of urban wastewater. EU member states have been implementing measures to comply with these standards, but the progress can vary. Microalgae-based technologies offer a feasible solution for small- and medium-sized cities. However, still, the economic feasibility of the process under practical operating conditions must be improved. To enhance the economic feasibility of microalgae-related wastewater treatment processes it is necessary to reduce labour costs, which implies that more automated designs need to be integrated into the design and operation to replace manpower. The selection of materials for the construction of reactors should focus on cheap and durable choices. The energy consumption for aeration, mixing and liquid conveying, microalgae harvesting and dewatering could be further reduced through the optimization of design and operation. Research on these topics will allow the development of robust and reliable technologies for distributed wastewater treatment.

5.4.3 Reuse of effluents in agriculture

Agriculture is a strategic sector for whatever society, the production of more sustainable and healthy foods being a demand of consumers, especially from developed countries. The EU has set forth various goals and initiatives to promote sustainable agriculture as part of its broader commitment to the United Nations' SDGs and the Agenda 2030. In this respect, the EU Commission imposes in the agenda for 2030 a 20% reduction in the use of fertilizers and pesticides, a 20% reduction in the use of land for food production and a 30% improvement in food production capacity. The recycling of water and nutrients contained in wastewater for the production of foods by agriculture is thus mandatory. Microalgae are an interesting option for this purpose because the environmental conditions required for their production are similar to those required for agriculture. Moreover, geographic areas devoted to agriculture are normally full of land, including non-arable land that can be used for microalgae-related processes. Finally, water and biomass obtained as products of the process are suitable for their utilization in food production by agriculture. Thus, as an example, for a population of 200,000 inhabitants (Almeria, Spain), it is estimated that up to 55,000 m^3/day of treated water can be obtained in a 10 ha facility by using microalgae-related processes for wastewater treatment. This amount of water corresponds to the water demand of up to 4,000 ha of tomato crops with a demand of 5,000 m^3/ha/year. This overall surface represents 14% of the overall surface of greenhouses in Almeria, producing 25% of horticulture crops consumed in Europe. Moreover, the overall biomass production capacity corresponding to these processes will represent up to 1,000 tons/year of dry matter, equivalent to 100 ton N/year and 20 ton P/year, allowing partially replace the consumption of mineral fertilizers in greenhouses, up to 10% of the current demand. However, the improvement of the performance of crops is more relevant, which allows increasing the food production by 10% by increasing the efficiency of nutrients uptake by the plants, which allows reducing up to 15% in the supply of mineral fertilizers, which means an overall reduction of 25% of the demand of mineral fertilizers. Thus, microalgae-based processes perfectly fit with the demand of the agriculture sector and consumers to produce and consume products from sustainable agriculture.

ACKNOWLEDGEMENTS

This research was funded by H2020 Research and Innovation Programme under the Marie Sklodowska-Curie (project: Digitalgaesation, 955520) and by the H2020 Research and Innovation Framework Programme (projects: PRODIGIO, 101007006; REALM, 101060991). Also, this study was supported by the Spanish Ministry of Science and Innovation (project: HYCO2BO, PID2020-112709RB-C21). All authors acknowledge the Institute for Agricultural and Fisheries Research and Training (IFAPA).

REFERENCES

Acién F. G. G., Gómez-Serrano C., Morales-Amaral M. M. M., Fernández-Sevilla J. M. M. and Molina-Grima E. (2016). Wastewater treatment using microalgae: how realistic a contribution might it be to significant urban wastewater treatment? *Applied Microbiology and Biotechnology*, **100**, 9013–9022, https://link.springer.com/article/10.1007/s00253-016-7835-7, https://doi.org/10.1007/s00253-016-7835-7

Acién Fernández F. G., Fernandez Sevilla J. M., Molina-Grima E., Fernandez F. G. A., Sevilla J. M. F. and Grima E. M. (2017). Microalgae: the basis of mankind sustainability. In: Case Study of Innovative Projects – Successful Real Cases, B. Llamas, F. L. Mazadiego and M. D. Storch De Gracia (eds.), InTech, Madrid, pp. 772–814, http://www.intechopen.com/books/case-study-of-innovative-projects-successful-real-cases/microalgae-the-basis-of-mankind-sustainability

Arashiro L. T. L. T., Montero N., Ferrer I., Acién F. G. F. G., Gómez C. and Garfí M. (2018). Life cycle assessment of high rate algal ponds for wastewater treatment and resource recovery. *Science of the Total Environment*, **622–623**, 1118–1130, https://www.sciencedirect.com/science/article/pii/S0048969717334708, https://doi.org/10.1016/j.scitotenv.2017.12.051

Arbib Z., de Godos Crespo I., Corona E. L. and Rogalla F. (2017). Understanding the biological activity of high rate algae ponds through the calculation of oxygen balances. *Applied Microbiology and Biotechnology*, **101**, 5189–5198, https://link.springer.com/article/10.1007/s00253-017-8235-3, https://doi.org/10.1007/s00253-017-8235-3

Arbib Z., Marín D., Cano R., Saúco C., Fernandez M., Lara E. and Rogalla F. (2022). Large-scale demonstration of microalgae-based wastewater biorefineries. In: Integrated Wastewater Management and Valorization Using Algal Cultures, Goksel N. Demirer, Sibel Uludag-Demirer (eds.), Elsevier, pp. 215–234, ISBN 9780323858595, https://doi.org/10.1016/B978-0-323-85859-5.00007-5

Arias D. M., García J. and Uggetti E. (2020). Production of polymers by cyanobacteria grown in wastewater: current status, challenges and future perspectives. *New Biotechnology*, **55**, 46–57, https://doi.org/10.1016/j.nbt.2019.09.001

Barceló-Villalobos M., Guzmán Sánchez J. L. L., Martín Cara I., Sánchez Molina J. A. A. and Acién Fernández F. G. G. (2018). Analysis of mass transfer capacity in raceway reactors. *Algal Research*, **35**, 91–97, https://doi.org/10.1016/j.algal.2018.08.017

Barceló-Villalobos M., Fernández-del Olmo P., Guzmán J. L. L., Fernández-Sevilla J. M. M., Acién Fernández F. G. G., Olmo P. F., Guzmán J. L. L., Fernández-Sevilla J. M. M. and Fernández F. G. A. (2019). Evaluation of photosynthetic light integration by microalgae in a pilot-scale raceway reactor. *Bioresource Technology*, **280**, 404–411, https://doi.org/10.1016/j.biortech.2019.02.032

Barceló-Villalobos M., Hoyo Á., Rodríguez-Miranda E., Guzmán J. L. and Acién F. G. (2022). A new control strategy to improve the mass transfer capacity and reduce air injection costs in raceway reactors. *New Biotechnology*, **70**, 49–56, https://doi.org/10.1016/j.nbt.2022.04.005

Cano R., Marín D., Fernández M., Lara E., Rogalla F. and Arbib Z. (2022). Transforming wastewater treatment plants in bioenergy factories based on microalgae technology. SSRN Electronic Journal, https://papers.ssrn.com/abstract=4109919 or https://doi.org/10.2139/ssrn.4109919

Casagli F., Rossi S., Steyer J. P., Bernard O. and Ficara E. (2021a). Balancing microalgae and nitrifiers for wastewater treatment: can inorganic carbon limitation cause an environmental threat? *Environmental Science & Technology*, **55**, 3940–3955, https://pubs.acs.org/doi/full/10.1021/acs.est.0c05264, https://doi.org/10.1021/acs.est.0c05264

Casagli F., Zuccaro G., Bernard O., Steyer J. P. and Ficara E. (2021b). ALBA: a comprehensive growth model to optimize algae-bacteria wastewater treatment in raceway ponds. *Water Research*, **190**, 116734, https://doi.org/10.1016/j.watres.2020.116734

Clagnan E., D'Imporzano G., Dell'Orto M., Sanchez-Zurano A., Acién-Fernandez F. G., Pietrangeli B. and Adani F. (2022). Profiling microalgal cultures growing on municipal wastewater and fertilizer media in raceway photobioreactors. *Bioresource Technology*, **360**, 127619, https://doi.org/10.1016/j.biortech.2022.127619

Collos Y. and Harrison P. J. (2014). Acclimation and toxicity of high ammonium concentrations to unicellular algae. *Marine Pollution Bulletin*, **80**, 8–23, https://doi.org/10.1016/j.marpolbul.2014.01.006

Colzi Lopes A., Valente A., Iribarren D. and González-Fernández C. (2018). Energy balance and life cycle assessment of a microalgae-based wastewater treatment plant: a focus on alternative biogas uses. *Bioresource Technology*, **270**, 138–146, https://doi.org/10.1016/j.biortech.2018.09.005

Costa J. A. V., Freitas B. C. B., Cruz C. G., Silveira J. and Morais M. G. (2019). Potential of microalgae as biopesticides to contribute to sustainable agriculture and environmental development. *Journal of Environmental Science and Health, Part B: Pesticides Food Contaminants and Agricultural Wastes*, **54**(5), 366–375.

Craggs R., Sutherland D. and Campbell H. (2012). Hectare-scale demonstration of high rate algal ponds for enhanced wastewater treatment and biofuel production. *Journal of Applied Phycology*, **24**:329–337, https://doi.org/10.1007/s10811-012-9810-8

Craggs R. J., Lundquist T. J. and Benemann J. R. (2013). Wastewater treatment and algal biofuel production. In: Algae for Biofuels and Energy, Dordrecht, Springer Netherlands, pp. 153–163.

de Godos I., Arbib Z., Lara E. and Rogalla F. (2016). Evaluation of high rate algae ponds for treatment of anaerobically digested wastewater: effect of CO_2 addition and modification of dilution rate. *Bioresource Technology*, **220**, 253–261, https://doi.org/10.1016/j.biortech.2016.08.056

De-Luca R., Bezzo F., Béchet Q. and Bernard O. (2019). Meteorological data-based optimal control strategy for microalgae cultivation in open pond systems. *Complexity*, **2019**, https://doi.org/10.1155/2019/4363895

Fernández F. G. A., Reis A., Wijffels R. H., Barbosa M., Verdelho V. and Llamas B. (2021). The role of microalgae in the bioeconomy. *New Biotechnology*, **61**, 99–107, https://doi.org/10.1016/j.nbt.2020.11.011

Garfí M., Flores L. and Ferrer I. (2017). Life cycle assessment of wastewater treatment systems for small communities: activated sludge, constructed wetlands and high rate algal ponds. *Journal of Cleaner Production*, **161**, 211–219, https://doi.org/10.1016/j.jclepro.2017.05.116

Grima E. M., Camacho F. G., Pérez J. A. S., Sevilla J. M. F., Fernández F. G. A. and Gómez A. C. (1994). A mathematical model of microalgal growth in light-limited chemostat culture. *Journal of Chemical Technology & Biotechnology*, **61**, 167–173, http://doi.wiley.com/10.1002/jctb.280610212, https://doi.org/10.1002/jctb.280610212

Guzmán J. L., Acién F. G. and Berenguel M. (2021). Modelling and control of microalgae production in industrial photobioreactors (Modelado y control de la producción de microalgas en fotobiorreactores industriales). *RIAI – Revista Iberoamericana de Automatica e Informatica Industrial*, **18**, 1–18.

Halim R., Harun R., Danquah M. K. and Webley P. A. (2012). Microalgal cell disruption for biofuel development. *Applied Energy*, **91**, 116–121, https://doi.org/10.1016/j.apenergy.2011.08.048

Hoyo Á., Rodríguez-Miranda E., Guzmán J. L., Acién F. G. F. G., Berenguel M. and Moreno J. C. J. C. (2022). A computer-based tool to simulate raceway photobioreactors for design, operation and control purposes. *Computers & Chemical Engineering*, **156**, 107572, https://doi.org/10.1016/j.compchemeng.2021.107572

Hreiz R., Sialve B., Morchain J. Ô., Escudié R., Steyer J. P. and Guiraud P. (2014). Experimental and numerical investigation of hydrodynamics in raceway reactors used for algaculture. *Chemical Engineering Journal*, **250**, 230–239, https://doi.org/10.1016/j.cej.2014.03.027

Inostroza C., Solimeno A., García J., Fernández-Sevilla J. M. and Acién F. G. (2021). Improvement of real-scale raceway bioreactors for microalgae production using computational fluid dynamics (CFD). *Algal Research*, **54**, 102207, https://doi.org/10.1016/j.algal.2021.102207

Inostroza C., Dávila J., Román S., Fernández-Sevilla J. M. and Acién F. G. (2023). Use of Airfoils for Enhancement of Photosynthesis Rate of Microalgae in Raceway, https://www.researchsquare.com

Kapoore R. V., Wood E. E. and Llewellyn C. A. (2021). Algae biostimulants: a critical look at microalgal biostimulants for sustainable agricultural practices. *Biotechnology Advances*, **49**, 107754, https://doi.org/10.1016/j.biotechadv.2021.107754

López Muñoz I. and Bernard O. (2021). Modeling the influence of temperature, light intensity and oxygen concentration on microalgal growth rate. *Processes*, **9**, 496, https://doi.org/10.3390/pr9030496

Lundquist T. J., Woertz I. and Benemann J. R. (2010). Microalgae for wastewater treatment and biofuels production. ACS National Meeting Book of Abstracts.

Mehrabadi A., Craggs R. and Farid M. M. (2017). Wastewater treatment high rate algal pond biomass for bio-crude oil production. *Bioresource Technology*, **224**, 255–264, https://doi.org/10.1016/j.biortech.2016.10.082

Mendoza J. L., Granados M. R., de Godos I., Acién F. G., Molina E., Banks C. and Heaven S. (2013a). Fluid-dynamic characterization of real-scale raceway reactors for microalgae production. *Biomass and Bioenergy*, **54**, 267–275, https://doi.org/10.1016/j.biombioe.2013.03.017

Mendoza J. L., Granados M. R., de Godos I., Acién F. G., Molina E., Heaven S. and Banks C. J. (2013b). Oxygen transfer and evolution in microalgal culture in open raceways. *Bioresource Technology*, **137**, 188–195, https://doi.org/10.1016/j.biortech.2013.03.127

Morillas-España A., Lafarga T., Sánchez-Zurano A., Acién-Fernández F. G., Rodríguez-Miranda E., Gómez-Serrano C. and González-López C. V. (2021). Year-long evaluation of microalgae production in wastewater using pilot-scale raceway photobioreactors: assessment of biomass productivity and nutrient recovery capacity. *Algal Research*, **60**, 102500, https://doi.org/10.1016/j.algal.2021.102500

Muga H. E. and Mihelcic J. R. (2008). Sustainability of wastewater treatment technologies. *Journal of Environmental Management*, **88**, 437–447, https://doi.org/10.1016/j.jenvman.2007.03.008

Murthy G. G. S. (2011). Overview and assessment of algal biofuels production technologies, pp. 415–437, https://books.google.es/books?hl=es&lr=&id=_oEe3J8GnC0C&oi=fnd&pg=PA415&dq=hou+microalgae+LCA+2011&ots=8kqUD6Q4M5&sig=j7ocXikyLLzRTLnRqseS78EY4BI

Nordio R., Delgado F. J., Sánchez-Zurano A., Hernandez J. G., Rodríguez-Miranda E., Guzmán J. L., Lafarga T. and Acién G. (2023). Long-term assessment of the nutrient recovery capacity and biomass productivity of *Scenedesmus almeriensis* in raceway reactors using unprocessed urban wastewater. *Bioresource Technology*, **369**, 128374, https://doi.org/10.1016/j.biortech.2022.128374

Olguín E. J. (2012). Dual purpose microalgae–bacteria-based systems that treat wastewater and produce biodiesel and chemical products within a biorefinery. *Biotechnology Advances*, **30**, 1031–1046, https://doi.org/10.1016/j.biotechadv.2012.05.001

Oswald W. J. and Golueke C. G. (1960). Biological transformation of solar energy. *Advances in Applied Microbiology*, **2**, 223–262, https://doi.org/10.1016/S0065-2164(08)70127-8

Otálora P., Guzmán J. L., Berenguel M. and Acién F. G. (2021). Dynamic model for the pH in a raceway reactor using deep learning techniques. *Lecture Notes in Electrical Engineering*, **695**, 190–199, https://doi.org/10.1007/978-3-030-58653-9_18

Otálora P., Guzmán J. L., Berenguel M. and Acién F. G. (2023). Data-driven pH model in raceway reactors for freshwater and wastewater cultures. *Mathematics*, **11**, 1–16, https://www.mdpi.com/2227-7390/11/7/1614/htm, https://doi.org/10.3390/math11071614

Park J. B. K., Craggs R. J. and Shilton A. N. (2013). Enhancing biomass energy yield from pilot-scale high rate algal ponds with recycling. *Water Research*, **47**, 4422–4432, https://doi.org/10.1016/j.watres.2013.04.001

Posadas E., Serejo M. L., Blanco S., Pérez R., García-Encina P. A. and Muñoz R. (2015). Minimization of biomethane oxygen concentration during biogas upgrading in algal-bacterial photobioreactors. *Algal Research*, **12**, 221–229, https://doi.org/10.1016/j.algal.2015.09.002

Robles Á., Capson-Tojo G., Galès A., Ruano M. V., Sialve B., Ferrer J. and Steyer J. P. (2020). Microalgae–bacteria consortia in high-rate ponds for treating urban wastewater: elucidating the key state indicators under dynamic conditions. *Journal of Environmental Management*, **261**, 110244, https://doi.org/10.1016/j.jenvman.2020.110244

Rodríguez-Miranda E., Sánchez-Zurano A., Guzmán J. L., Acién G. and Visioli A. (2022). A seasonal simulation approach for culture depth influence on the temperature for different characterized microalgae strains. *Biotechnology Journal*, **17**, 2100489, https://onlinelibrary.wiley.com/doi/full/10.1002/biot.202100489, https://doi.org/10.1002/biot.202100489

Rodríguez-Torres M. J., Morillas-España A., Guzmán J. L. and Acién F. G. (2021). Modelling and pH control in raceway and thin-layer photobioreactors for wastewater treatment. *Energies*, **14**, 1099, https://doi.org/10.3390/en14041099

Rossi S., Bellucci M., Marazzi F., Mezzanotte V. and Ficara E. (2018). Activity assessment of microalgal-bacterial consortia based on respirometric tests. *Water Science and Technology*, **78**, 207–215, https://doi.org/10.2166/wst.2018.078

Sánchez-Zurano A., Gómez-Serrano C., Acién-Fernández F. G., Fernández-Sevilla J. M. and Molina-Grima E. (2020). A novel photo-respirometry method to characterize consortia in microalgae-related wastewater treatment processes. *Algal Research*, **47**, 101858, https://doi.org/10.1016/j.algal.2020.101858

Sánchez-Zurano A., Guzmán J. L., Acién F. G. and Fernández-Sevilla J. M. (2021a). An interactive tool for simulation of biological models into the wastewater treatment with microalgae. *Frontiers in Environmental Science*, **9**, 298, https://doi.org/10.3389/fenvs.2021.721324

Sánchez-Zurano A., Lafarga T., del Morales-Amaral M. M., Gómez-Serrano C., Fernández-Sevilla J. M., Acién-Fernández F. G. and Molina-Grima E. (2021b). Wastewater treatment using *Scenedesmus almeriensis*: effect of operational conditions on the composition of the microalgae–bacteria consortia. *Journal of Applied Phycology*, **33**, 3885–3897, https://link.springer.com/article/10.1007/s10811-021-02600-2, https://doi.org/10.1007/s10811-021-02600-2

Sánchez-Zurano A., Morillas-España A., Gómez-Serrano C., Ciardi M., Acién G. and Lafarga T. (2021c). Annual assessment of the wastewater treatment capacity of the microalga *Scenedesmus almeriensis* and optimisation of operational conditions. *Science Reports*, **11**, 1–11, https://www.nature.com/articles/s41598-021-01163-z, https://doi.org/10.1038/s41598-020-79139-8

Sánchez-Zurano A., Rodríguez-Miranda E., Guzmán J. L., Acién-Fernández F. G., Fernández-Sevilla J. M. and Grima E. M. (2021d). Abaco: a new model of microalgae–bacteria consortia for biological treatment of wastewaters. *Applied Sciences*, **11**, 1–24, https://doi.org/10.3390/app11030998

Sánchez-Zurano A., Rossi S., Fernández-Sevilla J. M., Acién-Fernández G., Molina-Grima E. and Ficara E. (2022). Respirometric assessment of bacterial kinetics in algae-bacteria and activated sludge processes. *Bioresource Technology*, **352**, 127116, https://doi.org/10.1016/j.biortech.2022.127116

Sforza E., Pastore M., Barbera E. and Bertucco A. (2019). Respirometry as a tool to quantify kinetic parameters of microalgal mixotrophic growth. *Bioprocess and Biosystems Engineering*, **42**, 839–851, https://doi.org/10.1007/s00449-019-02087-9

Solimeno A. and García J. (2019). Microalgae and bacteria dynamics in high rate algal ponds based on modelling results: long-term application of BIO_ALGAE model. *Science of the Total Environment*, **650**, 1818–1831, https://doi.org/10.1016/j.scitotenv.2018.09.345

Solimeno A., Gómez-Serrano C. and Acién F. G. (2019). BIO_ALGAE 2: improved model of microalgae and bacteria consortia for wastewater treatment. *Environmental Science and Pollution Research*, **26**, 25855–25868, https://doi.org/10.1007/s11356-019-05824-5

Sompech K., Chisti Y. and Srinophakun T. (2012). Design of raceway ponds for producing microalgae. *Biofuels*, **3**, 387–397, https://doi.org/10.4155/bfs.12.39

Stirk W. A., Ördög V., Van Staden J. and Jäger K. (2002). Cytokinin- and auxin-like activity in *Cyanophyta* and microalgae. *Journal of Applied Phycology*, **14**, 215–221, https://link.springer.com/article/10.1023/A:1019928425569, https://doi.org/10.1023/A:1019928425569

Stirk W. A., Ördög V., Novák O., Rolčík J., Strnad M., Bálint P. and van Staden J. (2013). Auxin and cytokinin relationships in 24 microalgal strains. *Journal of Phycology*, **49**, 459–467, http://doi.wiley.com/10.1111/jpy.12061, https://doi.org/10.1111/jpy.12061

Sutherland D. L., Turnbull M. H. and Craggs R. J. (2014). Increased pond depth improves algal productivity and nutrient removal in wastewater treatment high rate algal ponds. *Water Research*, **53**, 271–281, https://doi.org/10.1016/j.watres.2014.01.025

Uggetti E., Passos F., Solé M., Garfí M. and Ferrer I. (2017). Recent achievements in the production of biogas from microalgae. *Waste and Biomass Valorization*, **8**, 129–139, https://doi.org/10.1007/s12649-016-9604-3

Umamaheswari J. and Shanthakumar S. (2016). Efficacy of microalgae for industrial wastewater treatment: a review on operating conditions, treatment efficiency and biomass productivity. *Reviews in Environmental Science and Bio/Technology*, **15**, 265–284, https://doi.org/10.1007/s11157-016-9397-7

United Nations. (2015). Transforming Our World: The 2030 Agenda for Sustainable Development. UN Doc. A/RES/70/1 (September 25, 2015), https://sdgs.un.org/publications/transforming-our-world-2030-agenda-sustainable-development-17981

Weissman J. C. and Goebel R. P. (1987). Design and analysis of microalgal open pond systems for the purpose of producing fuels: a subcontract report. SERI/STR-231-2840, http://www.osti.gov/energycitations/servlets/purl/6546458-qOBOVX/

Wu J., Liu J., Lin L., Zhang C., Li A., Zhu Y. and Zhang Y. (2015). Evaluation of several flocculants for flocculating microalgae. *Bioresource Technology*, **197**, 495–501, https://doi.org/10.1016/j.biortech.2015.08.094

Zhang X., Hu Q., Sommerfeld M., Puruhito E. and Chen Y. (2010). Harvesting algal biomass for biofuels using ultrafiltration membranes. *Bioresource Technology*, **101**, 5297–5304, https://doi.org/10.1016/j.biortech.2010.02.007

doi: 10.2166/9781789063547_0103

Chapter 6

Microalgae–methanotroph cocultures for carbon and nutrient recovery from wastewater

Jin Wang and Q. Peter He*

Department of Chemical Engineering, Auburn University, Auburn, Alabama, USA
**Corresponding author: wang@auburn.edu*

ABSTRACT

Wastewater produced from municipal, agricultural, and industrial processes has caused detrimental impacts on local communities and environments. In addition, wastewater is the fifth largest anthropogenic source of methane emissions globally. Although anaerobic digestion is a proven waste management technology with many environmental benefits, its application is limited to large-scale water resource recovery facilities. This is due to the poor return-on-investment caused by the contaminants present in raw biogas and nutrient-rich liquid digestate that require further treatments, some of which are costly. In this chapter, we discuss our recent development of a microalgae–methanotroph coculture-based platform for integrated biogas valorization and nutrient recovery. Development of coculture-based biotechnology faces many technical challenges, including tracking the growth of individual species in the coculture over time, quantifying and understanding inter-species metabolic interactions, and developing kinetic models for the coculture. There are also many practical considerations when applying coculture-based biotechnology for wastewater treatment. We discuss our proposed solutions to address these technical challenges and practical concerns. We also offer our perspective on future directions.

Keywords: anaerobic digestion, biogas valorization, microalgae–methanotroph coculture, nutrient recovery, photobioreactor

6.1 BACKGROUND

Municipal, agricultural, and industrial processes produce significant amounts of wastewater that contain organic carbon and high content of nitrogen, phosphorus, and other pollutants. If not adequately treated before discharge into waterways, wastewater can have detrimental impacts on local communities and the environment. Wastewater is the fifth largest anthropogenic source of methane (CH_4) emissions globally, contributing to worsening the greenhouse effect. Moreover, the chemicals and pollutants found in untreated wastewater, such as chlorofluorocarbons and other ozone-depleting substances, can have a harmful impact on the ozone layer. When these substances are released into the atmosphere, they can react with and destroy ozone molecules in the stratosphere, harming the protective ozone layer. This, in turn, can increase the amount of harmful ultraviolet radiation that reaches the Earth's surface, causing various health problems for humans and damaging ecosystems. Finally, wastewater also contributes to smog, acid rain, and drinking water contamination (Driscoll *et al.*, 2003; Galloway *et al.*, 2004).

Wastewater resource recovery facilities (WRRFs) typically use a combination of physical, chemical, and biological treatment techniques to remove pollutants and contaminants from wastewater. Among them, nitrification–denitrification is a biological process most commonly used in WRRFs to remove nitrogen from wastewater. It involves the conversion of ammonia (NH_3) to nitrate (NO_3^-) by nitrifying bacteria, followed by the conversion of nitrate to nitrogen gas (N_2) by denitrifying bacteria. A nitrification–denitrification process requires energy to operate and can be a significant source of operating costs. For example, nitrification requires oxygen (O_2), typically supplied using aeration systems that blow air into the wastewater. This is often the most energy-intensive process in the treatment plant, and the energy consumption can be in the range of 25–60% of the total energy use. Pumps and mixers are used to move and mix wastewater through the treatment process. These devices require energy to operate and can account for 10–25% of the total energy use (Siegrist *et al.*, 2008). A denitrification process often requires supplementing an organic carbon source (e.g., methanol) to support nitrate reduction (Tam *et al.*, 1992; Zhao *et al.*, 1999). Pumping air and supplying organic carbon sources are the primary contributors to high-operational costs for WRRFs (Drewnowski *et al.*, 2019).

On the contrary, the organic carbon stranded in wastewater is a valuable and often overlooked resource for producing fuels and chemicals. By transforming wastewater treatment processes to extract and utilize this carbon, it is possible to not only mitigate the negative environmental and societal consequences of wastewater, but also generate revenue to offset treatment costs and potentially create a profitable industry. This potential has spurred increasing research interest in waste-to-value (W2V) technologies, which include waste-to-energy, waste-to-fuel, waste-to-chemical, and other similar processes (Fei *et al.*, 2014; Haynes & Gonzalez, 2014; Henard *et al.*, 2016).

Anaerobic digestion (AD) is commonly regarded as the most effective waste management solution for wet organic waste and is currently the only widely commercialized W2V process at scale. In a controlled and contained manner, AD breaks down organic matter into biogas (consisting primarily of CH_4 and carbon dioxide (CO_2), with trace amounts of other gases such as hydrogen sulfide (H_2S)), and nutrient-rich digestate (containing valuable elements such as N, P, and K). Furthermore, AD is highly effective at mitigating odors and pathogens (Angelidaki & Ellegaard, 2003; Nasir *et al.*, 2012; Topper *et al.*, 2006). Given these benefits, AD has gained significant traction for converting stranded organic carbon in wastewater to biogas and is widely adopted by large-scale WRRFs. In the USA, 48% of total wastewater flows are treated by AD, with at least 1,238 WRRFs (out of 14,780 WRRFs in the USA) processing solids through AD (Qi *et al.*, 2013).

Many W2V technologies could utilize AD as a key process. The three most prominent ones are: (1) combined heat and power (CHP), which utilizes biogas produced from AD as a fuel source to generate heat and/or electricity; (2) renewable natural gas or biomethane (BM) where biogas is upgraded to meet pipeline quality standards, allowing it to be used as a transportation fuel or injected into natural gas pipelines; and (3) microalgae cultivation (MC) utilizing AD-generated digestate (and CO_2) for biofuels or bioplastics production (Ahmed *et al.*, 2021; Angelidaki *et al.*, 2018; Kapoor *et al.*, 2020). Despite their wide recognition, there are challenges associated with these technologies. For example, both CHP and BM require significant capital expenditure and operating expenses due to the impurities present in biogas (e.g., H_2S and NH_3). In addition, both technologies produce low-value products (electricity, heat, or CH_4), further deteriorating their economic viability. As a result of poor return on investment (ROI), both CHP and BM have limited success in the USA. For example, among the WRRFs that have AD installed, 15% (~184) of them flare the produced biogas, 64% (~791) of them burn the biogas for digester and building heating, whereas only 21% (~263) of them use biogas for power generation or driving machinery (Qi *et al.*, 2013). Although European Union countries have built technologically and economically mature CHP and BM industries, this has primarily been achieved through massive government subsidies that offset prices up to 10 times those of electricity and natural gas (Angelidaki *et al.*, 2018; Lombardi & Francini, 2020; Mishra *et al.*, 2021; Rosa, 2020; Scarlat *et al.*, 2018). As for integrated AD and microalgae cultivation (IADMC), the AD digestate provides the microalgae with nutrients, such as nitrogen and phosphorus, which are essential for their growth. The microalgae biomass can then be harvested and processed to produce biofuels, such as biodiesel and bioethanol, or other value-added products, such as nutraceuticals and cosmetics (Ho *et al.*, 2018; Rajagopal *et al.*, 2021; Vadiveloo *et al.*, 2021). IADMC achieves two goals simultaneously: AD provides a sustainable source of nutrients for MC, whereas MC helps reduce environmental impact of AD by recycling the nutrients produced. However, microalgae-based biofuels face several challenges that hinder their competitiveness with petroleum fuels, and sustainable production levels have not yet been achieved (Beckstrom *et al.*, 2020; Georgakopoulou, 2019; Richardson *et al.*, 2012; Sun *et al.*, 2011). These challenges include low biomass density, high costs associated with harvesting and downstream processing, and significant demands for water, energy, and land due to the limited light penetration in the liquid phase. Moreover, energetic and environmental viability of algae-based biofuels presents serious challenges that must be addressed (Choi *et al.*, 2021).

As mentioned earlier, wastewater is a significant contributor to anthropogenic CH_4 emissions, ranking as the fifth largest source globally and accounting for 7–9% of total global CH_4 emissions. The growing recognition of the prominent role of CH_4 emissions in climate change highlights the need for CH_4 remediation at WRRFs. This is especially crucial because CH_4 is 86 times more potent than CO_2 on a 20-year timescale and 34 times more potent on a 100-year timescale, making it a significant contributor to global warming. Multiple studies have demonstrated that mitigation of CH_4 can play an outsized role in limiting warming over the next few decades. It presents a potential for rapidly reducing climate warming, either in the near term to prevent temporary exceedance of the 1.5 or 2.0°C peak warming threshold, or later in the century to bring down temperatures after an overshoot to higher levels (Alvarez *et al.*, 2012; Collins *et al.*, 2018; Harmsen *et al.*, 2020; Ocko *et al.*, 2021; Rogelj *et al.*, 2015). Consequently, CH_4 remediation is a vital aspect of wastewater treatment and should be integrated into nutrient recovery.

6.2 OVERVIEW OF MICROALGAE–METHANOTROPH COCULTURES: A PROMISING W2V PLATFORM FOR WASTEWATER TREATMENT

In this chapter, we introduce a novel W2V biotechnology platform that employs a microalgae–methanotroph coculture for the integrated treatment of AD effluent (i.e., digestate) and conversion of biogas as shown in Figure 6.1. In microalgae–methanotroph coculture systems, a carefully selected microalgae–methanotroph coculture effectively converts both CO_2 and CH_4 in biogas to microbial biomass, while also recovering nutrients (e.g., N and P) from AD effluent to support microbial growth. We present some promising results from our laboratory experiments to demonstrate the potential of the microalgae–methanotroph coculture as a sustainable, profitable, and effective W2V platform.

The proposed microalgae–methanotroph coculture W2V platform is based on studies published in *Nature* (Raghoebarsing *et al.*, 2005) and *Nature Geoscience* (Kip *et al.*, 2010), which suggest that the

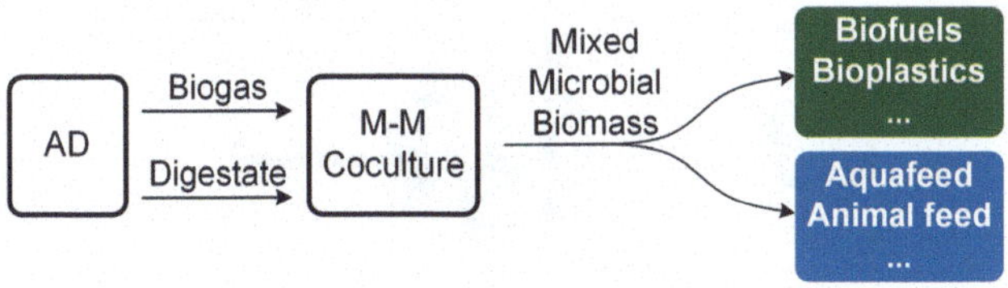

Figure 6.1 Overview of the proposed integrated AD and microalgae–methanotroph coculture W2V platform.

coupling of CH_4 oxidation (by methanotrophs) and oxygenic photosynthesis (by peat moss) is prevalent in wetlands to reduce CH_4 emissions and reuse CO_2. Another report in *Nature ISME* (Milucka *et al.*, 2015) shows that true aerobic oxidation by methanotrophs, fueled by *in-situ* O_2 produced from photosynthetic algae, is responsible for CH_4 removal in anoxic waters. These discoveries suggest that a microalgae–methanotroph coculture could be a novel and effective approach to recycling biogas and nutrients.

The coupling of methanotrophs and microalgae offers several advantages as shown in Figure 6.2: (1) dissolved O_2 produced by microalgae will be consumed by methanotrophs for cell growth, which not only eliminates the safety risk of mixed CH_4 and O_2, but also eliminates the O_2 inhibition to microalgae growth; (2) dissolved CO_2 produced by methanotrophs can further promote microalgae growth due to significantly reduced mass transfer resistance; and (3) microalgae growth can be significantly promoted by cofactors (e.g., biotin and thiamine), which are synthesized by methanotrophs (Cecchin *et al.*, 2018; Tandon *et al.*, 2017). By exploring the synergistic metabolic coupling of oxygenic photosynthesis and CH_4 oxidation, microalgae–methanotroph cocultures have demonstrated significantly increased biomass production and nutrient recovery. Compared to CHP, which only captures the energy contained in CH_4 while losing all carbon to CO_2, the coculture can recover 100% of the carbon in CH_4 and CO_2, resulting in zero emissions and close to 100% nutrient recovery (Roberts *et al.*, 2020).

To improve process economic feasibility, simple downstream processing that requires limited capital and operational cost is preferred. In addition, high-value products are needed to make the overall process profitable. With these considerations, coculture biomass produced from biogas and wastewater can serve as feedstock for animal feed supplements or bioplastics.

For wastewater produced from wineries and food-processing plants that are determined to be safe (e.g., low level of heavy metals and antibiotics), the coculture biomass could be used as single-cell protein (SCP) for aquafeed supplements. It is worth noting that both microalgae and methanotrophs have been studied extensively and tested as protein supplements for aquafeed. For example, trials in fish have shown that the protein meal derived from methanotrophs performs well as an alternative protein source to fish meal in feed formulations for Atlantic salmons (Aas *et al.*, 2006). Other studies on methanotroph-derived fish meal show improved growth performance and health benefits in aquatic and terrestrial animals (Øverland *et al.*, 2010; Romarheim *et al.*, 2010). For microalgae, positive testing results in fish and shrimp have suggested that a significantly higher dietary inclusion level of microalgal biomass in aquafeeds is expected (Becker, 2007; Gamboa-Delgado & Márquez-Reyes, 2018; Teimouri *et al.*, 2013). As a result, the coculture biomass of microalgae and methanotrophs could be a highly promising source for SCP.

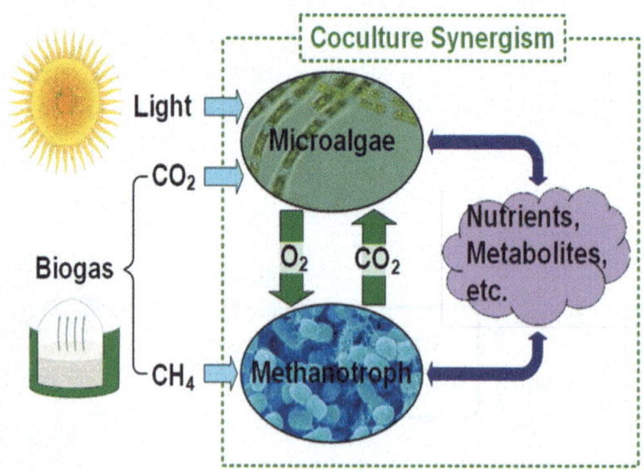

Figure 6.2 Synergistic interactions within microalgae–methanotroph cocultures.

For wastewater from municipal sewage and industrial processes, the produced coculture may not be safe for animal feed. In this case, they could be used as bioplastics feedstock (BPFS). Recent studies have shown that when simple and cost-effective processing of whole microalgal or bacterial cells is applied to produce bioplastics for packaging (which is in rapidly increasing demand), agricultural plastic products (e.g., horticultural containers), or apparel (e.g., footwear), these bioplastics products have been found to be immediately competitive without any government subsidy or incentive (Beckstrom *et al.*, 2020; Chia *et al.*, 2020; Choi *et al.*, 2021; Chong *et al.*, 2021; Coppola *et al.*, 2021; Karan *et al.*, 2019; Onen Cinar *et al.*, 2020; Tharani & Ananthasubramanian, 2020; Zeller *et al.*, 2013). This has been demonstrated by successful commercial applications at Algix – footwear brands, including Adidas and Merrell, who have launched commercial products containing bioplastics manufactured at Algix through thermo-mechanical molding of mixed whole-cell microbial biomass (Corcoran & Hunt, 2021). For microbial biomass to be used as BPFS, its protein content must be in the range of 35–60%; the protein content of the coculture biomass we produced from biogas and AD digestate was 43% (±4%) (Roberts *et al.*, 2020), ideal for BPFS.

In general, using mixed cultures as biocatalysts offers many advantages over conventional single or pure cultures. With mixed cultures, product yield and growth rate may be higher. In addition, mixed cultures usually enable better utilization of cheap, complex substrates. Mixed cultures often exhibit enhanced robustness and offer more protection against contamination. For example, studies have shown that, compared to a monoculture of microalgae, the cocultivation of microalgae and bacteria or fungi can deliver improved performance in terms of nutrient removal and biomass production. Zhang *et al.* (2020) systematically reviewed the recent progress in this area. However, using mixed cultures for bioconversion also presents many challenges. It is challenging to monitor and characterize a mixed culture in real time, which is the prerequisite for studying its growth kinetics. In addition, without real-time monitoring, it is almost impossible to maintain a mixed culture at its optimal state, such as controlling the optimal population of coculture to maximize substrate conversion. As a result, most existing research on mixed culture-based nutrient recovery from AD effluent mainly focused on the process performance (both biogas conversion and/or nutrient recovery), with few of them examining the inter-species interactions, and how such interactions would affect the overall performance of the process.

In this chapter, we present our recent progress on studying microalgae–methanotroph cocultures for biogas conversion and nutrient recovery, which includes the protocols for the coculture characterization in real time, kinetic modeling of the coculture, and many practical considerations for using microalgae–methanotroph cocultures for biogas conversion and wastewater treatment. The results presented in the following sections mainly utilized two-model coculture systems. One is *Arthrospira platensis–Methylomicrobium buryatense* 5GB1, which is a cyanobacterium–type I methanotroph coculture that prefers high-pH and high-salt medium. Using this model coculture, we developed experimental and computational tools for studying microalgae–methanotroph cocultures. The other model coculture is *Chlorella sorokiniana–Methylococcus capsulatus* (Bath), which is a microalgae–type X methanotroph coculture that prefers neutral pH and low-salt medium. We demonstrated integrated wastewater treatment and biogas valorization using the latter coculture pair.

6.3 EXPERIMENTAL AND COMPUTATIONAL TOOLS FOR REAL-TIME CHARACTERIZATION OF THE MICROALGAE–METHANOTROPH COCULTURES

Multispecies associations are prevalent in nature, providing essential ecosystem services such as carbon, nutrient, and metal cycling. This can be explained by natural selection – mixed cultures can offer numerous benefits over monocultures, as discussed previously. Despite these advantages, the use of mixed cultures for biotechnological applications in bioenergy and related fields has been limited. This is mainly due to the fact that the successful commercialization of potential biotechnologies requires a comprehensive understanding of the fundamental biological conversion steps within microorganisms.

Acquiring this knowledge requires accurate characterization of cell growth dynamics, substrate conversion, and product excretion rates. However, there is a lack of effective tools to accurately characterize the mixed culture in real time. In addition, the involvement of gas substrates (both CH_4 and CO_2) makes the characterization of the coculture more challenging, as obtaining accurate measurements of gas component uptake or production rates can be tricky due to their high sensitivity to system pressure or volume changes. In the following section, we will first discuss the experimental protocols that can deliver accurate gas-phase measurements, then the experimental–computational (E–C) protocols to characterize the coculture in real-time are discussed.

6.3.1 Accurate measurement of gas component uptake and production rates in bioconversion

The challenges with measuring gas component consumption and production rates are not caused by the precision of analytical equipment (e.g., gas chromatography (GC)). Instead, they are rooted in the fact that the consumption and/or production of gases alter the system headspace pressure (for batch processes) or gas-phase flow rate (for continuous processes). For batch experiments conducted in closed systems with constant volume, such as vials, the system pressure often experiences significant reduction when gas substrate(s) are involved. This can be caused by the overall gas consumption exceeding gas production, as well as gas and/or liquid sampling. On the contrary, vials can become over-pressurized if gas production dominates. When a headspace gas sample is obtained from an under- or over-pressurized system with a gas-tight syringe, after the syringe is withdrawn from the system, air can enter the syringe if under-pressurized, or the gas sample can escape from the syringe if over-pressurized. Our study demonstrated that such pressure differences could cause significant errors in the measured gas composition. For continuous chemostat operation, the system pressure should be constant. However, due to the imbalance between gas-phase substrate consumption and gaseous product excretion, the off-gas flow rate can significantly differ from the feeding gas flow rate. For continuous operation, accurate off-gas flow rate measurements are critical for determining the gas consumption and production rates through the mass balance of the system.

In Stone *et al.* (2019), we reported two easy-to-implement experimental protocols and associated computation procedures to obtain accurate measurements of gas component consumption and production rates: one for batch experiments, the other for continuous operation. For depressurized (i.e., system pressure below 1 atm) batch cultures, we use N_2 (or other inert gases that do not interfere with the measurements of other gases) to repressurize the system to 1 atm before obtaining samples, whereas for pressurized systems, accurate measurements can be obtained by measuring the system pressure and scale-up the measured composition by multiplying the ratio between the system pressure and the atmospheric pressure. For continuous cultures, we use helium (or other inert gases that can be accurately measured by GC) as an internal tracer to accurately measure the off-gas flow rate. Two abiotic and two biotic systems were used to conduct several case studies to validate the effectiveness and accuracy of the protocols and associated computational procedures.

The effectiveness of the developed repressurization protocol for batch systems is demonstrated in Figure 6.3a. Figure 6.3a clearly shows that the repressurization protocol can significantly reduce the measurement error in O_2 concentrations for a batch system with known gas composition. For a continuous system that cultivates a type I methanotroph (*M. buryatense* 5GB1), the significant difference between the feed and off-gas flow rates due to cell growth is demonstrated in Figure 6.3b. This result confirms the necessity to accurately measure the off-gas flow rate. In addition, the accuracy of the tracer protocol for a continuous system is validated through the total carbon balance. Details can be found in Stone *et al.* (2019).

6.3.2 Quantitative characterization of microalgae–methanotroph cocultures

One major challenge associated with characterizing mixed cultures is how to accurately determine the individual biomass concentration for each microorganism during the dynamic growth of the mixed culture. Existing characterization approaches can be categorized into molecular biological,

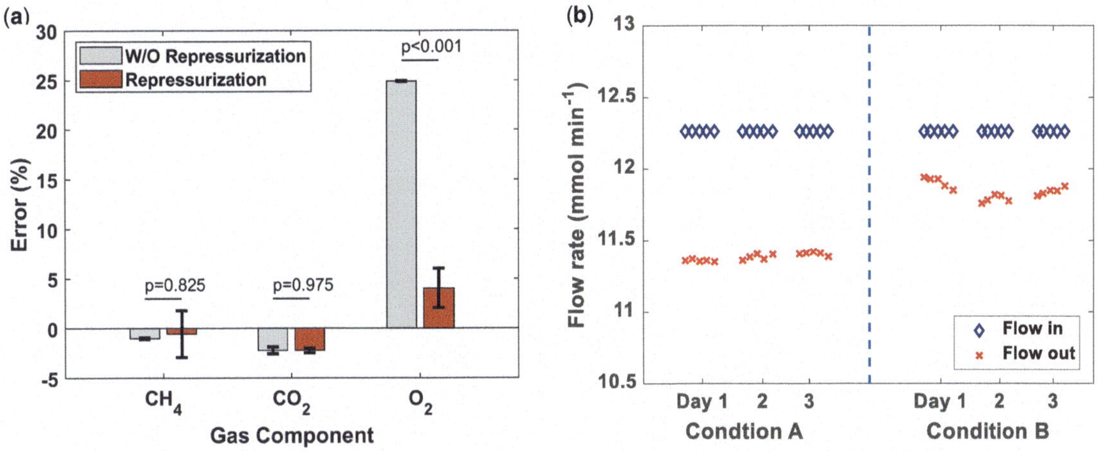

Figure 6.3 (a) Proposed experimental protocol significantly reduces measurement error in O_2 in a batch system; and (b) significant difference between the feed and off-gas flow rates due to cell growth in a continuous system.

biochemical, and microbiological methods (Sabra *et al.*, 2010; Spiegelman *et al.*, 2005). However, these methods require either expensive equipment, such as flow cytometry, community genome sequencing, or time-consuming and challenging techniques, such as ribonucleic acid (RNA)/deoxyribonucleic acid (DNA) extraction, isolation, or amplification. As a result, although these approaches deliver accurate offline characterization of a mixed culture, they are not suitable for frequent or real-time measurements desired for dynamic modeling of the mixed culture systems. As a result, among the published research on microalgae–methanotroph cocultures, only Hill *et al.* (2017) tracked the individual biomass concentration over time through cell counting using flow cytometry, whereas others only reported the total optical density (OD) of the coculture over time (Rasouli *et al.*, 2018; van der Ha *et al.*, 2012).

In addition to individual biomass concentration, the individual substrate consumption rates and product excretion rates of each organism over time are needed for the development of kinetic models. When there is cross-feeding in the coculture (i.e., any exchange of metabolite(s) between different organisms), the individual consumption/production rates cannot be measured directly. For the microalgae–methanotroph coculture, as shown in Figure 6.2, both O_2 and CO_2 are cross-feeding metabolites: O_2 is produced by microalgae while consumed by methanotrophs, whereas CO_2 is produced by methanotrophs and consumed by microalgae. The total consumption/production rates of O_2 and CO_2 by the coculture can be directly measured by GC, but accurate splitting of the total rates into two components is challenging. One could use labeled substrates, such as 13-CH_4, to determine the amount of cross-fed CO_2, but it requires expensive substrate and additional analysis, which is infeasible for real-time tracking.

To address these challenges and obtain real-time measurements for individual strains in the coculture, we have developed an E–C protocol to fully characterize the synthetic microalgae–methanotroph coculture that only requires commonly used analytical equipment. As shown in Figure 6.4, based on the measured total substrate consumption and production excretion rates, the E–C protocol computes the individual biomass concentration, individual substrate uptake rates, and product secretion rates based on the overall mass balance and growth stoichiometry of each organism. For microalgae–methanotroph cocultures, the substrate uptake rates include CH_4 and O_2 uptake rates for methanotrophs (can be computed or determined by Equations (6.1) and (6.2) in Figure 6.4, respectively), CO_2 uptake rate for microalgae (Equation (6.5)). The product secretion rates

(a)

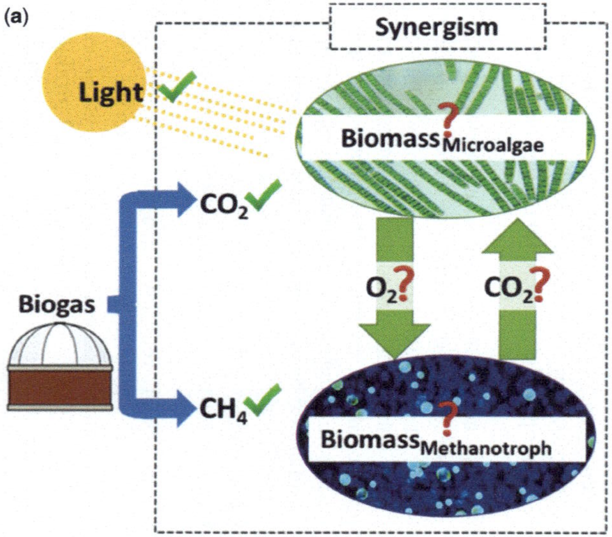

(b)

Measured
Estimated
Calculated

$$(\Delta CH_4)_M = (\Delta CH_4)_H \quad (1.1)$$

$$(\Delta O_2)_M = (\Delta CH_4)_M \cdot \left(Y_{\frac{O_2}{CH_4}}\right)_M \quad (1.2)$$

$$(\Delta CO_2)_M = -(\Delta CH_4)_M \cdot \left(Y_{\frac{CO_2}{CH_4}}\right)_M \quad (1.3)$$

$$(\Delta X)_M = (\Delta CH_4)_M \cdot \left(Y_{\frac{X}{CH_4}}\right)_M \quad (1.4)$$

$$(\Delta CO_2)_H = (\Delta CO_2)_M + (\Delta CO_2)_P + (\Delta CO_2)_L \quad (1.5)$$

$$(\Delta O_2)_H = (\Delta O_2)_M + (\Delta O_2)_P \quad (1.6)$$

$$(\Delta X)_P = \frac{1}{2}\left[(\Delta CO_2)_P \cdot \left(Y_{\frac{X}{CO_2}}\right)_P + (\Delta O_2)_P \cdot \left(Y_{\frac{X}{O_2}}\right)_P\right] \quad (1.7)$$

Subscripts:
M: methanotroph; P: microalgae/photoautotroph;
H: headspace ; L: liquid

Figure 6.4 Illustration of the E–C protocol for microalgae–methanotroph coculture characterization: (a) schematic diagram of the relationships among known variables (with checkmarks) and unknown variables (with crosses); and (b) equations used to compute the unknown variables based on the known measurements (in blue font) and yield parameters (in purple font).

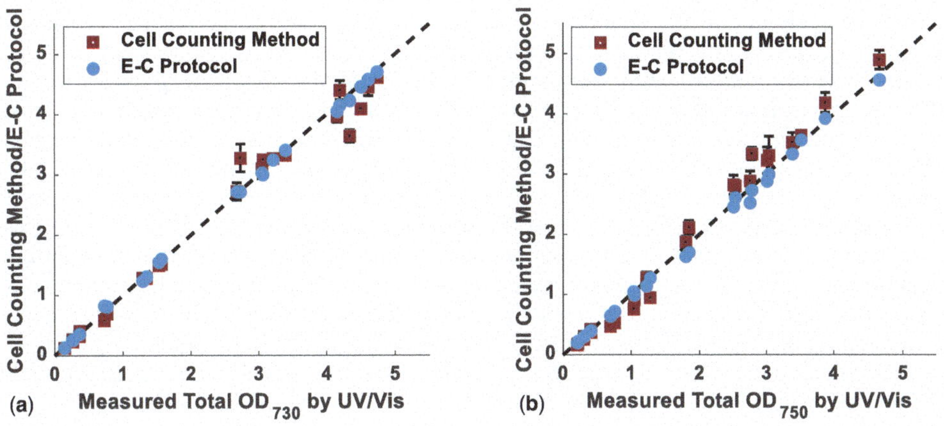

Figure 6.5 Comparison of the measured total OD vs. the total OD calculated using the individual biomass concentrations obtained through cell counting and the E–C protocol: (a) the salt water coculture pair *Synechococcus* sp. PCC7002–*M. alcaliphilum* 20ZR; and (b) the freshwater coculture pair *C. sorokiniana–M. capsulatus*. Points that lie closer to the diagonal dashed line have better accuracy than points that lie further away from the diagonal dashed line.

include CO_2 production rate by methanotrophs (Equation (6.3)), and O_2 production rate by microalgae (Equation (6.6)). Individual species growth rates include that of methanotrophs (Equation (6.4)) and microalgae (Equation (6.7)). More details can be found in Badr *et al.* (2021).

The E–C protocol was applied to characterize the growth of one cyanobacteria–methanotroph pair (*Synechococcus* sp. PCC7002–*Methylomicrobium alcaliphilum* 20ZR) and one microalgae–methanotroph pair (*C. sorokiniana–M. capsulatus*). The accuracy of the E–C protocol was validated by individual biomass concentrations measured through cell counting using flow cytometry. Moreover, we further showed that the E–C protocol provides better accuracy than the cell counting approach by comparing the predicted total OD from the individual biomass concentration by both approaches with the measured total OD and statistical testing. This is illustrated in Figure 6.5. For both cases, the E–C protocol provides noticeably better estimates of total OD than those based on cell counting – points that lie closer to the diagonal dashed line have better accuracy than points that lie further away from the diagonal dashed line.

The E–C protocol only requires the commonly used analytical equipment, including GC, ultraviolet-visible spectrometry, and total carbon analyzer. It does not require any special sample preparation such as deoxyribonucleic acid (DNA) or ribonucleic acid (RNA) extraction or cell fixation. Therefore, it is suitable for real-time characterization of microalgae–methanotroph cocultures. In addition, it is shown that the E–C protocol is more accurate than cell counting such as flow cytometry. The significance of the E–C protocol is that it provides the real-time quantitative coculture characterizations that are required for the kinetic modeling of the coculture. We expect that the E–C protocol and its adaptations to other systems will become convenient and valuable tools for developing coculture or mixed culture-based biotechnologies.

6.4 SEMI-STRUCTURED KINETIC MODELING OF THE COCULTURE

A key enabler for the development of any biotechnology is a high-quality kinetic model that can predict the growth of biocatalysts under different conditions. Such a kinetic model can enable optimal design and scale up of the bioreactor. It also provides the foundation for model-based control of the bioreactor. For the microalgae–methanotroph coculture, there is a lack of effort in developing kinetic models. This is largely due to the challenges in obtaining real-time characterization of microalgae–methanotroph cocultures and the lack of understanding of the inherently complex interactions in the coculture. For example, the time-series measurements of the coculture growth over time are prerequisite for the estimation of the kinetic parameters, which has not been available until recently. In addition, the *in-situ* exchange of O_2 and CO_2 between the microalgae and methanotroph, as well as additional unknown interspecies interactions make the kinetic modeling of the system highly challenging.

Enabled by the E–C protocol discussed in the previous section, we have published the very first kinetic model that captures the known cross-feeding mechanism within a microalgae–methanotroph coculture. Our kinetic model is a semi-structured model, as it explicitly models the exchange of *in-situ* produced O_2 and CO_2 between the two species. The other unknown interactions are captured in model parameters. For example, if the coculture promotes the growth of the methanotroph, we would expect the maximum methanotroph growth rate in the coculture model to have a greater value than that of the methanotroph in its single-culture model.

Figure 6.6 illustrates the flow chart and key equations for the semi-structured kinetic model. The model consists of four components: (1) microalgae growth; (2) methanotroph growth; (3) mass balance in the liquid phase; and (4) mass balance in the gas phase. Note that the growth of each organism in the coculture is coupled with the changes in the gas-phase composition through the mass balances in the liquid and gas phases. The equations that model each component are listed in Figure 6.6 as well, with the terms that coupling different components highlighted in dotted boxes. These terms include the O_2 produced by the microalgae, CO_2 produced by the methanotroph, and their contributions to the liquid phase mass balances.

Several sets of wet-lab experiments were conducted to test the accuracy of the semi-structured kinetic modeling using a cyanobacteria–methanotroph coculture (*A. platensis–M. buryatense* 5GB1). It was shown that the semi-structured kinetic model accurately predicted the individual growth

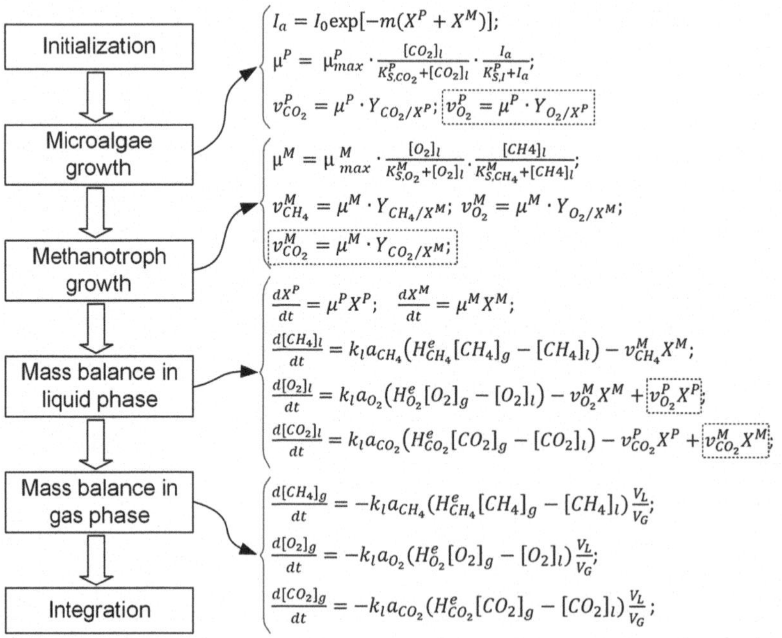

Figure 6.6 Flow chart of the semi-structured kinetic model and the associated model equations.

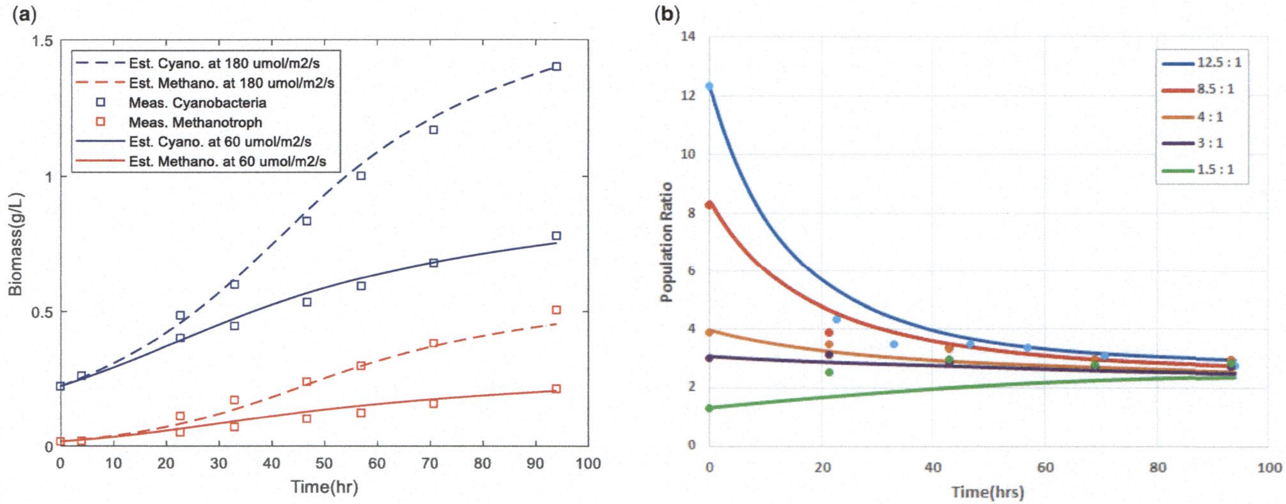

Figure 6.7 Experimental validation of the semi-structured kinetic model using a methanotroph–cyanobacteria coculture: (a) comparison of the biomass concentration in the coculture predicted by the model (lines and dashed lines) with measurements (squares) and (b) both model simulation (lines) and experimental measurements (dots) confirm that despite different inoculum ratios (cyanobacterium:methanotroph), after achieving steady state, the population ratio of the two strains converge to the same value.

rates, and the individual consumption/production rates of O_2 and CO_2 for the methanotroph and cyanobacteria in the coculture (Badr *et al.*, 2022). Specifically, experiments were conducted to examine the effect of several factors on the growth of the coculture, including light intensity, gas-phase composition, and inoculation ratio. The model predictions showed excellent agreement with the experimental data under all conditions examined, as illustrated in Figure 6.7.

With the details provided by the model that are often impossible to obtain through experiments, we revealed that for the cyanobacteria–methanotroph coculture cultivated on biogas without external O_2 supply, light limitation (due to self-shading effects) becomes the growth-limiting factor before mass transfer limitation (due to the small solubility of CH_4 and O_2). In addition, the semi-structured kinetic model captures the effect of other emergent metabolic interactions through the maximum growth rate. In fact, the maximum growth rates for the model coculture showed a 48 and 42% increase compared to their corresponding monocultures for cyanobacteria and methanotrophs, respectively. The significant increase in the maximum growth rates confirms the existence of the emergent metabolic interactions within the model coculture, although their identities are not yet known. More details can be found in Badr *et al.* (2022).

6.5 INTEGRATED NUTRIENT RECOVERY AND MITIGATION OF GREENHOUSE GAS EMISSIONS FROM WASTEWATER USING MICROALGAE–METHANOTROPH COCULTURES

As discussed in Section 6.1, due to many environmental benefits of AD and production of a valuable fuel (CH_4), it has been commonly deployed in WRRFs to convert organic wastes contained in wastewater into biogas, particularly in large-scale WRRFs. However, the installations of AD at mid- and small-scale WRRFs have been very limited. This is because the utilization of the AD-generated biogas has been limited to heating and electricity generation. The low value of heat/electricity is the main reason for poor ROI that prevents AD installations at mid- and small-scale WRRFs. Another drawback of AD is that its nutrient-rich liquid effluent is usually required to go through a nitrification–denitrification process prior to discharge, which is energy intensive and costly as discussed before.

The microalgae–bacteria or microalgae–fungi mixed culture has been studied for wastewater treatment and biogas upgrading (Kleerebezem & van Loosdrecht, 2007). In addition, inspired by how nature recycles biogas and nutrient, microalgae–methanotroph cocultures have been examined for integrated nutrient recovery and/or biogas valorization. For example, van der Ha *et al.* (2012) reported that a coculture of *Methylocystis parvus–Scenedesmus* sp. could completely convert a synthetic biogas (60% CH_4, 40% CO_2) into microbial biomass without external O_2 supply. Hill *et al.* (2017) demonstrated that *M. alcaliphilum–Synechococcus* PCC 7002 could maintain stable growth on gas mixtures with a wide range of compositions, including raw biogas and synthetic biogas. Rasouli *et al.* (2018) demonstrated the application of using a microalgae–methanotroph (*C. sorokiniana–M. capsulatus*) coculture for nutrient recovery from a potato plant wastewater with synthetic biogas.

Most recently, through an ongoing collaboration with Columbus Water Works (CWW), our research (Roberts *et al.*, 2020) showed that *C. sorokiniana–M. capsulatus* can efficiently recover nutrients (N and P) in wastewater while converting biogas into microbial biomass. Located in Columbus (GA, USA), CWW is the fourth largest WRRF in Georgia and has been a leader among WRRFs in technology innovation. In 2012, CWW implemented and demonstrated a new technology that integrates green power and class-A biosolids production with wastewater treatment. The technology, known as the Columbus Biosolids Flow-Through Thermophilic Treatment and Cogeneration System, was the first thermophilic AD process in the USA that runs entirely off the heat derived from the AD-biogas-fueled power generation. Using *C. sorokiniana–M. capsulatus* (Bath) as the model coculture, we demonstrated that the coculture achieved 100% recovery of NH_3 (80% recovery of total nitrogen (TN)) and 100% recovery of orthophosphate (98% recovery of total phosphorous) when biogas supply is unlimited. In addition, the coculture achieved 100% CH_4 and CO_2 conversion into microbial biomass when nutrient supply is unlimited. Also, biomass production, TN recovery, and total phosphorus (TP) recovery performance of the microalgae–methanotroph coculture were significantly better than those of the microalgae monoculture, achieving 120, 71, and 164% improvements, respectively, when the same amount of biogas was used.

The practical implementation of the microalgae–methanotroph coculture-based W2V technology requires consideration of several critical factors. In this section, we examine four major issues that could affect the performance of the microalgae–methanotroph coculture for nutrient recovery and biogas valorization, along with our preliminary results on how to address them. These factors include the selection of the biocatalyst, the tolerance of the coculture to contaminants in raw biogas, the freshwater consumption rate, and the pretreatment of AD effluent. Furthermore, in order to attract more attention to coculture or mixed culture W2V technology from researchers in the W2V field, and to achieve wide acceptance of the platform in the field, we need to provide convincing and experimentally verifiable results to demonstrate advantages of cocultures over single cultures or sequential single cultures.

6.5.1 Choice of a suitable biocatalyst

One key factor that has significant impact on the performance of the coculture W2V platform is the choice of biocatalyst. The ideal candidates should tolerate various inhibitors in the AD digestate and raw biogas, while delivering robust and stable growth under various disturbances, such as light intensity, light–dark cycle, and availability of macro- and micro-nutrients (e.g., N and P).

Our research has shown that not all microalgae–methanotroph pairs form a synergistic partnership. This is true even if their preferred growth condition matches, which include pH and salinity. In addition, different AD digestates contain different inhibitors and stressors depending on the organic waste fed to the AD, and the AD types and operations. Therefore, to optimize the performance of the coculture system in terms of carbon and nutrient recovery, it is important to screen different microalgae and methanotrophic strains to identify the best candidates.

The screening of different coculture pairs is a time- and labor-intensive process. Given the number of microalgae and methanotrophic species n_1 and n_2, there are n_1n_2 potential coculture candidates to be screened. To speed up the screening process, we have developed a specialized equipment, termed species screening station (S3), as shown in Figure 6.8. S3 consists of nine parallel-fed batch reactors that control temperature, pH, agitation rate, light intensity, gas composition, and feed rate. It is important to note that abiotic tests are necessary to verify that the equipment provides consistent or uniform growth conditions among different reactors, with the only difference being the biocatalysts in each reactor. Besides abiotic tests, biotic tests should be conducted by cultivating the same species in

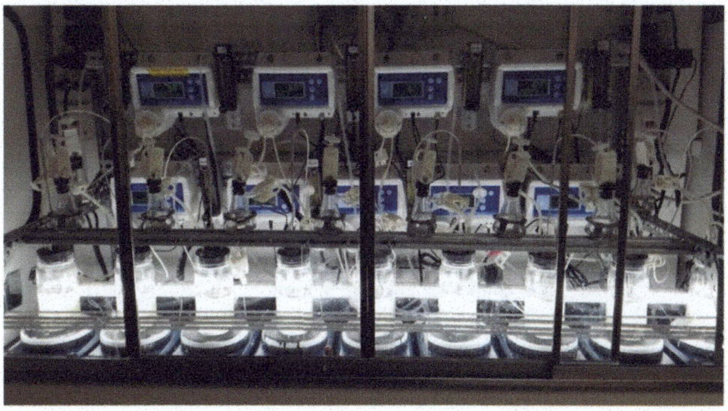

Figure 6.8 Photograph of S3 that can run nine parallel-fed batch bioreactors simultaneously.

all nine reactors under the same conditions. The growth rates measured from different rectors should be comparable to confirm they are indeed nine replicates.

Another important issue to consider for strain screening is to provide a consistent adaptation process for all strains to be examined. Because of different inhibitors in the AD digestate, some strains may not adapt as fast as others. Without any adaptation, the fast-adaptive strains may exhibit better growth performance on the wastewater in short terms, whereas others that exhibit better growth performance in longer terms may appear to perform worse during the screening. To avoid missing out the slower adapters, we have developed a standard protocol to provide a consistent adaptation period as described below. Before the screening, each strain will first be introduced to a transition medium (5% AD digestate, 45% clarifier water, both from CWW, plus 50% defined medium). The strains will be harvested during their mid-exponential growth phase on the transition medium and used as inoculation for screening in the S3. With the added adaptation period, we can ensure that slower adapters would have sufficient time to adjust to the AD digestate. The strategy will reduce the risk of missing out any promising strains. Our results show that the S3 can greatly expedite the screening process. For example, we screened 12 monocultures and six cocultures in triplicate – a total of 54 individual runs in less than 3 months (Murphy *et al.*, 2022).

After the screening, we chose *C. sorokiniana*–*M. capsulatus* (Bath) as the model coculture pair for the following reasons: (1) the pair showed one of the best growth performances on raw biogas and diluted digestate provided by CWW; (2) both *M. capsulatus* and *C. sorokiniana* have long served as model organisms to understand CH_4 oxidation and phototrophy, respectively. And both strains have complete and expert-annotated genomic information, (draft) genome-scale metabolic models and genetic tools; (3) *C. sorokiniana* has been extensively studied for wastewater treatment, particularly for digestate treatment; and (4) *M. capsulatus* (Bath) is the only industrial methanotroph strain for commercial applications due to its robustness and stability. For example, it has been commercialized for the production of SCP as animal feed (e.g., FeedKind® from Calysta, Menlo Park, CA).

6.5.2 Coculture tolerance to contaminants in raw biogas

The major contaminants in raw biogas are NH_3 and H_2S. Although many microalgae–methanotroph coculture pairs (including the model coculture pair *C. sorokiniana*–*M. capsulatus*) prefer NH_3 as the nitrogen source, a high concentration of H_2S may inhibit the coculture growth. To confirm the tolerance of the model pair to H_2S in raw biogas, we have tested the pair by adding different concentrations of H_2S (1,000–5,000 ppm) to the synthetic feeding gas (70% CH_4 and 30% CO_2), as the raw biogas from CWW only contains ~1,000 ppm H_2S. The experiments were conducted in serum bottles and lasted for 8 days. Figure 6.9 shows the coculture OD over time, which suggests that 1,000 and 2,000 ppm H_2S had little inhibition to the coculture; whereas 3,000 ppm H_2S lengthened the lag phase, the growth rate was similar to that of 1,000 and 2,000 ppm after the lag phase. In addition, the

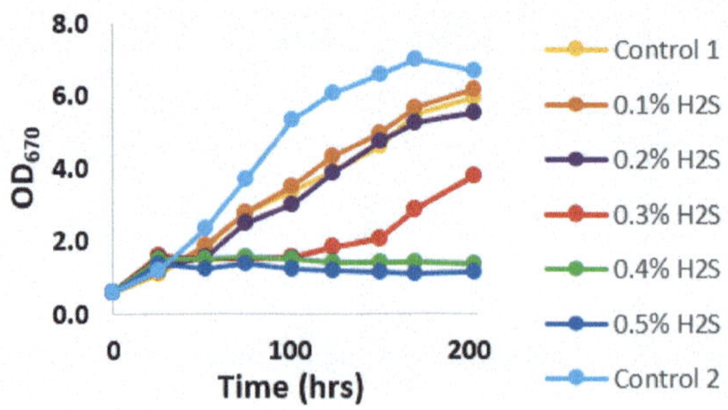

Figure 6.9 H_2S tolerance test shows healthy coculture growth up to 3,000 ppm (i.e., 0.3% H_2S).

measured individual biomass concentrations confirmed the steady growth of both strains. This result confirms that the model pair can tolerate biogas with 3,000 ppm H_2S, which is the upper limit of H_2S concentrations in most AD-generated biogases.

If the biogas feed contains higher H_2S concentrations that the coculture pair cannot tolerate, one could either add a pre-cleaning column to use chemical solutions (such as an alkaline solution) to remove part of the H_2S, or introduce a compatible sulfide-oxidizing bacterium to the coculture to achieve simultaneous H_2S removal. The coculture of sulfide-oxidizing bacteria with microalgae has been examined for biogas upgrading, which can effectively remove CO_2, NH_3, and H_2S from raw biogas, so that the upgraded biogas can be injected into existing natural gas pipelines (Muñoz *et al.*, 2015).

6.5.3 Freshwater consumption required by wastewater treatment

AD effluent often contains various inhibitors, including volatile fatty acids and antibiotics that may severely inhibit the growth of both microalgae and methanotrophs in the coculture. For microalgae-based wastewater treatment, the digestate is usually diluted 10 or 20 times to achieve sustained growth of microalgae and enable sufficient nutrient recovery rates (Wang *et al.*, 2018; Wen *et al.*, 2017; Xia & Murphy, 2016). However, using fresh water to dilute AD effluent is not practical because fresh water is a limited resource in most locations.

To eliminate or reduce freshwater usage, we have examined the possibility of using secondary clarifier water (the treated water at a WRRF before the final cleanup) as a diluent. In Roberts *et al.* (2020), we compared three diluents to examine their effects on coculture growth. The three diluents were: (1) tap water, (2) secondary clarifier water, and (3) a modified ammonium mineral salt medium (modified AMS), which is the standard AMS medium (Whittenbury *et al.*, 1970) without inorganic nitrogen (NH_4^+-N) and inorganic phosphorus (PO_4^{3-}-P).

The effect of different diluents on coculture growth was evaluated by biomass production, biogas utilization, and nutrient recovery. As reported in Roberts *et al.* (2020), three diluents had negligible effects on coculture growth and biomass productivity. A comparison of the coculture biomass growth and nutrient recovery of the coculture on AD effluent diluted with different diluents is presented in Figure 6.10. Clearly, the coculture performed similarly on the three diluents. This indicates that the secondary clarifier water can be used to replace fresh water to dilute AD effluent, thereby avoiding

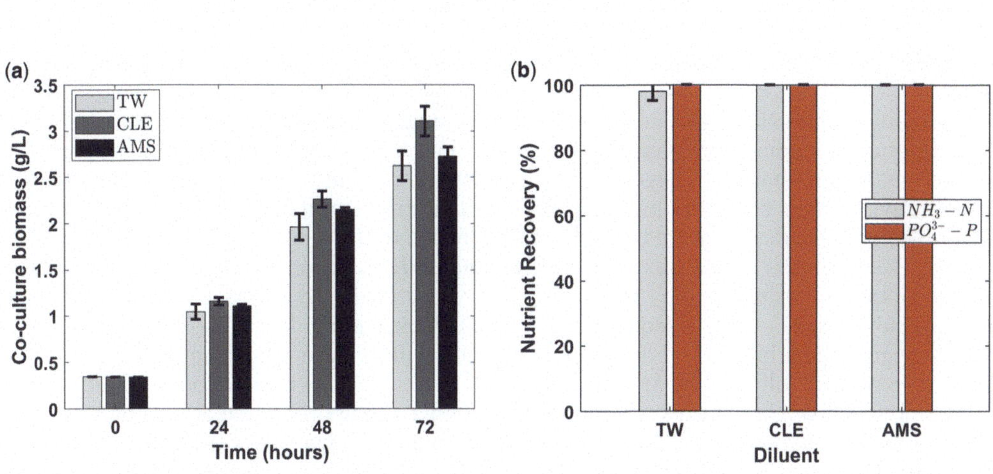

Figure 6.10 Comparison of (a) biomass growth and (b) nutrient recovery of coculture on AD effluent diluted with tap water (TW), secondary clarifier water (CLE), and modified AMS (AMS).

the need of fresh water for the proposed technology. In addition, the model coculture demonstrated effective nutrient recovery with nearly 100% recovery of ammonia nitrogen and orthophosphates.

6.5.4 Pretreatment of AD effluent

Besides macro- and micro-nutrients, AD effluent also contains solid contents, which require pretreatment before being fed to the coculture. Suspended solids often contain native microbial communities (e.g., methanogens) in the AD effluent, which could potentially compete with or cause contamination of the microalgae–methanotroph coculture. Liquid medium sterilization can completely eliminate potential competition or contamination. However, the process would incur significant cost, making the coculture-based W2V technology impractical for wastewater treatment from a financial perspective. To ensure the economic viability of the technology for wastewater treatment, it is necessary to lower or minimize the AD effluent pretreatment requirement. This is in fact possible because the native microbial communities in AD effluent thrive under anaerobic conditions. As a result, we argue that they do not compete well with the microalgae–methanotroph coculture due to the aerobic conditions in the bioreactor. This is supported by a study in which it was found that the fast-growing methanotrophs in AD effluent were significantly enriched and became the dominant species after prolonged cultivation on CH_4 and O_2 (Kim *et al.*, 2018). Specifically, in the enriched culture, *Methylosarcina fibrata* accounts for 94.1% of the methanotroph population, and 53.8% of the total microbial population. This result agrees with another recent study by Perera *et al.* (2022) where the synergistic interactions in a defined microalgae–bacteria consortium were not perturbed by the native heterotrophic bacteria in the wastewater, and a community shift occurred, which balanced the interactions and resulted in enhanced wastewater treatment (Perera *et al.*, 2022).

To further validate our argument and demonstrate the feasibility of using cocultures for wastewater treatment, we have conducted experiments to examine the effects of different AD effluent pretreatment methods on coculture growth. Three different pretreatment methods of AD effluent were examined: settling, filtering, and autoclaving. In addition, we tested coculture growth on a sterilized AMS medium, which served as the control. As reported in Roberts *et al.* (2020), there were no statistically significant differences among different pretreatment methods. In other words, the coculture grew similarly on the three differently treated AD effluents. Figure 6.11 shows the biomass productivity of individual strains in the coculture on differently pretreated AD effluents. This result further confirms the robustness of the model coculture, and that minimum treatment by settling (hence incurring minimum cost) would suffice the AD effluent pretreatment requirement.

6.5.5 Advantage of the coculture over sequential single cultures in carbon and nutrient recovery

There are many studies that have demonstrated advantages of mixed cultures over single cultures in a variety of applications. However, for the application of both microalgae and methanotrophs to wastewater treatment, it has been suggested that sequentially cultivating microalgae and methanotrophs is an alternative to their coculture and may achieve similar performance as the coculture (personal communication). We believe that this is highly unlikely due to the missing synergism when cultivated sequentially. Nevertheless, there was no experimental study to support one or the other. Therefore, we designed experiments to compare the two options. Specifically, to examine whether the coculture exhibits any advantages over the sequential single cultures, we conducted experiments to compare the coculture growth on biogas and diluted AD effluent, with the growth of sequential single culture, that is, *C. sorokiniana* followed by *M. capsulatus*. To achieve biogas conversion without external O_2 supply by sequential single cultures, we first cultivate microalgae on diluted AD using raw biogas as the carbon substrate, which fixes CO_2 and produces O_2, then the spent gas is fed to the methanotrophs which convert CH_4 and CO_2 into microbial biomass. As nutrient limitation would limit the growth of both strains in the coculture, in these experiments 20 ml of undiluted AD effluent was added to the vessel 48 h after the inoculation to ensure there is unlimited supply of macronutrients.

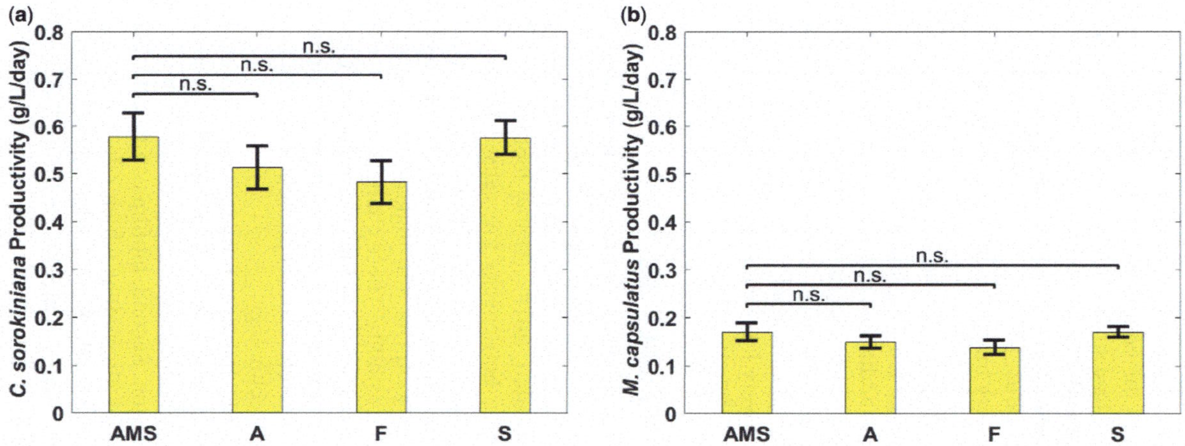

Figure 6.11 Biomass productivity of individual coculture strains on differently pretreated AD effluent diluted by CLE: (A) *C. sorokiniana* and (B) *M. capsulatus*. The Dunnett's test, with AMS as the control, indicates that these differences are not statistically significant (n.s.) at the significance level $\alpha = 0.05$.

Our study (Roberts *et al.*, 2020) shows that both microalgae (*C. sorokiniana*) and methanotroph (*M. capsulatus*) in the coculture exhibited significantly improved growth compared to the sequential single cultures. With the same amount of biogas supply, the biomass production of *C. sorokiniana* and *M. capsulatus* in the coculture showed a 64 and 58% increase compared to the sequential single-culture counterparts. We believe this significantly improved biomass production can be attributed to the metabolic coupling of CH_4 oxidation and oxygenation through photosynthesis. Enabled by the metabolic coupling, *C. sorokiniana* in the coculture showed a significantly higher CO_2 fixation rate and O_2 evolution rate than the microalgae single culture; and *M. capsulatus* in the coculture showed a much higher CH_4 assimilation rate and CO_2 production rate than the methanotroph single culture. In other words, there was no negative impact such as inhibition on *C. sorokiniana* growth due to the extra CO_2 produced by *M. capsulatus*. In addition, the results showed that CH_4 had no effect on microalgae. This is likely due to the very limited solubility of CH_4 in the liquid medium. More details can be found in Roberts *et al.* (2020).

To determine whether the model coculture offers an improvement in nutrient recovery compared to the sequential single cultures, we conducted experiments similar to the growth experiment described above. The only difference was that no additional nutrients were added after 48 h. For all cultures, TN, NH_4^+-N, TP, and PO_4^{3-}-P were measured to assess the nutrient recovery by different cultures. These experiments confirmed that the coculture offered significantly enhanced nutrient recovery than the single cultures, which was mainly due to enhanced biomass production as the correlation between the biomass production and nutrient recovery for the coculture was almost the same as that of the single cultures (Roberts *et al.*, 2020).

6.6 NEXT-GENERATION PHOTOBIOREACTORS

To achieve broad adoption of the microalgae–methanotroph coculture-based W2V technology in real applications, we must address the relevant engineering challenges associated with microbial cell cultivation. Traditional suspended or planktonic MC has two critical bottlenecks that limit its commercial application (Georgakopoulou, 2019): (1) low footprint areal biomass productivity and (2) high-cost biomass recovery. These bottlenecks are caused by the light attenuation in culture broth and mass transfer resistance of gaseous substrate into the liquid broth, both of which severely limit the achievable cell density in the liquid medium and the scale-up of the biotechnology. The low cell density further results in a large footprint and high energy cost for biomass harvesting which drastically diminishes the economic feasibility. These challenges are the main reasons for the limited commercialization of microalgae-based waste-to-fuel technologies (Cheah *et al.*, 2016). The same challenges apply to microalgae–methanotroph cocultures, with additional challenge of low solubility of CH_4 and O_2 in aqueous solutions. Therefore, novel photobioreactors are needed for the microalgae–methanotroph coculture-based W2V technology.

In the past decade, biofilm-based MC has drawn increasing attention (Gross & Wen, 2014; Gross *et al.*, 2013, 2015). By cultivating microalgae in biofilms on a supporting substratum, biofilm-based cultivation offers many advantages: first, biomass harvest is made easy and energy-efficient – biomass can be simply scraped off the substratum. Second, biofilm can be cultivated on a moving belt to be in contact with the gas and liquid phases alternately, which offers additional benefits of reduced mass transfer resistance of gaseous substrate reaching microalgae cells. Exploring the idea of biofilm-based cultivation, we have developed a patented circulating coculture biofilm photobioreactor (CCBP) for the microalgae–methanotroph coculture (He *et al.*, 2022). The schematic diagram of the CCBP is shown in Figure 6.12, in which a conveyor belt (substratum) is stretched around shafts to form a zig-zag configuration that supports microalgae–methanotroph biofilm growth on it. The lower part of the configuration is submerged in the diluted AD effluent, while the upper part is exposed directly to biogas and sunlight. The movement of the conveyor belt is driven by connecting one shaft with a motor, enabling the attached biofilm to alternately access nutrients in the liquid phase and carbon

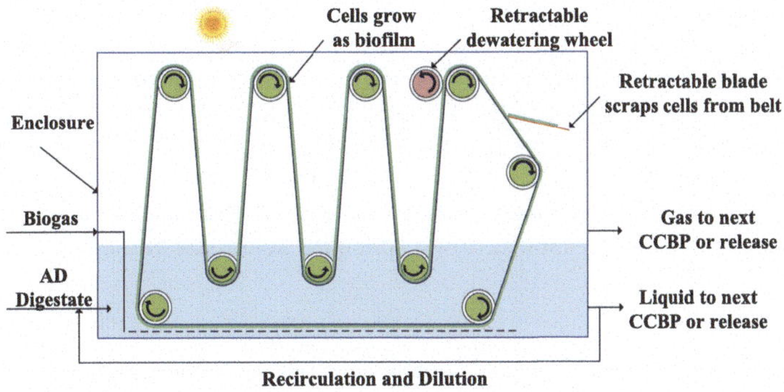

Figure 6.12 Schematic diagram of a patented CCBP (He *et al.*, 2022).

substrate (CH_4 and CO_2) in the gas phase. With the biocatalysts (coculture biofilm) directly exposed to the gas phase, and significantly increased surface area available for sunlight, the CCBP has achieved an aerial biomass productivity of 144.6 g Dry Cell Weight/m^2/day.

6.7 OUTLOOK AND CONCLUSION

Wastewater, if not properly treated, causes acute economic and environmental problems. In addition, wastewater is the fifth largest anthropogenic source of CH_4 emission globally. Therefore, mitigation of CH_4 is an imperative task in wastewater treatment. AD converts wet organic wastes into biogas and inorganic nutrients in a controlled and contained fashion, thereby offering many environmental benefits while producing a valuable fuel (CH_4). However, the poor ROI, caused by the low value of biogas (due to contaminants) and nutrient-rich AD effluent that requires further treatment, limits the application of AD to large-scale WRRFs. To broaden the adoption of AD, it is necessary to convert biogas into high-value products and reduce the cost of nutrient recovery.

Recently, microalgae–bacterial cultures have drawn increasing research interest for their application in integrated biogas upgrading and wastewater treatment. In particular, microalgae–methanotroph cocultures have been demonstrated to provide a highly promising platform for integrated nutrient recovery from wastewater and biogas valorization. In this chapter, we showed that microalgae–methanotroph cocultures have the potential to play an important role in reducing greenhouse gas emissions and energy consumption from wastewater treatment processes and producing value-added products (e.g., animal feed or aquafeed). Existing research, including ours, has established the scientific foundations for using microalgae–methanotroph cocultures for practical applications. These foundations include demonstrated robustness in tolerating the contaminants/inhibitors in raw biogas, confirmed feasibility of minimum pretreatment of the AD effluent, demonstrated operation with treated wastewater instead of fresh water, and quantified high-protein content suitable for animal feed or bioplastics production.

To fully explore the significant potential of microalgae–methanotroph cocultures for biogas valorization and nutrient recovery, there are many research questions that remain to be answered. Currently, all existing research on microalgae–methanotroph cocultures utilizes wild strains, and the product is microbial biomass. This is due to the lack of understanding on the biological foundation for the synergistic effect within the coculture. The molecular mechanism of the inter-species interactions is still largely unknown. Understanding the molecular mechanism for the synergistic interactions between the microalgae and methanotroph will pave the road for many more applications, such as engineering the coculture for the production of desired biochemicals, instead of just producing biomass. Another unexplored area is to construct mixed cultures with other species, such as sulfide-oxidizing bacteria, to handle challenging situations such as extremely high concentrations of H_2S or other contaminants. Such research will require the development of new analytical procedures to monitor the mixed culture in real time. Last but not least is the development of photobioreactors. Currently, there is little understanding on the growth of the microalgae and methanotrophs in biofilms. There has not been any research that examines the mass transfer of gaseous substrates and macronutrients in the biofilm, and whether the enhanced growth observed in the biofilm-based cultivation of the coculture is caused only by the increased surface area or if the biofilm (i.e., extracellular matrix) contributed in any way. There is a vast unknown space for scientists and engineers to explore.

REFERENCES

Aas T. S., Grisdale-Helland B., Terjesen B. F. and Helland S. J. (2006). Improved growth and nutrient utilisation in Atlantic salmon (*Salmo salar*) fed diets containing a bacterial protein meal. *Aquaculture*, **259**, 365–376, https://doi.org/10.1016/j.aquaculture.2006.05.032

Ahmed S. F., Mofijur M., Tarannum K., Chowdhury A. T., Rafa N., Nuzhat S., Kumar P. S., Vo D.-V. N., Lichtfouse E. and Mahlia T. M. I. (2021). Biogas upgrading, economy and utilization: a review. *Environmental Chemistry Letters*, **19**, 4137–4164, https://doi.org/10.1007/s10311-021-01292-x

Alvarez R. A., Pacala S. W., Winebrake J. J., Chameides W. L. and Hamburg S. P. (2012). Greater focus needed on methane leakage from natural gas infrastructure. *Proceedings of the National Academy of Sciences*, **109**, 6435–6440, https://doi.org/10.1073/pnas.1202407109

Angelidaki I. and Ellegaard L. (2003). Codigestion of manure and organic wastes in centralized biogas plants. *Applied Biochemistry and Biotechnology*, **109**, 95–106, https://doi.org/10.1385/ABAB:109:1-3:95

Angelidaki I., Treu L., Tsapekos P., Luo G., Campanaro S., Wenzel H. and Kougias P. G. (2018). Biogas upgrading and utilization: current status and perspectives. *Biotechnology Advances*, **36**, 452–466, https://doi.org/10.1016/j.biotechadv.2018.01.011

Badr K., Whelan W., He Q. P. and Wang J. (2021). Fast and easy quantitative characterization of methanotroph–photoautotroph cocultures. *Biotechnology and Bioengineering*, **118**, 703–714, https://doi.org/10.1002/bit.27603

Badr K., He Q. P. and Wang J. (2022). A novel semi-structured kinetic model of methanotroph–photoautotroph cocultures for biogas conversion. *Chemical Engineering Journal*, **431**, 133461, https://doi.org/10.1016/j.cej.2021.133461

Becker E. W. (2007). Micro-algae as a source of protein. *Biotechnology Advances*, **25**, 207–210, https://doi.org/10.1016/j.biotechadv.2006.11.002

Beckstrom B. D., Wilson M. H., Crocker M. and Quinn J. C. (2020). Bioplastic feedstock production from microalgae with fuel co-products: a techno-economic and life cycle impact assessment. *Algal Research*, **46**, 101769, https://doi.org/10.1016/j.algal.2019.101769

Cecchin M., Benfatto S., Griggio F., Mori A., Cazzaniga S., Vitulo N., Delledonne M. and Ballottari M. (2018). Molecular basis of autotrophic vs mixotrophic growth in *Chlorella sorokiniana*. *Scientific Reports*, **8**, 1, https://doi.org/10.1038/s41598-018-24979-8

Cheah W. Y., Ling T. C., Show P. L., Juan J. C., Chang J.-S. and Lee D.-J. (2016). Cultivation in wastewaters for energy: a microalgae platform. *Applied Energy*, **179**, 609–625, https://doi.org/10.1016/j.apenergy.2016.07.015

Chia W. Y., Tang D. Y. Y., Khoo K. S., Lup A. N. K. and Chew K. W. (2020). Nature's fight against plastic pollution: algae for plastic biodegradation and bioplastics production. *Environmental Science and Ecotechnology*, **4**, 100065.

Choi H. i., Sung Y. J., Hong M. E., Han J., Min B. K. and Sim S. J. (2021). Reconsidering the potential of direct microalgal biomass utilization as end-products: a review. Renewable and Sustainable Energy Reviews, **155**, 111930.

Chong J. W. R., Khoo K. S., Yew G. Y., Leong W. H., Lim J. W., Lam M. K., Ho Y.-C., Ng H. S., Munawaroh H. S. H. and Show P. L. (2021). Advances in production of bioplastics by microalgae using food waste hydrolysate and wastewater: a review. *Bioresource Technology*, **342**, 125947, https://doi.org/10.1016/j.biortech.2021.125947

Collins W. J., Webber C. P., Cox P. M., Huntingford C., Lowe J., Sitch S., Chadburn S. E., Comyn-Platt E., Harper A. B. and Hayman G. (2018). Increased importance of methane reduction for a 1.5 degree target. *Environmental Research Letters*, **13**, 054003, https://doi.org/10.1088/1748-9326/aab89c

Coppola G., Gaudio M. T., Lopresto C. G., Calabro V., Curcio S. and Chakraborty S. (2021). Bioplastic from renewable biomass: a facile solution for a greener environment. *Earth Systems and Environment*, **5**, 231–251.

Corcoran A. A. and Hunt R. W. (2021). Capitalizing on harmful algal blooms: from problems to products. *Algal Research*, **55**, 102265, https://doi.org/10.1016/j.algal.2021.102265

Drewnowski J., Remiszewska-Skwarek A., Duda S. and Łagód G. (2019). Aeration process in bioreactors as the main energy consumer in a wastewater treatment plant. Review of solutions and methods of process optimization. *Processes*, **7**(5), p. 311, https://doi.org/10.3390/pr7050311

Driscoll C., Whitall D., Aber J., Boyer E., Castro M., Cronan C., Goodale C., Groffman P., Hopkinson C., Lambert K., Lawrence G. (2003). Nitrogen pollution: sources and consequences in the US northeast. *Environment: Science and Policy for Sustainable Development*, **45**, 8–22, https://doi.org/10.1080/00139150309604553

Fei Q., Guarnieri M. T., Tao L., Laurens L. M. L., Dowe N. and Pienkos P. T. (2014). Bioconversion of natural gas to liquid fuel: opportunities and challenges. *Biotechnology Advances*, **32**, 596–614, https://doi.org/10.1016/j.biotechadv.2014.03.011

Galloway J. N., Dentener F. J., Capone D. G., Boyer E. W., Howarth R. W., Seitzinger S. P., Asner G. P., Cleveland C. C., Green P. A. and Holland E. A., others (2004). Nitrogen cycles: past, present, and future. *Biogeochemistry*, **70**, 153–226, https://doi.org/10.1007/s10533-004-0370-0

Gamboa-Delgado J. and Márquez-Reyes J. M. (2018). Potential of microbial-derived nutrients for aquaculture development. *Reviews in Aquaculture*, **10**, 224–246, https://doi.org/10.1111/raq.12157

Georgakopoulou M. (2019). The use of microalgae in biotechnological applications: case studies.

Gross M. and Wen Z. (2014). Yearlong evaluation of performance and durability of a pilot-scale revolving algal biofilm (RAB) cultivation system. *Bioresource Technology*, **171**, 50–58, https://doi.org/10.1016/j.biortech.2014.08.052

Gross M., Henry W., Michael C. and Wen Z. (2013). Development of a rotating algal biofilm growth system for attached microalgae growth with in situ biomass harvest. *Bioresource Technology*, **150**, 195–201, https://doi.org/10.1016/j.biortech.2013.10.016

Gross M., Mascarenhas V. and Wen Z. (2015). Evaluating algal growth performance and water use efficiency of pilot-scale revolving algal biofilm (RAB) culture systems. *Biotechnology and Bioengineering*, **112**, 2040–2050, https://doi.org/10.1002/bit.25618

Harmsen M., van Vuuren D. P., Bodirsky B. L., Chateau J., Durand-Lasserve O., Drouet L., Fricko O., Fujimori S., Gernaat D. E. H. J. and Hanaoka T. (2020). The role of methane in future climate strategies: mitigation potentials and climate impacts. *Climatic Change*, **163**, 1409–1425, https://doi.org/10.1007/s10584-019-02437-2

Haynes C. A. and Gonzalez R. (2014). Rethinking biological activation of methane and conversion to liquid fuels. *Nature Chemical Biology*, **10**, 331–339, https://doi.org/10.1038/nchembio.1509

He Q., Wang J. and Hilliard M. (2022). Culture systems and methods of using same. 11,339,360 B2.

Henard C. A., Smith H., Dowe N., Kalyuzhnaya M. G., Pienkos P. T. and Guarnieri M. T. (2016). Bioconversion of methane to lactate by an obligate methanotrophic bacterium. *Scientific Reports*, **6**, 21585, https://doi.org/10.1038/srep21585

Hill E. A., Chrisler W. B., Beliaev A. S. and Bernstein H. C. (2017). A flexible microbial co-culture platform for simultaneous utilization of methane and carbon dioxide from gas feedstocks. *Bioresource Technology*, **228**, 250–256, https://doi.org/10.1016/j.biortech.2016.12.111

Ho S.-H., Nagarajan D., Ren N. and Chang J.-S. (2018). Waste biorefineries – integrating anaerobic digestion and microalgae cultivation for bioenergy production. *Current Opinion in Biotechnology*, **50**, 101–110, https://doi.org/10.1016/j.copbio.2017.11.017

Kapoor R., Ghosh P., Tyagi B., Vijay V. K., Vijay V., Thakur I. S., Kamyab H., Nguyen D. D. and Kumar A. (2020). Advances in biogas valorization and utilization systems: a comprehensive review. *Journal of Cleaner Production*, **273**, 123052, https://doi.org/10.1016/j.jclepro.2020.123052

Karan H., Funk C., Grabert M., Oey M. and Hankamer B. (2019). Green bioplastics as part of a circular bioeconomy. *Trends in Plant Science*, **24**, 237–249, https://doi.org/10.1016/j.tplants.2018.11.010

Kim J., Kim D. D. and Yoon S. (2018). Rapid isolation of fast-growing methanotrophs from environmental samples using continuous cultivation with gradually increased dilution rates. *Applied Microbiology and Biotechnology*, **102**, 5707–5715, https://doi.org/10.1007/s00253-018-8978-5

Kip N., van Winden J. F., Pan Y., Bodrossy L., Reichart G.-J., Smolders A. J. P., Jetten M. S. M., Damsté J. S. S. and den Camp H. J. M. O. (2010). Global prevalence of methane oxidation by symbiotic bacteria in peat-moss ecosystems. *Nature Geoscience*, **3**, 617–621, https://doi.org/10.1038/ngeo939

Kleerebezem R. and van Loosdrecht M. C. M. (2007). Mixed culture biotechnology for bioenergy production. *Current Opinion in Biotechnology*, **18**, 207–212, https://doi.org/10.1016/j.copbio.2007.05.001

Lombardi L. and Francini G. (2020). Techno-economic and environmental assessment of the main biogas upgrading technologies. *Renewable Energy*, **156**, 440–458, https://doi.org/10.1016/j.renene.2020.04.083

Milucka J., Kirf M., Lu L., Krupke A., Lam P., Littmann S., Kuypers M. M. M. and Schubert C. J. (2015). Methane oxidation coupled to oxygenic photosynthesis in anoxic waters. The *ISME Journal*, **9**(9), 1991–2002.

Mishra A., Kumar M., Bolan N. S., Kapley A., Kumar R. and Singh L. (2021). Multidimensional approaches of biogas production and up-gradation: opportunities and challenges. *Bioresource Technology*, **338**, 125514, https://doi.org/10.1016/j.biortech.2021.125514

Muñoz R., Meier L., Diaz I. and Jeison D. (2015). A review on the state-of-the-art of physical/chemical and biological technologies for biogas upgrading. *Reviews in Environmental Science and Bio/Technology*, **14**, 727–759, https://doi.org/10.1007/s11157-015-9379-1

Murphy L., Badr K., O'Gwynn R., He Q. P. and Wang J. (2022). Fed-batch screening of methanotroph–algae cocultures for wastewater treatment and valorization. AIChE Annual Meeting, Phoenix, AZ.

Nasir I. M., Mohd Ghazi T. I. and Omar R. (2012). Anaerobic digestion technology in livestock manure treatment for biogas production: a review. *Engineering in Life Sciences*, **12**, 258–269, https://doi.org/10.1002/elsc.201100150

Ocko I. B., Sun T., Shindell D., Oppenheimer M., Hristov A. N., Pacala S. W., Mauzerall D. L., Xu Y. and Hamburg S. P. (2021). Acting rapidly to deploy readily available methane mitigation measures by sector can immediately slow global warming. *Environmental Research Letters*, **16**, 054042, https://doi.org/10.1088/1748-9326/abf9c8

Onen Cinar S., Chong Z. K., Kucuker M. A., Wieczorek N., Cengiz U. and Kuchta K. (2020). Bioplastic production from microalgae: a review. *International Journal of Environmental Research and Public Health*, **17**, 3842, https://doi.org/10.3390/ijerph17113842

Øverland M., Tauson A.-H., Shearer K. and Skrede A. (2010). Evaluation of methane-utilising bacteria products as feed ingredients for monogastric animals. *Archives of Animal Nutrition*, **64**, 171–189, https://doi.org/10.1080/17450391003691534

Perera I. A., Abinandan S., Panneerselvan L., Subashchandrabose S. R., Venkateswarlu K., Naidu R. and Megharaj M. (2022). Co-culturing of microalgae and bacteria in real wastewaters alters indigenous bacterial communities enhancing effluent bioremediation. *Algal Research*, **64**, 102705, https://doi.org/10.1016/j.algal.2022.102705

Qi Y., Beecher N. and Finn M. (2013). Biogas Production and Use at Water Resource Recovery Facilities in the United States. Water Environment Federation, Alexandria, VA.

Raghoebarsing A. A., Smolders A. J. P., Schmid M. C., Rijpstra W. I. C., Wolters-Arts M., Derksen J., Jetten M. S. M., Schouten S., Damsté J. S. S. and Lamers L. P. M., others (2005). Methanotrophic symbionts provide carbon for photosynthesis in peat bogs. *Nature*, **436**, 1153–1156, https://doi.org/10.1038/nature03802

Rajagopal R., Mousavi S. E., Goyette B. and Adhikary S. (2021). Coupling of microalgae cultivation with anaerobic digestion of poultry wastes: toward sustainable value added bioproducts. *Bioengineering*, **8**, 57, https://doi.org/10.3390/bioengineering8050057

Rasouli Z., Valverde-Pérez B., D'Este M., De Francisci D. and Angelidaki I. (2018). Nutrient recovery from industrial wastewater as single cell protein by a co-culture of green microalgae and methanotrophs. *Biochemical Engineering Journal*, **134**, 129–135, https://doi.org/10.1016/j.bej.2018.03.010

Richardson J. W., Johnson M. D. and Outlaw J. L. (2012). Economic comparison of open pond raceways to photo bio-reactors for profitable production of algae for transportation fuels in the southwest. *Algal Research*, **1**, 93–100, https://doi.org/10.1016/j.algal.2012.04.001

Roberts N., Hilliard M., He Q. P. and Wang J. (2020). A microalgae–methanotroph coculture is a promising platform for fuels and chemical production from wastewater. *Frontiers in Energy Research*, **8**, 230, https://doi.org/10.3389/fenrg.2020.563352

Rogelj J., Meinshausen M., Schaeffer M., Knutti R. and Riahi K. (2015). Impact of short-lived non-CO_2 mitigation on carbon budgets for stabilizing global warming. *Environmental Research Letters*, **10**, 075001, https://doi.org/10.1088/1748-9326/10/7/075001

Romarheim O. H., Øverland M., Mydland L. T., Skrede A. and Landsverk T. (2010). Bacteria grown on natural gas prevent soybean meal-induced enteritis in Atlantic salmon. *The Journal of Nutrition*, **141**, 124–130, https://doi.org/10.3945/jn.110.128900

Rosa M. d. (2020). Economic assessment of producing and selling biomethane into a regional market. *Energy & Environment*, **31**, 60–76, https://doi.org/10.1177/0958305X18762581

Sabra W., Dietz D., Tjahjasari D. and Zeng A.-P. (2010). Biosystems analysis and engineering of microbial consortia for industrial biotechnology. *Engineering in Life Sciences*, **10**, 407–421, https://doi.org/10.1002/elsc.201000111

Scarlat N., Dallemand J. F. and Fahl F. (2018). Biogas: developments and perspectives in Europe. *Renewable Energy*, **129**, 457–472, https://doi.org/10.1016/j.renene.2018.03.006

Siegrist H., Salzgeber D., Eugster J. and Joss A. (2008). Anammox brings WWTP closer to energy autarky due to increased biogas production and reduced aeration energy for N-removal. Water Science and Technology, **57**(3), 383–388, https://doi.org/10.2166/wst.2008.048

Spiegelman D., Whissell G. and Greer C. W. (2005). A survey of the methods for the characterization of microbial consortia and communities. *Canadian Journal of Microbiology*, **51**, 355–386, https://doi.org/10.1139/w05-003

Stone K. A., He Q. P. and Wang J. (2019). Two experimental protocols for accurate measurement of gas component uptake and production rates in bioconversion processes. *Scientific Reports*, **9**, 5899, https://doi.org/10.1038/s41598-019-42469-3

Sun A., Davis R., Starbuck M., Ben-Amotz A., Pate R. and Pienkos P. T. (2011). Comparative cost analysis of algal oil production for biofuels. *Energy*, **36**, 5169–5179, https://doi.org/10.1016/j.energy.2011.06.020

Tam N. F. Y., Wong Y. S. and Leung G. (1992). Effect of exogenous carbon sources on removal of inorganic nutrient by the nitrification–denitrification process. Water Research, **26**(9), 1229–1236, https://doi.org/10.1016/0043-1354(92)90183-5

Tandon P., Jin Q. and Huang L. (2017). A promising approach to enhance microalgae productivity by exogenous supply of vitamins. *Microbial Cell Factories*, **16**, 1, https://doi.org/10.1186/s12934-017-0834-2

Teimouri M., Amirkolaie A. K. and Yeganeh S. (2013). The effects of *Spirulina platensis* meal as a feed supplement on growth performance and pigmentation of rainbow trout (*Oncorhynchus mykiss*). *Aquaculture*, **396–399**, 14–19, https://doi.org/10.1016/j.aquaculture.2013.02.009

Tharani D. and Ananthasubramanian M. (2020). Microalgae as sustainable producers of bioplastic. In: Microalgae Biotechnology for Food, Health and High Value Products, Springer, Singapore, pp. 373–396.

Topper P. A., Graves R. E. and Richard T. (2006). The Fate of Nutrients and Pathogens during Anaerobic Digestion of Dairy Manure. Penn State University, Lehman (PA). College of Agricultural Science, Cooperative Extension Bulletin G 71.

Vadiveloo A., Foster L., Kwambai C., Bahri P. A. and Moheimani N. R. (2021). Microalgae cultivation for the treatment of anaerobically digested municipal centrate (ADMC) and anaerobically digested abattoir effluent (ADAE). *Science of the Total Environment*, **775**, 145853, https://doi.org/10.1016/j.scitotenv.2021.145853

van der Ha D., Nachtergaele L., Kerckhof F.-M., Rameiyanti D., Bossier P., Verstraete W. and Boon N. (2012). Conversion of biogas to bioproducts by algae and methane oxidizing bacteria. *Environmental Science & Technology*, **46**, 13425–13431, https://doi.org/10.1021/es303929s

Wang Q., Higgins B., Ji H. and Zhao D. (2018). Improved microalgae biomass production and wastewater treatment: Pre-treating municipal anaerobic digestate for algae cultivation. ASABE 2018 Annual International Meeting. https://doi.org/10.13031/aim.201801333

Wen Y., He Y., Ji X., Li S., Chen L., Zhou Y., Wang M. and Chen B. (2017). Isolation of an indigenous *Chlorella vulgaris* from swine wastewater and characterization of its nutrient removal ability in undiluted sewage. *Bioresource Technology*, **243**, 247–253, https://doi.org/10.1016/j.biortech.2017.06.094

Whittenbury R., Phillips K. C. and Wilkinson J. F. (1970). Enrichment, isolation and some properties of methane-utilizing bacteria. *Journal of General Microbiology*, **61**(2), 205–218, https://doi.org/10.1099/00221287-61-2-205

Xia A. and Murphy J. D. (2016). Microalgal cultivation in treating liquid digestate from biogas systems. *Trends in Biotechnology*, **34**(4), 264–275, https://doi.org/10.1016/j.tibtech.2015.12.010

Zeller M. A., Hunt R., Jones A. and Sharma S. (2013). Bioplastics and their thermoplastic blends from *Spirulina* and *Chlorella* microalgae. *Journal of Applied Polymer Science*, **130**(5), 3263–3275, https://doi.org/10.1002/app.39559

Zhang B., Li W., Guo Y., Zhang Z., Shi W., Cui F., Lens P. N. L. and Tay J. H. (2020). Microalgal–bacterial consortia: from interspecies interactions to biotechnological applications. *Renewable and Sustainable Energy Reviews*, **118**, 109563, https://doi.org/10.1016/j.rser.2019.109563

Zhao H. W., Mavinic D. S., Oldham W. K. and Koch F. A. (1999). Controlling factors for simultaneous nitrification and denitrification in a two-stage intermittent aeration process treating domestic sewage. *Water Research*, **33**(4), 961–970, https://doi.org/10.1016/S0043-1354(98)00292-9

Part 3

Integration with Other Technologies

doi: 10.2166/9781789063547_0129

Chapter 7

Microalgae cultivation in bio-electrochemical systems

A. Mishra and M. Chhabra*

Department of Biosciences and Bioengineering, Indian Institute of Technology Jodhpur, Jodhpur, India
**Correspondence author: meenuchhabra@iitj.ac.in*

ABSTRACT

Biofuels from algae have the potential to completely replace fossil-based fuels and provide energy security for the future. However, the cost of algae biofuels is still too high for commercial application. In this context, producing algae and electrical energy using photosynthetic microbial fuel cells (PMFCs) is an attractive option. PMFCs utilize the natural process of photosynthesis for algae generation or algae degradation at the anode. In the former system, the process of organic matter degradation complements the process of algae biomass production with concomitant power generation. Electrogenic bacteria oxidize organic matter at the anode anaerobically. The anode transfers the electrons released through oxidation to the cathode, where photosynthetic organisms produce oxygen (O_2) as a cathodic electron acceptor. The suitability of bio-electrochemical systems such as microbial fuel cells for algae cultivation can be assessed by comparing them with the conventional method of cultivation, namely open ponds and photobioreactors. PMFCs offer a process that can provide high carbon dioxide concentrations for algal growth, has a mechanism to prevent high inhibitory O_2 concentrations and can meet a fraction of the process electricity requirements. The algae biomass can go as high as 4–5 g/L in a PMFC and power output doubles due to activity of algae at the cathode compartment. This chapter discusses the algal growth in bio-electrochemical systems, the factors that influence them and directions for future research.

Keywords: bioelectricity, biofuel, carbon dioxide, dissolved oxygen, electrogenic bacteria, energy, microalgae, photobioreactor, PMFC, wastewater.

7.1 INTRODUCTION

The International Energy Agency (IEA) reported that biofuels have a high potential to meet a substantial fraction of the world's energy demands. As per a 2015 report, biofuel constitutes 10% of the world's cumulative energy source (IEA, 2015). Third-generation biofuels derived from algae are a prominent part of bioenergy initiatives. However, several bottlenecks hinder bioenergy generation from algae. This includes algae cultivation, harvesting, oil extraction and algae-to-fuel conversion efficiency (Reddy *et al.*, 2019). The processes that bring significant enhancement of efficiency in any of these factors enable algae technology for bioenergy production. A sustainable technology for algae cultivation is the need of the hour. The definition of sustainability involves environmental, economic and social aspects. Algae cultivation is an environmental friendly process and an attractive greenhouse effect mitigation method.

Photosynthetic microbial fuel cells (PMFCs) offer a process that can provide high carbon dioxide (CO_2) concentrations for algal growth, has a mechanism to prevent high inhibitory oxygen (O_2) concentrations and can meet a fraction of the process electricity requirements. PMFCs are a modified form of microbial fuel cells (MFCs) and considered promising systems for wastewater treatment and power generation. In such systems, microorganisms convert organic matter into electricity (Khandelwal *et al.*, 2018). A conventional two-chamber MFC design contains an anodic chamber and a cathodic chamber separated by a proton exchange membrane (PEM) or an ion-selective membrane (Logan *et al.*, 2006). Microorganisms oxidize the organic matter in the anodic chamber to produce electrons, protons and CO_2. The electrons are transferred onto the anode surface, from where they flow across an external circuit to the cathode, constituting the flow of current. The protons migrate to the cathode through the PEM. At the cathode, molecular O_2 reduction to water depletes electrons and protons. In PMFCs, the CO_2 produced at the anode is fed to the cathode for fixation by algae, which in turn produce O_2 as an electron acceptor to complete the MFC circuit (Wang *et al.*, 2010).

Wang *et al.* (2010) reported proof of using PMFCs for algae cultivation at the cathode with simultaneous anodic CO_2 fixation and generation of O_2 as an electron acceptor. The power output from the device was 5.6 W/m^3, with algal growth in the cathode chamber linked directly to power generation. The cathode dissolved oxygen (DO) concentration was found to be 6.6 mg/L. The power generation from the MFC depends on the O_2 concentration at the cathode and a concentration level of 6.6 mg/L is considered suitable (Kang *et al.*, 2003). The device was demonstrated to be an efficient carbon capture device for algae biomass production. However, they used glucose as a carbon source at the anode. Pandit *et al.* (2012) used a similar technology with wastewater to generate cyanobacterial biomass at the cathode. A number of reports then published on either wastewater treatment using these devices (Zhang *et al.*, 2011) or comparing the efficiency of these devices with mechanically aerated devices (Juang *et al.*, 2012).

Schamphelaire and Verstraete (2009) reported an integrated system consisting of an algal biomass production unit, an anaerobic digester to convert algae biomass to biogas and an MFC to generate electricity while treating digester effluent. They estimated their system results in a ~9 kW/ha capacity power plant, with 23 kW/h prospects. Khandelwal *et al.* (2018) showed that the algae biomass degradation at the anode was coupled with algae biomass production at the cathode. The algae biomass harvested from the cathode was used for oil extraction and fed back into the system at the anode for degradation by a microbial consortium. The replacement of costly fuels such as glucose/acetate at the anode with algae biomass and capture of O_2 generated by the photosynthesis process via cathodic reduction to water was achieved, as high O_2 concentrations are inhibitory to the growth of algae, particularly in closed systems.

Considering the advantages of using PMFCs for algae cultivation, it is essential to look at the process with respect to some key points such as the feasibility of the process in dual-chambered systems/open pond reactors, the highest algae productivity obtainable in such systems, highest power/energy output, scalability and sustainability of the process. The points mentioned above require discussion on two different aspects of MFC operation: the first one being electrochemical and design parameters such as MFC configuration, electrodes, separators, electrode catalysts and power management systems (Figure 7.1). The second one is biological parameters such as the microbial community at the anode, choice of algae species at the cathode, interaction of algae cells with the cathode and optimal conditions for microbial growth. This chapter discusses algal growth with respect to these aspects. In addition, a brief comparison of algal growth in bio-electrochemical systems and conventional systems is presented.

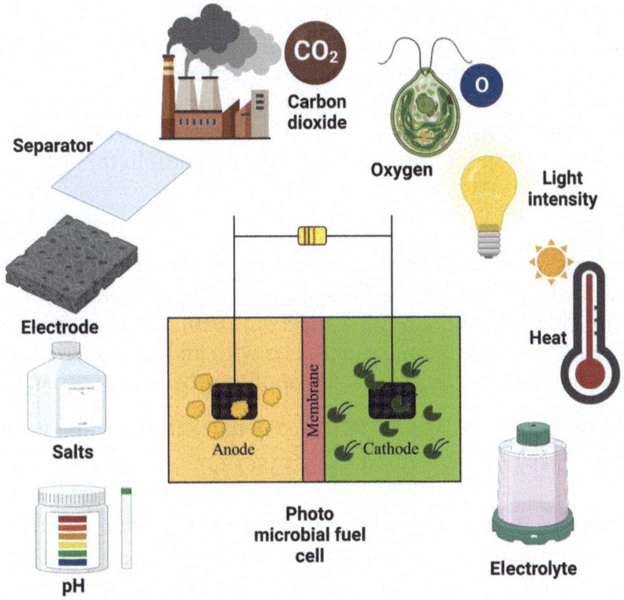

Figure 7.1 Overview of the factors affecting the efficiency of PMFCs.

7.2 USE OF ALGAE IN MFCS

7.2.1 Algae as primary producers

Microalgae are one of the primitive organisms on Earth that carry out oxygenic photosynthesis (Hopkins, 1999). The water splits with the help of light energy at the photosynthetic reaction center housed in the eukaryote's chloroplast and the cyanobacterial cell membrane. The splitting of water is accompanied by electron flow, adenosine triphosphate (ATP) and nicotinamide adenine dinucleotide phosphate hydrogen (NADPH) generation. The ATP and NADPH then reduce CO_2 through dark reactions (Hopkins, 1999). Algae are at the bottom of the aquatic food chain and are called autotrophs, which means they get their food from themselves via CO_2 fixation. Cyanobacteria, also known as blue-green algae (BGA), are considered living fossils of the planet. The algae class Euglenophyceae is thought to be the oldest lineage of algae that includes zooflagellates.

Algae, unlike other photosynthetic organisms, can grow in barren lands flooded with wastewater, exhibit much higher biomass productivity and effectively purify wastewater of nitrogen, phosphorus and chemical oxygen demand (COD). Microalgae (microscopic algae) are not only used as a source of bioenergy (Chapter 9) but also as food and dietary supplements (Chapter 10). Because of these tiny photosynthetic microorganisms known as phytoplankton, the ocean is a massive ecosystem that acts as a reservoir of fixed atmospheric CO_2.

Only a small group of algal genera have been explored to date. About 30,000 species have been studied and classified (Enzing *et al.*, 2014). Algae are a diverse group of organisms that range from single-celled to filamentous, solitary to colonial, micro to macro and prokaryotes to eukaryotes, with shorter life cycles and the ability to survive extreme conditions. They are divided into distinct classes based on pigment composition, cell membrane (single-, bi- or multi-layered), mode of reproduction and food reserve. There are a total of 11 classes of algae classified by Fritsch (1945), namely Chlorophyceae, Xanthophyceae, Chrysophyceae, Bacillariophyceae, Cryptophyceae, Dinophyceae, Chloromonadineae, Euglenineae, Pheophyceae, Rhodophyceae and Myxophyceae (Cyanophyceae).

The most recent study by Gololobova and Belyakova (2022) shows how chloroplasts were crucial to the evolution of algae. They divided the algae into five monophyletic supergroups named mainly Archaeplastida (Glaucocystophyta, Rhodophyta, Prasinodermophyta, Chlorophyta and Charophyta), TSAR defined as telonemids, stramenopiles, alveolates and rhizaria (Ochrophyta, Dinophyta, Chlorarachniophyta and photosynthetic species of the genera Chromera, Vetrella and Paulinella), Haptista (Prymnesiophyta and Rappemonads), Cryptista (Cryptophyta) and Discoba (Euglenophyta).

7.2.2 Algae metabolism

Algae can grow in various cultural media and environmental conditions (Sharma *et al.*, 2011). Physicochemical factors such as sunlight, CO_2 concentration, pH, temperature, salinity and stress influence algal growth and biomass production (Mata *et al.*, 2010). The photoperiod shows ascendancy by observing the growth of the algae. The biomass increases with an increase in photoperiod from 6 to 12 h (Ip *et al.*, 1982). The following reaction represents the photosynthesis process of algae (Wang *et al.*, 2017):

$$\begin{aligned} &106\,CO_2 + 16\,NO_3^- + HPO_4^{2-} + 112\,H_2O + 18H^+ + \text{energy} + \text{trace elements} \\ &\quad \rightarrow (CH_2O)_{106}(NH_3)_{16}(H_3PO_4)(\text{algae biomass}) + 138O_2(g) \end{aligned} \tag{7.1}$$

Temperature, light, dissolved CO_2, DO and pH affect each other, and their interplay determines the algal growth. Biomass tends to increase with a rise in temperature. The optimal temperature for algal growth is between 20 and 30°C (Ip *et al.*, 1982). The temperature also influences the solubility of CO_2/O_2 and the mass transfer rate in the culture medium (Vale *et al.*, 2020). Depending on the algal species, growth is affected by the concentration of CO_2. High levels of DO are detrimental to microalgae due to the formation of reactive oxygen species, which causes oxidative stress. As far as CO_2 is concerned, gaseous CO_2 has limited solubility in water and is not readily assimilated by algae. CO_2 first dissolves in the solution as carbonic acid (Equation (7.2)), which eventually dissociates to bicarbonate or carbonate (Equation (7.3)), depending on the pH. Carbonic acid formation is accelerated by carbonic anhydrase, an enzyme all algal species produce (Shukla & Kumar, 2018). Microalgae supplied with bicarbonates grow better than the ones provided with gaseous CO_2. The pH of the medium is another critical parameter that affects algal growth as it determines the predominant carbon species and nutrient bioavailability. An alkaline pH favors CO_2 dissolution in water and the formation of bicarbonates. The reaction between water and CO_2 dissolved in water can be represented by the below equation (Hopkins, 1999):

$$CO_2(g) + H_2O(l) \leftrightarrow H_2CO_3(aq.) \tag{7.2}$$

Depending on the pH of the solution, carbonic acid can dissociate to produce HCO_3^- and carbonate (CO_3^{2-}). The below equation depicts the equilibrium between the three species (Hopkins, 1999):

$$H_2CO_3(aq.) \leftrightarrow HCO_3^-(aq.) + H^+(aq.) \leftrightarrow CO_3(aq.) + 2H^+(aq.) \tag{7.3}$$

At high pH values, the reaction proceeds in favor of carbonate formation. Carbonate is the predominant species in solution under highly alkaline conditions. Bicarbonate is the most common species when the pH is close to neutral. Microalgae prefer near neutral pH for optimal growth and readily assimilate bicarbonates from solution. The utilization of bicarbonates regenerates alkaline water that needs to be controlled by supplying CO_2.

An adequate concentration of CO_2 is mandatory for photosynthesis (Creswell, 2010). Algae exhibit low production rates at ambient CO_2 concentration (0.035%). A CO_2 concentration of 2–6% supports high photosynthetic activity and biomass growth (Chinnasamy *et al.*, 2009). To supply a high CO_2 concentration, flue gas from power generation plants is an attractive source. A high flue gas temperature lowers the process efficiency and requires an additional step to cool the gas. Economically, algae cultivation is still a costly affair. The mechanical stirring for mixing in photo-bioreactors/open ponds and degasification to remove high inhibitory O_2 concentrations in photo-bioreactors adds to the cost of the process. Nutrients such as nitrogen and phosphorus must be present in adequate quantities to support algal growth. A lack of nitrogen slows respiration rates and affects amino acids and protein synthesis. Limiting phosphorus concentrations affect protein synthesis, nucleic acid synthesis and cell metabolism. This restriction also impacts the photosynthetic energy conversion lowering microalgal growth.

7.2.3 Large-scale microalgae cultivation

There are several ways to grow microalgae, from batch cultures, which are simple and do not require any inputs or outputs, to continuous systems, in which fresh medium is added to the culture and the spent medium is withdrawn from the system at the same rate promoting steady biomass growth. Microalgae can be grown indoors or outdoors in closed or open systems. Open systems for growing microalgae are less expensive. Still, they are more likely to get contaminated or affected by environmental factors. This makes it harder to maintain microalgae growing for long periods. Regardless of the species used as an inoculum, several parameters, such as temperature, light intensity and pH, can significantly affect open pond growth. Also, outdoor cultures only work for a few species and are prone to crashes because the parameters cannot be controlled and sometimes differ from batch to batch. In contrast, closed cultivation systems such as photobioreactors (PBRs) are more expensive but enable strict control of the cultivation parameters that are suitable for microalgal growth. As closed systems, PBRs prevent direct gas exchange between the atmosphere and the algal culture. PBR tubes must be transparent to allow light to pass through and are typically made of glass or acrylic. These systems provide the cultures with a controlled environment (pH, temperature, light intensity and dissolved O_2 and CO_2) and prevent bacterial contamination. Algae productivity is higher in PBRs, but the productivities are not good enough to cover the operational costs of a PBR (Table 7.1).

Table 7.1 Comparison of the specific growth rate of microalgae in different operating systems.

Microorganism	Operational Condition	Specific Growth Rate (/day)	References
Chlorella sp.	Bubble column PBRs	0.3	Naira *et al.* (2019)
Chlorella ellipsoidea	Bubble column PBRs	Indoor; $V = 2$ L – 0.168 (±0.006) 20 L – 0.168 (±0.011) Outdoor; $V = 200$ L – 0.145 (±0.026)	Wang *et al.* (2014)
Spirulina sp.	Plastic rectangular tanks	Indoor; $V = 50$ L – 0.42 (±0.030) Outdoor; $V = 500$ L – 0.1 (±0.02)	Krishnamoorthy *et al.* (2019)
Chlorella sp.	Batch-scale PBRs for indoor Pilot-scale PBRs for outdoor	Indoor; $V = 500$ mL – 0.3225 (±0.0039) Outdoor; $V = 40$ L RDW 5% – 0.978 RDW 10% – 0.2012 RDW 25% – 0.1652	Lu *et al.* (2015)
C. vulgaris	PMFCs (indoor; $V = 100$ mL)	LEA – 0.275 (±0.02) FP – 0.208 (±0.015)	Khandelwal *et al.* (2018)
C. vulgaris	LDPE PBR PMFCs (outdoor; $V = 10$ L)	RPMFC 5% – 0.63 (±0.056) 10% – 0.59 (±0.049) CW-MFC – 0.54 (±0.035)	Khandelwal *et al.* (2020)

PBR, photobioreactor; RDW, raw dairy waste; RP-MFC, rock phosphate-microbial fuel cell; CW-MFC, clayware-microbial fuel cell; LDPE, low-density polythene; LEA, lipid extracted algae; FP, fruit pulp.

7.3 ROLE OF ALGAE IN PMFCS

Algae are O_2 producers at the PMFC cathode; the cathodic reactions require a continuous supply of electron acceptors, which are provided by the algae by generating O_2. In a PMFC, the need of O_2 supply is met by utilizing algae's ability to produce O_2 through photosynthesis. The below reactions describe the activity of microflora at the anode and cathode of PMFCs:

At the anode:

$$\begin{gathered}\text{Organic matter} \rightarrow \text{Acetate,} \\ C_2H_4O_2 + 2H_2O \rightarrow 8H^+ + 2CO_2 + 8e^-\end{gathered} \tag{7.4}$$

At the cathode:

(1) Photosynthetic reaction:

$$6CO_2 + 6H_2O + \text{light energy} \rightarrow C_6H_{12}O_6 + 6O_2 \tag{7.5}$$

(2) Reduction reaction:

$$2O_2 + 8H^+ + e^- \rightarrow 4H_2O \tag{7.6}$$

In PMFCs, *Chlorella vulgaris* microalgae produce 750 mmol/L of DO daily; the values exceed by a factor of 3 from the DO concentration of 250 mmol/L that was attained by bubbling air in the cathode (Commault *et al.*, 2017). Algae also fix CO_2, making it a carbon capture system. Continuous O_2 quenching via cathodic oxygen reduction further enhances CO_2 capture rates and algae biomass production. The rate of CO_2 fixation by algae is ~6.24 kg/m³/day through photosynthesis (Elmekawy *et al.*, 2014). Several researchers have explored algae cultivation in PMFCs with promising results obtained in terms of algal growth, water treatment and power generation.

7.3.1 Algal species tested in MFC cathode compartment

Different species of algae present different growth rates, O_2 production rates, carbon capture rates and nutrient assimilation rates. Algae differ in terms of their intracellular biomolecular content and composition as well. Depending on the intended application, a desirable species of algae can be grown in a PMFC. The role of algal species is to replace unsustainable chemical acceptors at the PMFC cathode with photosynthetic O_2. Algae species also impact the power output of a PMFC significantly. *C. vulgaris* is one of the many algal strains frequently used in the cathode compartment of a PMFC due to its high photosynthetic efficiency. The cathode of PMFCs was also tested with several other pure algal species, such as *Dunaliella salina*, *Chlamydomonas reinhardtii*, *Scenedesmus obliquus*, *Desmodesmus* sp. and *Microcystis aeruginosa* to capture solar energy for photosynthesis and generate bioelectricity (Table 7.2).

7.3.2 Mechanism of bioelectricity generation in PMFCs

An energy-generating redox reaction is separated into two chambers of a PMFC, with oxidation taking place at the anode and reduction at the cathode. The oxidation generates electrons that travel from the anode to the cathode constituting the current. The protons diffuse through the PEM to balance the moving charges. The electrons, protons and O_2 combine at the cathode surface to produce water. The main steps in PMFCs are: (1) photosynthesis by algae at the cathodic chamber, (2) oxidation of organic matter under anaerobic conditions by electrogenic bacteria at the anode chamber, (3) transfer of protons and electrons from the anode to the cathode chamber and (4) reduction of O_2 at the cathode.

Table 7.2 Performance assessment of PMFCs using different algal strains for power generation.

Class of Algae	Algae Used at the Cathode	Culture Medium	Light Intensity	Types of MFC	Electrode Used	MFC Material	Power Output	Biomass Produced	COD Removal Efficiency (%)	Reference
Cyanophyceae	BGA	Waste water		Single-chambered tubular	Graphite felts	Polyvinyl chloride	114 mW/m^2		78.9	Yuan *et al.* (2011)
Chlorophyceae	*Chlorella pyrenoidosa*	TAP	3,500 lux	Dual-chambered	Carbon paper	Glass	30.15 (±0.02) mW/m^2	5.94×10^6 cells/mL		Xu *et al.* (2015)
Chlorophyceae	Mixed algae	Modified Hoagland		Dual-chambered	Graphite felts	Glass	14 mW/m^2			Strik *et al.* (2008)
Chlorophyceae	*C. vulgaris*	Modified Bold's	26 W	Dual-chambered	Carbon graphite	Glass	0.6 mW/m^2	437–2,140 mg/L		Mitra and Hill (2012)
Chlorophyceae	*S. obliquus*	Bold basal	1,600 lux	Dual-chambered	Carbon paper		102 mW/m^2		74	Kondaveeti *et al.* (2014)
Chlorophyceae	*C. vulgaris*	Acetate-free phosphate-buffered solution		Dual-chambered	Carbon felts	Glass	187 mW/m^2			Liu *et al.* (2015)
Chlorophyceae	*C. vulgaris*	Modified Zehnder		Dual-chambered	Graphite plates	Polycarbonate	15.0 (±0.1)/722 (±62) mW/m^2		17.6	Lakaniemi *et al.* (2012)
Chlorophyceae	*Dunaliella tertiolecta*	Modified Zehnder		Dual-chambered	Graphite felts	Polycarbonate	5.3 (±2.6)/277 (±133) mW/m^2		7.7	Lakaniemi *et al.* (2012)
Pheophyceae	*Laminaria saccharina*	Minimal		Dual-chambered	Graphite felts	Glass with rubber septa	250 mW/m^2		60–85	Gadhamshetty *et al.* (2013)
Cyanophyceae	*Arthrospira maxima*	Basal		Dual-chambered	Graphite granules	Perspex frames, six rubber gaskets, stainless steel	5,800 mW/m^3		67	Inglesby and Fisher (2013)
Cyanophyceae	*A. maxima*	Zarrouk's		Dual-chambered	Graphite granules	Perspex frames, six rubber gaskets, stainless steel	10.2 (±1.3) W/m^3		41–63	Inglesby and Fisher (2012)

(Continued)

Table 7.2 Performance assessment of PMFCs using different algal strains for power generation (*Continued*).

Class of Algae	Algae Used at the Cathode	Culture Medium	Light Intensity	Types of MFC	Electrode Used	MFC Material	Power Output	Biomass Produced	COD Removal Efficiency (%)	Reference
Chlorophyceae	*Scenedesmus* sp.	Open pond (fresh water)		Dual-chambered	Carbon brushes	Acrylic	1.78 W/m^2			Rashid *et al.* (2013)
Cyanophyceae	*M. aeruginosa*	Lake water		Dual-chambered	Graphite brushes	Polymethyl methacrylate	4.14 (±0.05) W/m^3		81 (±6) to 23 (±4)	Wang *et al.* (2012)
Chlorophyceae	*C. vulgaris*	Lake water		Dual-chambered	Graphite brushes	Polymethyl methacrylate	3.70 (±0.02) W/m^3		73 (±3) to 30 (±5)	Wang *et al.* (2012)
Cyanophyceae, Chlorophyceae, Bacillariophyceae	*Ankistrodesmus*, *Chlorella*, *Oscillatoria*, *Scenedesmus*, *Diatom*, *Cosmarium*	Synthetic wastewater	3,000 (±200) lux	Single-chambered	Graphite plates	Perspex	3.55 $\mu W/m^2$	2.87 g/L		Subhash *et al.* (2013)
Chlorophyceae	*Scenedesmus abundans*	BBM media	94.6 $\mu mol/m^2/s$	Dual-chambered	Graphite rods	Plexiglass	838.68 mW/m^2	0.94 (±0.01) g/L	97.24	Nayak and Ghosh (2019)
Cyanophyceae	*Synechococcus leopoliensis*	BG-11	40 (±5) $\mu E/m^2/s$	Dual-chambered	Carbon fiber veils	Acrylic	42.5 W/m^3	6×10^5 cells/mL		Walter *et al.* (2015)
Cyanophyceae	*M. aeruginosa*	BG-11	7,000 lux	Dual-chambered	Carbon paper	Glass	58.4 mW/m^3			Cai *et al.* (2013)
Chlorophyceae	*Desmodesmus* sp.	BG-11	3,000 lux	Dual-chambered	Graphite felts	Plexiglass	99.09 mW/m^2			Wu *et al.* (2014)
Chlorophyceae	*C. vulgaris*	Bold's basal		Dual-chambered	Toray carbon cloths with 10% Teflon		13.5 mW/m^2	300 mg/dm^3	80	Campo *et al.* (2013)
Chlorophyceae	*C. reinhardtii* transformation F5	TAP	900 lux	Dual-chambered	Graphite electrodes	Poly-acrylic plastic	12.947 mW/m^2			Lan *et al.* (2013)
Chlorophyceae	*Scenedesmus quadricauda* SDEC-8	Domestic wastewater	135 $\mu mol/m^3/s$	Single-chambered	Carbon cloth with titanium wire	Plexiglas cylinder	62.93 mW/m^2		62	Yang *et al.* (2018)
Chlorophyceae	*C. vulgaris* strain CS-42	SWW 4% and phosphate buffer	200 $mmol/m^2/s$	Dual-chambered	Graphite plates		34.2 (±10) mW/m^2			Commault *et al.* (2017)

(*Continued*)

Table 7.2 Performance assessment of PMFCs using different algal strains for power generation (*Continued*).

Class of Algae	Algae Used at the Cathode	Culture Medium	Light Intensity	Types of MFC	Electrode Used	MFC Material	Power Output	Biomass Produced	COD Removal Efficiency (%)	Reference
Chlorophyceae	*C. vulgaris*	Effluent water from a chocolate factory	75,000 lux	Dual-chambered	Graphite plates	Glass	23.17–327.67 mW/m²	5.2 mg/mL	78.6	Huarachi-Olivera *et al.* (2018)
Chlorophyceae	*Chlorella sorokiniana*	BG-11	2,000 lux	Dual-chambered	Carbon felts	Poly-acrylic	3.2 W/m³		65.97 (±0.83)	Neethu *et al.* (2018)
Chlorophyceae	*C. vulgaris*	BG-11	Solar radiation	Dual-chambered	Graphite felts	RP-mixed CW	11.5318 kWh/m³	0.307 kg/m³/day		Khandelwal *et al.* (2020)
Chlorophyceae	*C. vulgaris*	BG-11	PBRs	Dual-chambered	Graphite felts	Acrylic	2.7 W/m³	0.028 kg/m³/day		Khandelwal *et al.* (2018)
Unspecified	Microalgae (unspecified)	Wastewater		Dual-chambered	Carbon veils	Terracotta cylinder	44 μW			Salar-García *et al.* (2016)
Chlorophyceae	*Dunaliella salina*	Modified Johnson's media	9800 lx	Dual chambered	Graphite felt	Acrylic material	213.38 mW/m²	$4.02 \pm 6 \times 10^6$ cells/mL	59.32%	Mishra and Chhabra (2022)

BBM, Bold's basal medium; BG-11, Bluegreen-11; BGA, blue-green algae; TAP, Tris-acetate-phosphate; RP, rock phosphate; CW, clayware; SWW, Synthetic wastewater.

In the case of anodic electrogenic bacteria, oxidation of NADH to NAD^+ during the mitochondrial respiration process plays a principal role in voltage generation (Logan *et al.*, 2006). Many dehydrogenase reactions are an integral part of the plasma membrane of algae and have an essential role in the electron transport chain (Shukla & Kumar, 2018):

$$CH_2O\,(\text{Organic matter}) + H_2O \rightarrow 4e^- + 4H^+ + CO_2 \tag{7.7}$$

In algae, photosystems (PSs) comprise several pigments, including chlorophyll. The pigments are arranged in a fashion that harvests maximum solar energy and transmits that to the reaction center. The carotenoids are at the outer surface and chlorophyll forms the reaction center. Algal photosynthesis is completed in two steps: light and dark reactions. The light mediates electron ejection from the two reaction centers of the P680 and P700 PSs during the light reaction. The oxidized P680 splits water into O_2, protons and electrons. The electrons fill the hole and stabilize the reaction center for the next cycle. The splitting molecule breaks water molecules into O_2, protons and electrons. Meanwhile, the ejected electron transports through the Z-scheme from PS II (P680) to PS I (P700), resulting in a proton gradient that generates ATP. In non-cyclic photophosphorylation, $NADP^+$ accepts the electron generated through the oxidation of PS I, producing NADPH. In cyclic photophosphorylation, only ATP is generated as the electron flows in a cycle coming back to the same reaction center:

$$2NADP + 2H_2O + 2ADP + 2P_i \xrightarrow{\text{light energy/chlorophyll}} 2NADPH_2 + 2ATP + O_2 \tag{7.8}$$

During the dark reaction, NADPH and ATP reduce CO_2 through the Calvin–Benson cycle (C_3 cycle). The first step of this cycle is catalyzed by ribulose-1,5-bisphosphate carboxylase/oxygenase (Rubisco) (Taiz *et al.*, 2015) as shown in the below equation:

$$CO_2 + 4H^+ + 4e^- \xrightarrow{2NADPH_2;3ATP} CH_2O + H_2O \tag{7.9}$$

The Rubisco enzyme shows a higher affinity for O_2, thereby reducing the rate of CO_2 fixation and algal growth (Hopkins, 1999). As O_2 inhibits the process of photosynthesis, continuous/periodic deoxygenation of the algal growth medium is essential to sustain healthy algal growth.

7.4 PMFC DESIGN PARAMETERS

7.4.1 Dual chambers vs sediment MFCs

PMFCs can be single- or dual-chambered. In both types, an anode and a cathode are immersed in two redox environments. In sediment MFCs, the top layers are suitable for cathode placement as these are exposed to light and support photosynthesis and O_2 generation (Shukla & Kumar, 2018). Anaerobiosis prevails in the bottom-most layers of the water column and sediment where the anode can be placed. In dual chambers, the electrodes can be on the same plane, but separated into different compartments via a selectively permeable membrane. Both systems can be operated in batch or continuous mode. The shape of the chambers can vary. Dual-chambered PMFCs are difficult for industrial scale-up due to their complex design and high-operational costs. Hence, single-chambered PMFCs are preferred over the dual-chambered design.

7.4.2 Construction materials, electrolytes, electrodes and separators

MFCs can be made of glass, concrete, acrylic, polyvinyl chloride, polycarbonate, perspex, stainless steel, polymethyl methacrylate, plastic and ceramics (Shukla & Kumar, 2018). In the case of PMFCs, the cathodic chamber has to be transparent for light penetration. Ceramics are the most sustainable for anodic compartments among the said materials as these are cost-effective, use renewable materials and do not create pollution. The most common ceramics are earthenware pots, terracotta pots and cylinders and goethite cylinders (Behera & Ghangrekar, 2011).

7.4.3 Electrode materials

An MFC electrode should have high conductivity, higher surface area, non-toxic, non-polluting and have a low charge transfer resistance. The latter often requires the modification of electrodes with electrode catalysts that accelerate the rate of charge transfer reactions. Researchers attempt to enhance power output of MFCs by changing the shape and material of electrodes. Carbon-based electrodes such as carbon cloth, carbon felts, carbon brushes, carbon paper, graphite plates, graphite rods and graphite foils are commonly used as base electrodes (Shukla & Kumar, 2018). Electrode catalysts can be platinum black or metal oxides as well as their nanoparticles coated on/dispersed in carbon materials. Similarly, carbon nanotubes, graphene and N-doped graphene/graphene oxide find applications as electrode materials. Other less commonly used electrode materials include gold, silver, copper, nickel, cobalt, aluminum, titanium and stainless steel (Zhao *et al.*, 2008). Baudler *et al.* (2015) found that carbon electrodes are better than other electrodes as they offer low internal resistance, are cheap, recyclable, non-corrosive and usable for a long time. The drawback of carbon electrodes, particularly for PMFC applications, is the dark and opaque appearance that blocks light or has a shading effect. In addition, these have low thermal conductivity, resulting in low power output. Therefore, transparent metal electrodes are suitable for PMFC applications (Baudler *et al.*, 2015).

The disadvantages of metal electrodes include low surface roughness that reduces charge transfer rates and the absence of well-defined pores that prevents proper microbial attachment (Logan *et al.*, 2007). Moreover, metal electrodes may corrode over time resulting in background currents and lower conductivity. Literature studies suggest that the nanoscale surface morphology of electrodes decides the conductivity and ability to interface with microbes (Bacakova *et al.*, 2011; Cao *et al.*, 2009; Legeay *et al.*, 2010; Sekar *et al.*, 2004). Even with electrodes made of similar carbon material, differences in conductivity, surface area, space and size of carbon microfiber can result in variable MFC performance (Sanchez *et al.*, 2015).

7.4.4 Separators

The electrolyte and separator should have high conductivity. The anions and cations flow in opposite directions in closed circuits with flow rates proportional to the magnitude of the current. The separator has two roles: (1) to prevent the mixing of anodic and cathodic substrates and (2) to allow the movement of ions across the chambers. Separators such as electrodes should be non-toxic, have high ionic conductivity and be impermeable to O_2. The most commonly used separators are PEMs that wear negatively charged groups such as -PO_3^-, -PO_3^{2-}, -COO^-, -$C_6H_4O^-$ and -SO^{3-} that successively protonate and deprotonate (Rodenas *et al.*, 2015). The lower ion/proton conductivity increases the internal resistance, lowers conductivity and creates a pH gradient with low pH at the anode. Low pH disrupts microbial growth and lowers the power output (Winfield *et al.*, 2013). MFCs can be operated without any separator, but this decreases the Coulombic efficiency of the system due to the diffusion of O_2 to the anode and electron donor to the cathode (Ghangrekar & Shinde, 2007). The anion exchange membranes with positively charged groups, such as -PR^+, -SR^+ and -NH_3^+ help exchange negative ions creating an ionic balance (Zhuang *et al.*, 2012). Separators such as salt bridges, glass fibers, glass wool, clayware (CW) membranes and ceramic membranes can exchange cations and anions primarily due to their high water-holding capacity. Separators need periodic replacement/cleaning as they tend to clog in long-term operations lowering the MFC performance. Separators made of non-biodegradable materials are undesirable. Thus, ceramic separators such as CW hold promise for scale-up of MFC systems. Behera and Ghangrekar (2011) demonstrated the efficiency of the terracotta separator that has a wall thickness of 4 mm. Similarly, rock phosphate (RP)-blended CW can be used as a separator showing maximum power density, 5% RP blend yields 890 (±95), 10% RP-blend yields 960 (±120) and CW yields 1200 (±152) mW/m^3 of power density (Khandelwal *et al.*, 2018).

7.4.5 Effect of light intensity, temperature, DO, CO_2, pH and salts

Light is essential and the duration of the photoperiod determines the power output. High light intensity increases the temperature, enhancing the reaction rate and the substrate utilization rate up to a certain temperature (Shukla & Kumar, 2018). A light intensity of 3,500–100,000 lux with a light/dark regime of 18/6, 12/12 and 16/8 h was reported as optimum for PMFC operation (Reddy *et al.*, 2019). The optimum temperature for PMFC operation is 30°C but may vary from organism to organism. DO also increases with light intensity. However, as mentioned in the previous sections, high DO inhibits algal growth. The DO content also depends on the types of algal species used at the cathode. PMFCs can provide a DO content as high as 18–19 mg/L for cathodic reactions (Taskan & Taskan, 2022). A minimum DO of 2.2 mg/L is required for the continuous operation of PMFCs (Jang *et al.*, 2013). The DO content also decreases with salinity (Kim & Chung, 1984).

A neutral pH range of 6.8–7.5 is optimal for algal growth. CO_2 dissolves well in alkaline pH and algae absorb better bicarbonates (Reddy *et al.*, 2019). The CO_2 concentration at the cathode plays an essential role in MFC performance, with an optimum value of ~5% CO_2–air mixture (Reddy *et al.*, 2019). There are two methods for supplying CO_2 to the PMFCs: *in-situ* and *ex-situ*. In the *ex-situ* method, the catholyte is purged or supplied with pure CO_2 gas or bicarbonate generated elsewhere. In an *in-situ* way, CO_2 generated through anaerobic digestion of organic matter at the anode is used. There are benefits and drawbacks of both strategies. Microalgae grow faster with the *ex-situ* method, but the system requires additional units for the transport of CO_2. The *in-situ* method relies on the metabolic rate of microbes at the cathode, the partial pressure and pH of the catholyte. Generally, anodic off-gas supports effective algal growth (Khandelwal *et al.*, 2018). It directly affects how well the PMFC works, making the material more conductive.

7.5 ECONOMIC IMPORTANCE OF PMFCS

Algae accumulate a high oil content in their cells and the oil composition is suitable for biodiesel production. The residual algae biomass is rich in proteins and carbohydrates. In a biorefinery, the residual algae are often subjected to biogas generation (Figure 7.2). However, the same can be used as an anodic substrate in MFCs. Algae biomass is also bioconvertible to ethanol, biogas and biohydrogen (Figure 7.2). Algae harvest and drying prior to lipid extraction is challenging. Therefore, wet algal biomass transesterification has been tested. Direct or wet transesterification is the same as dry transesterification except that the extraction step is skipped and the whole biomass is used as the feedstock of the reaction (Shukla & Kumar, 2018). Microalgae only require a little pre-treatment before fermentation because their cell walls have only a thin cellulosic fence and lack lignin.

Conversely, macroalgae must be treated first to get the stored carbohydrates out, whereas microalgal fermentation produces ethanol, acetate, hydrogen and CO_2. Hydrogen production from algal biomass can be accomplished through photobiological and fermentative processes (Shaishav *et al.*, 2013). Hydrogenase is the main enzyme that speeds up reactions that lead to the production of biohydrogen. Some microbes can break down the organic compounds in algae without O_2 to generate methane and CO_2. Algal biomass can avoid the requirement of harvesting, dewatering, drying or oil extraction to produce biogas (Shukla & Kumar, 2018).

Algae biomass is rich in carotenoids, terpenoids, xanthophylls, chlorophylls, phycobilins, polyunsaturated fatty acids, polysaccharides, vitamins, sterols, tocopherols and phycocyanins (Ammar *et al.*, 2022). Algae are frequently utilized as dietary supplements because of their nutritional value. *Spirulina* and *Chlorella* are the two well-known algal species used as food sources. Algae aid in treating diseases such as diabetes, rheumatic disorders and high blood pressure in the arteries. Algae also benefit memory and concentration by providing omega 3 and omega 6 polyunsaturated fatty acids required for brain development. Additionally, they fight off bacterial, fungal and viral infections. Algae are also rich in natural antioxidants and antimicrobials that improve shelf life and circumvent artificial preservatives (Gonçalves, 2021).

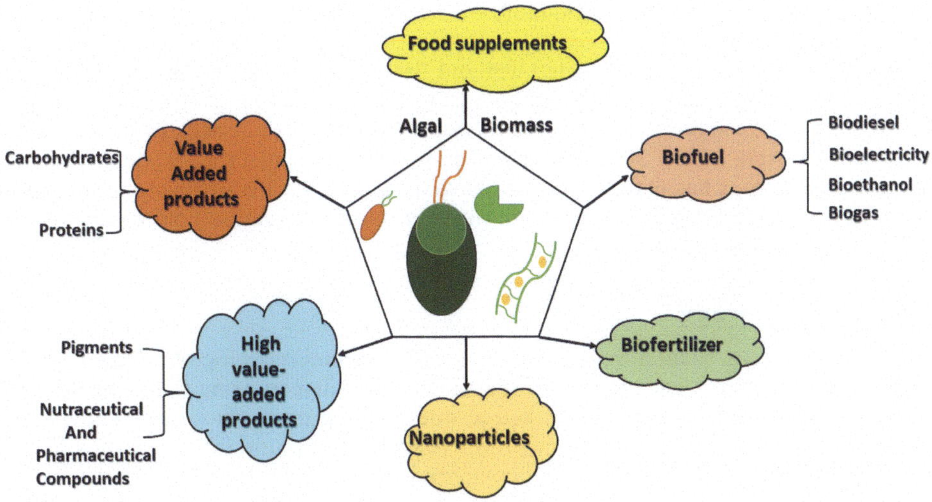

Figure 7.2 Cultivation, value-added product generation, power generation, biofuel generation and resource recovery of algae biomass.

Algae, which can be unicellular, multicellular, filamentous or saponaceous, are examples of organisms that produce their food through photosynthesis. With over 200,000 species, they are also the most prominent primary producers in the world. Microalgal production requires mass cultivation, biomass recovery and downstream operations to ensure a consistent yield for food, chemicals, feed, biofuel and biofertilizers, as shown in Figure 7.3 (Balasubramaniam *et al.*, 2021). BGA (microalgae) can yield plant growth hormones, polysaccharides, chemicals that kill bacteria and other metabolites.

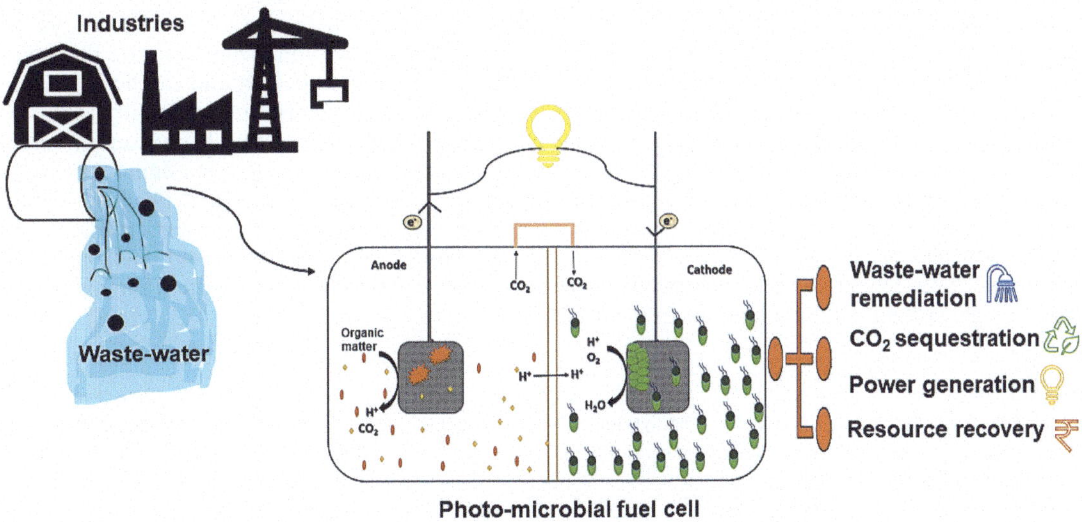

Figure 7.3 PMFCs for wastewater remediation, algae cultivation, biomass recovery, power generation and downstream operations to ensure a consistent yield for food, chemicals, feed and biofuel.

They can also improve the fertility and quality of the soil (Ronga *et al.*, 2019). According to Guo *et al.* (2020), the primary sources of organic matter of the agroecosystem are cyanobacteria and green microalgae. In the case of constructed wetland PMFCs, algae can make a big difference in the amount of organic carbon in the soil by taking up CO_2. Cyanobacterial heterocyst cells (e.g., *Nostoc*, *Anabaena* and *Aulosira*) may fix atmospheric nitrogen to satisfy the needs of the microbiota and plants of the soil (Fritsch, 1945). Several studies have shown that injecting cyanobacteria or groups of cyanobacteria into crops has made a big difference in the nitrogen content of the soil. Inoculating the soil with cyanobacteria can reduce the soil nitrogen fertilizer requirement by 25–40% (Ammar *et al.*, 2022).

7.6 FUTURE PERSPECTIVES

Microalgae use CO_2 as a carbon source in their metabolism, reducing the load of CO_2 in the environment. Because microalgae can grow independently while capturing CO_2, this process can be combined with other techniques already linked to microalgal growth, such as cleaning wastewater, generating biofuels and producing high-value products. Open ponds and PBRs are suitable techniques for algal growth in conventional methods, but compared with PMFCs, the latter showed better results for algal growth, where O_2 degassing is natural because it acts as the terminal electron acceptor.

Compared to heterotrophic MFCs or photovoltaic cells, PMFCs have some observable advantages. They can generate electricity solely from natural resources such as sunlight, water and CO_2. As a result, it is not necessary to load MFCs with organic compounds and the utilization of CO_2 also contributes to carbon sequestration, resulting in a clean environment. Regardless of day or night, PMFCs can produce power continuously. The procedure can turn the algal biomass into proteins, pigments and biofuels such as biodiesel, biogas and bioethanol. It can also be used with PBRs to supply O_2 to MFCs, thus allowing clean wastewater by MFCs. The performance of the system is influenced by light and DO, making it challenging to produce electricity continuously and sustainably. When assessing how sustainable these systems are, it is also essential to look at the relationship between the amount of electricity used and the removal of pollutants and how well the system is set up and run.

The challenge lies in the process scale-up. The scale-up studies of PMFCs have revealed higher capital costs ranging from \$735/m^3 to \$36,000/m^3 (Liang *et al.*, 2018; Wang *et al.*, 2020). Low-cost PMFC systems depend on CW separators, and low-cost PBRs have also been tested and the system worked well for power generation (Khandelwal *et al.*, 2020). Electrodes and membranes are the major causes for high cost of an MFC system. Sediment PMFCs are relatively cheaper and can be applied to real-world applications. The operating parameters, such as temperature, pH, organic loading, salinity, conductivity, start-up and hydraulic retention time have been optimized for a number of contaminated soils or artificial wetlands. Although promise associated with the process has been demonstrated, a need to study the process for large-scale *in-situ* bioremediation, bioaugmentation and algae cultivation remains. This involves a close understanding of anodic and cathodic microenvironments, mass transfer efficiencies, soil/sediment characteristics in the case of sediment MFCs and environmental conditions. The impact of these factors on algae biomass growth and composition is important for reproducible commercial applications of PMFCs.

REFERENCES

Ammar E. E., Aioub A. A., Elesawy A. E., Karkour A. M., Mouhamed M. S., Amer A. A. and El-Shershaby N. A. (2022). Algae as bio-fertilizers: between current situation and future prospective. *Saudi Journal of Biological Sciences*, **29**(5), 3083–3096, https://doi.org/10.1016/j.sjbs.2022.03.020

Bacakova L., Filova E., Parizek M., Ruml T. and Svorcik V. (2011). Modulation of cell adhesion, proliferation and differentiation on materials designed for body implants. *Biotechnology Advances*, **29**(6), 739–767, https://doi.org/10.1016/j.biotechadv.2011.06.004

Balasubramaniam V., Gunasegavan R. D. N., Mustar S., Lee J. C. and Mohd Noh M. F. (2021). Isolation of industrial important bioactive compounds from microalgae. *Molecules*, **26**(4), 943, https://doi.org/10.3390/molecules26040943

Baudler A., Schmidt I., Langner M., Greiner A. and Schröder U. (2015). Does it have to be carbon? Metal anodes in microbial fuel cells and related bioelectrochemical systems. *Energy & Environmental Science*, **8**(7), 2048–2055, https://doi.org/10.1039/C5EE00866B

Behera M. and Ghangrekar M. M. (2011). Electricity generation in low-cost microbial fuel cell made up of earthenware of different thickness. *Water Science and Technology*, **64**(12), 2468–2473, https://doi.org/10.2166/wst.2011.822

Cai P. J., Xiao X., He Y. R., Li W. W., Zang G. L., Sheng G. P. and Yu H. Q. (2013). Reactive oxygen species (ROS) generated by cyanobacteria act as an electron acceptor in the biocathode of a bio-electrochemical system. *Biosensors and Bioelectronics*, **39**(1), 306–310, https://doi.org/10.1016/j.bios.2012.06.058

Campo A. G., Cañizares P., Rodrigo M. A., Fernández F. J. and Lobato J. (2013). Microbial fuel cell with an algae-assisted cathode: a preliminary assessment. *Journal of Power Sources*, **242**, 638–645, https://doi.org/10.1016/j.jpowsour.2013.05.110

Cao J., Yuan W., Pei Z. J., Davis T., Cui Y. and Beltran M. (2009). A preliminary study of the effect of surface texture on algae cell attachment for a mechanical–biological energy manufacturing system. *Journal of Manufacturing Science and Engineering*, **131**(6), 064505.

Chinnasamy S., Ramakrishnan B., Bhatnagar A. and Das K. C. (2009). Biomass production potential of a wastewater alga *Chlorella vulgaris* ARC 1 under elevated levels of CO_2 and temperature. *International Journal of Molecular Sciences*, **10**(2), 518–532.

Commault A. S., Laczka O., Siboni N., Tamburic B., Crosswell J. R., Seymour J. R. and Ralph P. J. (2017). Electricity and biomass production in a bacteria-*Chlorella* based microbial fuel cell treating wastewater. *Journal of Power Sources*, **356**, 299–309, https://doi.org/10.1016/j.jpowsour.2017.03.097

Creswell L. (2010). Phytoplankton Culture for Aquaculture Feed. Southern Regional Aquaculture Center. Department of Agriculture, US, PP. 14.

Elmekawy A., Hegab H. M., Vanbroekhoven K. and Pant D. (2014). Techno-productive potential of photosynthetic microbial fuel cells through different configurations. *Renewable and Sustainable Energy Reviews*, **39**, 617–627, https://doi.org/10.1016/j.rser.2014.07.116

Enzing C., Ploeg M., Barbosa M. and Sijtsma L. (2014). Microalgae-based products for the food and feed sector: an outlook for Europe. *JRC Scientific and Policy Reports*, Eur 26255, Publications Office of the European Union: JRC85709, Luxembourg, 19–37.

Fritsch F. E. (1945). The Structure and Reproduction of the Algae, London, Cambridge University Press.

Gadhamshetty V., Belanger D., Gardiner C. J., Cummings A. and Hynes A. (2013). Evaluation of *Laminaria*-based microbial fuel cells (LbMs) for electricity production. *Bioresource Technology*, **127**, 378–385, https://doi.org/10.1016/j.biortech.2012.09.079

Ghangrekar M. M. and Shinde V. B. (2007). Performance of membrane-less microbial fuel cell treating wastewater and effect of electrode distance and area on electricity production. *Bioresource Technology*, **98**(15), 2879–2885, https://doi.org/10.1016/j.biortech.2006.09.050

Gololobova M. A. and Belyakova G. A. (2022). Position of algae on the tree of life. *Doklady Biological Sciences*, **507**(1), 12–326.

Gonçalves A. L. (2021). The use of microalgae and cyanobacteria in the improvement of agricultural practices: a review on their biofertilising, biostimulating and biopesticide roles. *Applied Sciences*, **11**(2), 871, https://doi.org/10.3390/app11020871

Guo S., Wang P., Wang X., Zou M., Liu C. and Hao J. (2020). Microalgae as biofertilizer in modern agriculture. *Microalgae Biotechnology for Food, Health and High Value Products*, **7**, 397–411, https://doi.org/10.1007/978-981-15-0169-2_12

Hopkins W. G. (1999). Introduction to Plant Physiology, 2nd edn, John Wiley and Sons, New York, USA.

Huarachi-Olivera R., Dueñas-Gonza A., Yapo-Pari U., Vega P., Romero-Ugarte M., Tapia J. and Esparza M. (2018). Bioelectrogenesis with microbial fuel cells (MFCs) using the microalga *Chlorella vulgaris* and bacterial communities. *Electronic Journal of Biotechnology*, **31**, 34–43, https://doi.org/10.1016/j.ejbt.2017.10.013

Inglesby A. E. and Fisher A. C. (2012). Enhanced methane yields from anaerobic digestion of *Arthrospira maxima* biomass in an advanced flow-through reactor with an integrated recirculation loop microbial fuel cell. *Energy & Environmental Science*, **5**(7), 7996, https://doi.org/10.1039/c2ee21659k

Inglesby A. E. and Fisher A. C. (2013). Downstream application of a microbial fuel cell for energy recovery from an *Arthrospira maxima* fed anaerobic digester effluent. *RSC Advances*, **3**(38), 17387, https://doi.org/10.1039/c3ra42277a

International Energy Agency (2015). Key World Energy Statistics, https://iea.blob.core.windows.net/assets/5a314029-69c2-42a9-98ac-d1c5deeb59b3/WEO2015.pdf, 81

Ip S. Y., Bridger J. S., Chin C. T., Martin W. R. B. and Raper W. G. C. (1982). Algal growth in primary settled sewage: the effects of five key variables. *Water Research*, **16**(5), 621–632, https://doi.org/10.1016/0043-1354(82)90083-5

Jang J. K., Kan J., Bretschger O., Gorby Y. A., Hsu L., Kim B. H. and Nealson K. H. (2013). Electricity generation by microbial fuel cell using microorganisms as catalyst in cathode. *Journal of Microbiology and Biotechnology*, **23**(12), 1765–1773, https://doi.org/10.4014/jmb.1310.10117

Juang D. F., Lee C. H. and Hsueh S. C. (2012). Comparison of electrogenic capabilities of microbial fuel cell with different light power on algae grown cathode. *Bioresource Technology*, **123**, 23–29.

Kang K. H., Jang J. K., Pham T. H., Moon H., Chang I. S. and Kim B. H. (2003). A microbial fuel cell with improved cathode reaction as a low biochemical oxygen demand sensor. *Biotechnology Letters*, **25**, 1357–1361.

Khandelwal A., Vijay A., Dixit A. and Chhabra M. (2018). Microbial fuel cell powered by lipid extracted algae: a promising algal lipids and power generation system. *Bioresource Technology*, **247**, 520–527, https://doi.org/10.1016/j.biortech.2017.09.119

Khandelwal A., Chhabra M. and Yadav P. (2020). Performance evaluation of algae-assisted microbial fuel cell under outdoor conditions. *Bioresource Technology*, **310**, 123418, https://doi.org/10.1016/j.biortech.2020.123418

Kim K. and Chung J. Y. (1984). On the salinity-minimum and dissolved oxygen-maximum layer in the East Sea (Sea of Japan). *Elsevier Oceanography Series*, **39**, 55–65, https://doi.org/10.1016/S0422-9894(08)70290-3

Kondaveeti S., Choi K. S., Kakarla R. and Min B. (2014). Microalgae *Scenedesmus obliqus* as renewable biomass feedstock for electricity generation in microbial fuel cells (MFCs). *Frontiers of Environmental Science & Engineering*, **8**(5), 784–791, https://doi.org/10.1007/s11783-013-0590-4

Krishnamoorthy S., Manickam P. and Muthukaruppan V. (2019). Evaluation of distillery wastewater treatability in a customized photobioreactor using blue-green microalgae–laboratory and outdoor study. *Journal of Environmental Management*, **234**, 412–423, https://doi.org/10.1016/j.jenvman.2019.01.014

Lakaniemi A. M., Tuovinen O. H. and Puhakka J. A. (2012). Production of electricity and butanol from microalgal biomass in microbial fuel cells. *Biology Energy Research*, **5**(2), 481–491.

Lan J. C. W., Raman K., Huang C. M. and Chang C. M. (2013). The impact of monochromatic blue and red LED light upon performance of photo microbial fuel cells (PMFCs) using *Chlamydomonas reinhardtii* transformation F5 as biocatalyst. *Biochemical Engineering Journal*, **78**, 39–43, https://doi.org/10.1016/j.bej.2013.02.007

Legeay G., Coudreuse A., Poncin-Epaillard F., Herry J. M. and Bellon-Fontaine M. N. (2010). Surface engineering and cell adhesion. *Journal of Adhesion Science and Technology*, **24**(13–14), 2301–2322, https://doi.org/10.1163/016942410X508037

Liang P., Duan R., Jiang Y., Zhang X., Qiu Y. and Huang X. (2018). One-year operation of 1000-L modularized microbial fuel cell for municipal wastewater treatment. *Water Research*, **141**, 1–8, https://doi.org/10.1016/j.watres.2018.04.066

Liu T., Rao L., Yuan Y. and Zhuang L. (2015). Bioelectricity generation in a microbial fuel cell with a self-sustainable photocathode. *The Scientific World Journal*, **2015**, 2356–6140.

Logan B. E., Hamelers B., Rozendal R., Schröder U., Keller J., Freguia S. and Rabaey K. (2006). Microbial fuel cells: methodology and technology. *Environmental Science & Technology*, **40**(17), 5181–5192, https://doi.org/10.1021/es0605016

Logan B., Cheng S., Watson V. and Estadt G. (2007). Graphite fiber brush anodes for increased power production in air-cathode microbial fuel cells. *Environmental Science & Technology*, **41**(9), 3341–3346, https://doi.org/10.1021/es062644y

Lu W., Wang Z., Wang X. and Yuan Z. (2015). Cultivation of *Chlorella* sp. using raw dairy wastewater for nutrient removal and biodiesel production: characteristics comparison of indoor bench-scale and outdoor pilot-scale cultures. *Bioresource Technology*, **192**, 382–388, https://doi.org/10.1016/j.biortech.2015.05.094

Mata T. M., Martins A. A. and Caetano N. S. (2010). Microalgae for biodiesel production and other applications: a review. *Renewable and Sustainable Energy Reviews*, **14**(1), 217–232, https://doi.org/10.1016/j.rser.2009.07.020

Mishra A. and Chhabra M. (2022). Performance of photo-microbial fuel cell with *Dunaliella salina* at the saline cathode. *Bioresource Technology Reports*, **19**, 101199, https://doi.org/10.1016/j.biteb.2022.101199

Mitra P. and Hill G. A. (2012). Continuous microbial fuel cell using a photoautotrophic cathode and a fermentative anode. *The Canadian Journal of Chemical Engineering*, **90**(4), 1006–1010, https://doi.org/10.1002/cjce.20605

Naira V. R., Das D. and Maiti S. K. (2019). Real-time light intensity-based carbon dioxide feeding for high cell-density microalgae cultivation and biodiesel production in a bubble column photobioreactor under outdoor natural sunlight. *Bioresource Technology*, ***284***, 43–55, https://doi.org/10.1016/j.biortech.2019.03.102

Nayak J. K. and Ghosh U. K. (2019). Post-treatment of microalgae treated pharmaceutical wastewater in a photosynthetic microbial fuel cell (PMFC) and biodiesel production. *Biomass and Bioenergy*, **131**, 105415, https://doi.org/10.1016/j.biombioe.2019.105415

Neethu B., Bhowmick G. D. and Ghangrekar M. M. (2018). Enhancement of bioelectricity generation and algal productivity in microbial carbon-capture cell using low-cost coconut shell as membrane separator. *Biochemical Engineering Journal*, **133**, 205–213, https://doi.org/10.1016/j.bej.2018.02.014

Pandit S., Nayak B. K. and Das D. (2012). Microbial carbon capture cell using cyanobacteria for simultaneous power generation, carbon dioxide sequestration and wastewater treatment. *Bioresource Technology*, **107**, 97–102.

Rashid N., Cui Y. F., Rehman M. S. U. and Han J. I. (2013). Enhanced electricity generation by using algae biomass and activated sludge in microbial fuel cell. *Science of the Total Environment*, **456–457**, 91–94, https://doi.org/10.1016/j.scitotenv.2013.03.067

Reddy C. N., Nguyen H. T., Noori M. T. and Min B. (2019). Potential applications of algae in the cathode of microbial fuel cells for enhanced electricity generation with simultaneous nutrient removal and algae biorefinery: current status and future perspectives. *Bioresource Technology*, **292**, 122010, https://doi.org/10.1016/j.biortech.2019.122010

Rodenas M. P., Ter Heijne A., van der Weijden R., Saakes M., Buisman C. J. and Sleutels T. H. (2015). High-rate copper and energy recovery in microbial fuel cells. *Frontiers in Microbiology*, **6**, 527, https://doi.org/10.3389/fmicb.2015.00527

Ronga D., Biazzi E., Parati K., Carminati D., Carminati E. and Tava A. (2019). Microalgal biostimulants and biofertilisers in crop productions. *Agronomy*, **9**(4), 192, https://doi.org/10.3390/agronomy9040192

Salar-García M. J., Gajda I., Ortiz-Martínez V. M., Greenman J., Hanczyc M. M., de Los Ríos A. P. and Ieropoulos I. A. (2016). Microalgae as substrate in low-cost terracotta-based microbial fuel cells: novel application of the catholyte produced. *Bioresource Technology*, **209**, 380–385, https://doi.org/10.1016/j.biortech.2016.02.083

Sanchez D. V., Jacobs D., Gregory K., Huang J., Hu Y., Vidic R. and Yun M. (2015). Changes in carbon electrode morphology affect microbial fuel cell performance with *Shewanella oneidensis* MR-1. *Energies*, **8**(3), 1817–1829, https://doi.org/10.3390/en8031817

Schamphelaire De L. and Verstraete W. (2009). Revival of the biological sunlight-to-biogas energy conversion system. *Biotechnology and Bioengineering* **103**(2), 296–304.

Sekar R., Venugopalan V. P., Satpathy K. K., Nair K. V. K. and Rao V. N. R. (2004). Laboratory studies on adhesion of microalgae to hard substrates. Asian Pacific Phycology in the 21st Century: Prospects and Challenges, Dordrecht, Springer, pp. 109–116.

Shaishav S., Singh R. N. and Satyendra T. (2013). Biohydrogen from algae: fuel of the future. *International Research Journal of Environmental Sciences*, **2**(4), 44–47.

Sharma Y. C., Singh B. and Korstad J. (2011). A critical review on recent methods used for economically viable and eco-friendly development of microalgae as a potential feedstock for synthesis of biodiesel. *Green Chemistry*, **13**(11), 2993, https://doi.org/10.1039/c1gc15535k

Shukla M. and Kumar S. (2018). Algal growth in photosynthetic algal microbial fuel cell and its subsequent utilization for biofuels. *Renewable and Sustainable Energy Reviews*, **82**, 402–414, https://doi.org/10.1016/j.rser.2017.09.067

Strik D. P., Terlouw H., Hamelers H. V. and Buisman C. J. (2008). Renewable sustainable biocatalyzed electricity production in a photosynthetic algal microbial fuel cell (PAMFC). *Applied Microbiology and Biotechnology*, **81**(4), 659–668, https://doi.org/10.1007/s00253-008-1679-8

Subhash G. V., Chandra R. and Mohan S. V. (2013). Microalgae mediated bio-electrocatalytic fuel cell facilitates bioelectricity generation through oxygenic photomixotrophic mechanism. *Bioresource Technology*, **136**, 644–653, https://doi.org/10.1016/j.biortech.2013.02.035

Taiz L., Zeiger E., Møller I. M. and Murphy A. (2015). Plant Physiology and Development. Sinauer Associates, Inc., Sunderland.

Taşkan B. and Taşkan E. (2022). Sustainable bioelectricity generation using *Cladophora* sp. as a biocathode in membrane-less microbial fuel cell. *Bioresource Technology*, **347**, 126704, https://doi.org/10.1016/j.biortech.2022.126704

Vale M. A., Ferreira A., Pires J. C. and Gonçalves A. L. (2020). CO_2 capture using microalgae. In: Advances in Carbon Capture. Woodhead Publishing, Swaston, UK, pp. 381–405.

Walter X. A., Greenman J., Taylor B. and Ieropoulos I. A. (2015). Microbial fuel cells continuously fuelled by untreated fresh algal biomass. *Algal Research*, **11**, 103–107, https://doi.org/10.1016/j.algal.2015.06.003

Wang X., Feng Y., Liu J., Lee H., Li C., Li N. and Ren N. (2010). Sequestration of CO_2 discharged from anode by algal cathode in microbial carbon capture cells (MCCs). *Biosensors and Bioelectronics*, **25**(12), 2639–2643.

Wang H., Liu D., Lu L., Zhao Z., Xu Y. and Cui F. (2012). Degradation of algal organic matter using microbial fuel cells and its association with trihalomethane precursor removal. *Bioresource Technology*, **116**, 80–85, https://doi.org/10.1016/j.biortech.2012.04.021

Wang S. K., Hu Y. R., Wang F., Stiles A. R. and Liu C. Z. (2014). Scale-up cultivation of *Chlorella ellipsoidea* from indoor to outdoor in bubble column bioreactors. *Bioresource Technology*, **156**, 117–122, https://doi.org/10.1016/j.biortech.2014.01.023

Wang L., Wang X., Jin X., Xu J., Zhang H., Yu J. and Wang L. (2017). Analysis of algae growth mechanism and water bloom prediction under the effect of multi-affecting factor. *Saudi Journal of Biological Sciences*, **24**(3), 556–562, https://doi.org/10.1016/j.sjbs.2017.01.026

Wang A. J., Wang H. C., Cheng H. Y., Liang B., Liu W. Z., Han J. L. and Wang S. S. (2020). Electrochemistry-stimulated environmental bioremediation: development of applicable modular electrode and system scale-up. *Environmental Science and Ecotechnology*, **3**, 100050, https://doi.org/10.1016/j.ese.2020.100050

Winfield J., Greenman J., Huson D. and Ieropoulos I. (2013). Comparing terracotta and earthenware for multiple functionalities in microbial fuel cells. *Bioprocess and Biosystems Engineering*, **36**(12), 1913–1921, https://doi.org/10.1007/s00449-013-0967-6

Wu Y. C., Wang Z. J., Zheng Y., Xiao Y., Yang Z. H. and Zhao F. (2014). Light intensity affects the performance of photo microbial fuel cells with *Desmodesmus* sp. A8 as cathodic microorganism. *Applied Energy*, **116**, 86–90, https://doi.org/10.1016/j.apenergy.2013.11.066

Xu C., Poon K., Choi M. M. and Wang R. (2015). Using live algae at the anode of a microbial fuel cell to generate electricity. *Environmental Science and Pollution Research*, **22**(20), 15621–15635, https://doi.org/10.1007/s11356-015-4744-8

Yang Z., Pei H., Hou Q., Jiang L., Zhang L. and Nie C. (2018). Algal biofilm-assisted microbial fuel cell to enhance domestic wastewater treatment: nutrient, organics removal and bioenergy production. *Chemical Engineering Journal*, **332**, 277–285, https://doi.org/10.1016/j.cej.2017.09.096

Yuan Y., Chen Q., Zhou S., Zhuang L. and Hu P. (2011). Bioelectricity generation and microcystins removal in a blue-green algae powered microbial fuel cell. *Journal of Hazardous Materials*, **187**(1–3), 591–595, https://doi.org/10.1016/j.jhazmat.2011.01.042

Zhang Y., Noori J. S. and Angelidaki I. (2011). Simultaneous organic carbon, nutrients removal and energy production in a photomicrobial fuel cell (PFC). *Energy & Environmental Science*, **4**(10), 4340–4346.

Zhao F., Rahunen N., Varcoe J. R., Chandra A., Avignone-Rossa C., Thumser A. E. and Slade R. C. (2008). Activated carbon cloth as anode for sulfate removal in a microbial fuel cell. *Environmental Science & Technology*, **42**(13), 4971–4976, https://doi.org/10.1021/es8003766

Zhuang L., Zheng Y., Zhou S., Yuan Y., Yuan H. and Chen Y. (2012). Scalable microbial fuel cell (MFC) stack for continuous real wastewater treatment. *Bioresource Technology*, **106**, 82–88, https://doi.org/10.1016/j.biortech.2011.11.019

doi: 10.2166/9781789063547_0147

Chapter 8

Integrated anaerobic digestion and algae cultivation

Maria Lúcia Calijuri[*1], Paula Assemany[2], Juliana F. Lorentz[1], Jackeline S. Castro[1], Alexia S. A. P. Pereira[1], Bianca B. Marangon[1], Iara B. Magalhães[1], Thiago A. Silva[1], Jéssica Ferreira[1], Rafael C. N. Gama[1] and Matheus Q. Braga[2]*

[1]*Federal University of Viçosa (Universidade Federal de Viçosa/UFV), Post-Graduate Program in Civil Engineering, Department of Civil Engineering, Campus Universitário, 36570-900 Viçosa, MG, Brazil*
[2]*Federal University of Lavras (Universidade Federal de Lavras/UFLA), Post-Graduate Program in Environmental Engineering, Department of Environmental Engineering, Campus Universitário, 37203-202 Lavras, MG, Brazil*
**Correspondence author: lucia.calijuri@gmail.com*

ABSTRACT

Anaerobic digestion is considered a versatile process that for years has been used to treat various types of waste. Besides being a low-cost technology applicable in rural and urban locations, anaerobic digestion produces multiple by-products that can be integrated into a biorefinery scenario. Similarly, microalgae biotechnology can adequately complement anaerobic digestion by improving resource recovery through a closed-loop process and contributing to a biobased green circular economy model. Therefore, this chapter aims to address current perspectives on the topic. It covers algae cultivation from anaerobic digestion residues as a post-treatment option and digestate as a potential medium for microalgae growth. Moreover, anaerobic digestion is presented as an energetic valorization route of algae biomass, including strategies to overcome main challenges, such as pre-treatment of microalgae biomass and anaerobic co-digestion. Biogas upgrading during algae cultivation is also discussed. Finally, it presents biorefinery models based on integrated microalgae and anaerobic digestion, reporting the technologies' sustainability and environmental impacts. Future perspectives on the subject are highlighted, encouraging further studies to improve microalgae biomass production, nutrient recovery, wastewater treatment, and biogas upgrading.

8.1 INTRODUCTION

Anaerobic digestion (AD) is one of the most diffused biotechnologies for converting organic biomass to bioenergy (Chen *et al.*, 2018). In this oxygen-deprived process, organic substrates are submitted to microbial conversion in reactors to produce biogas (Greene, 2019). In addition, the process also results in liquid or solid residues containing residual nutrients and microbiota (Zicari *et al.*, 2019). Moreover, organic waste streams' use as a substrate is highly attractive from the economic and environmental perspectives, with food wastes, agricultural, municipal solids, animal manure, poultry, and microalgae as reported substrates (Khan *et al.*, 2021).

Research on algal substrates for AD dates back to the late 1950s (Golueke *et al.*, 1957), based on its potential for biofuel production through biomass valorization. In addition, algal cultivation can be employed as photosynthetic biogas upgrading technology, recovering CO_2, and purifying CH_4

(Franco-Morgado *et al.*, 2021). The mobilization of nutrients (nitrogen (N) and phosphorus (P)) and the CO_2 availability made the AD process highly attractive for microalgae applications (Solé-Bundó *et al.*, 2019b). Anaerobically treated effluent or digestate can thus be useful as a microalgae culture medium, allowing effluent polishing and providing nutrients for algae growth (Barreiro-Vescovo *et al.*, 2020). In this context, several pathways can be employed to integrate AD technology with algae cultivation to promote resource recovery and waste treatment.

Different types of biomass can be obtained in microalgae biotechnology according to the wastewater used for growth medium or culture conditions. Multiple products can be produced due to the great versatility of microalgae biomass and AD process. In a circular economy biorefinery concept, AD and algal biotechnology integration occurs via nutrient recycling (Chen *et al.*, 2018), being an interesting multi-arrangement alternative for environmental, social, and economic development. In addition to compensating for each process's limitations, coupling two or more waste treatment technologies can act as an engine to improve resource recovery through a closed-loop process (Sikarwar *et al.*, 2021).

Bearing this context in mind, this chapter includes current perspectives on the topic, covering algae cultivation from AD residues, AD as an energetic valorization route of algae biomass, algae cultivation for biogas upgrading, and coupling technologies for sustainable biorefineries.

8.2 ALGAE CULTIVATION FROM AD RESIDUES

8.2.1 Liquid effluent

Considering biological systems for wastewater treatment, several technologies based on anaerobic treatment are available, such as high-rate anaerobic systems, namely up-flow anaerobic sludge blanket reactor (UASB), anaerobic contact process, anaerobic filter or fixed film reactors and fluidized-bed reactors (Khan *et al.*, 2011). Among them, the UASB reactor is widely used in countries with hot climates, such as Brazil, Colombia, and India (Chernicharo *et al.*, 2018). It offers economic and operating benefits and less area demand than conventional treatments like activated sludge and stabilization ponds (Vassalle *et al.*, 2020a).

Although these anaerobic reactors are used for organic matter removal, it is widely accepted that their performance needs to be improved to meet many wastewater discharge standards. It results in an effluent that may still contain high concentrations of organic matter, suspended solids, nutrients (Khan *et al.*, 2011), and pathogenic organisms, thus requiring post-treatment steps.

The chemical oxygen demand (COD) in domestic wastewater treated in different conventional anaerobic systems commonly presents concentrations ranging from 70 to 160 mg/L (Chernicharo, 2006). The typical COD removal efficiency in a UASB reactor varies around 55% and 70% (Chernicharo, 2006). Table 8.1 presents studies using anaerobic effluents for microalgae cultivation. Concentrations ranging from 141 to 45,875 mg COD/L can be present in anaerobic effluents, mainly explained by several factors, such as wastewater type, reactor scale, and treatment parameters. Notably, in some works, wastewater dilution was necessary to adapt the culture medium organic matter load to the ideal conditions for microalgae development (de Godos *et al.*, 2016; Kimura *et al.*, 2019; Xie *et al.*, 2018; Zhen *et al.*, 2022).

The greatest COD values (Table 8.1) do not meet the release standards of Brazil ($150 < COD < 225$ mg/L) (Morais & Santos, 2019) and the European Union ($COD = 125$ mg/L). Regarding nutrients, the Brazilian standard establishes a limit of 20 mg/L for NH_4^+ (CONAMA, 2011), while the European Commission requires a more rigorous standard of 15 mg/L for total nitrogen. Thus, a post-treatment is needed not only at a secondary level but also at a tertiary level to maintain balanced discharge into ecosystems.

Some technologies have been suggested for UASB reactor effluent post-treatment. Among them are trickling filters, submerged aerated biofilters, rotating biological contactors, wetlands, sequencing batch reactors, chemically enhanced sedimentation, zeolite columns, and dissolved air flotation (Khan *et al.*, 2011). In recent decades, microalgae-based technologies, such as the high-rate algal pond (HRAP), have also been evaluated and shown to be promising (Assemany *et al.*, 2018; Benett

Table 8.1 Concentrations of organic matter and nutrients from effluents treated in an anaerobic system followed by microalgae-based treatment.

Type of Effluent	Anaerobic/Aerobic Treatment Unit	Raw Effluent (mg/L)	Microalgae	Treated Effluent (mg/L)	Biomass Productivity	References
Domestic sewage	UASB/Hybrid system composed of an HRAP and biofilm reactor	CODs = 116.0 $N\text{-}NH_4^+ = 37.3$ $N\text{-}NO_3^- = 1.6$ Ps = 5.2	Mixed culture (autochthonous species)	CODs = 78.0 (33%) $N\text{-}NH_4^+ = 6.1$ (84%) $N\text{-}NO_3^- = 29.9$ (−1.769%) Ps = 4.1 (21%)	6.79 g/m^2day	Assis *et al.* (2017)
Raw domestic wastewater (screened)	Anaerobic pond/ HRAP	$BOD_5 = 94$ $N\text{-}NH_4^+ = 36$ $N\text{-}NO_3^- = 0.1$	Mixed culture (autochthonous species)	$BOD_5 = 52$ $N\text{-}NH_4^+ = 15.2$ (33–76%) $N\text{-}NO_3^- = 0.2$	1.281–4.112 mg/L (chlorophyll-a)	Sutherland *et al.* (2017)
Domestic sewage	Septic tank/HRAP fed with gas from the combustion of gasoline	CODs = 174.5 TKN = 87.8 Ps = 12.3	Mixed culture (autochthonous species)	CODs = 110.8 (30%) TKN = 36.2 (37.9%) Ps = 13.5 (−11.3%)	6.12 g/m^2day	Assis *et al.* (2019)
Anaerobically digested distillery (diluted)	UASB/ Photobioreactor (rectangular tank with submerged mixer aerator)	COD = 45,875	*Spirulina* sp.	COD = (60–70%)	0.08–0.094 g dry biomass per L/d	Krishnamoorthy *et al.* (2019)
Domestic sewage	Septic tank/Hybrid system composed of a HRAP and biofilm reactor	COD = 329.2 $N\text{-}NH_4^+ = 87.4$ $NO_3^- = 1.1$ TP = 9.1	Mixed culture (autochthonous species)	COD = 135.5 (58.8%) $N\text{-}NH_4^+ = 19.9$ (77.3%) $NO_3^- = 40.3$ TP = 7.7 (16.2%)	6.13 g/m^2day	Assis *et al.* (2020)
Municipal wastewater	UASB/HRAP	COD = 232.69 (55%) TN = 54.33 $N\text{-}NH_4^+ = 34.21$	Mixed culture (autochthonous species)	COD = 146.08 (38%) TN = 24.31 (30%) $N\text{-}NH_4^+ = 14.31$ (44%)	1.01 g/L (volatile solids)	Vassalle *et al.* (2020b)
Domestic sewage	UASB/HRAP	COD = 141 ± 48 TKN = 41.1 ± 12.0 TP = 4.4 ± 0.8	Mixed culture	COD = 63.7 ± 11.3 TKN = 9 ± 4.2 TP = 3.4 ± 0.6	NA	Oss *et al.* (2022)

Note: Average (±standard deviation) final concentration values of water quality variables found in the literature were reported, and the values within parentheses refer to the removal efficiency. NA = not available. UASB: upflow anaerobic sludge blanket; HRAP: high-rate algal pond; BOD_5: biochemical oxygen demand; COD: chemical oxygen demand; CODs: soluble chemical oxygen demand; TKN: total Kjeldahl nitrogen; TN: total nitrogen; $N\text{-}NH_4^+$: ammoniacal nitrogen; $N\text{-}NO_3^-$: nitrate nitrogen; PO_4^{3-}: phosphorus; Ps: soluble phosphorus; TP: total phosphorus.

et al., 2008; Couto *et al.*, 2020; Magalhães *et al.*, 2022; Santiago *et al.*, 2013; Vassalle *et al.*, 2020a, 2020b; Villar-Navarro *et al.*, 2018) (Figure 8.1). Systems that use algal–bacterial symbiosis represent a wastewater treatment technology (Zhen *et al.*, 2022) with the advantages of reduced energy consumption during aeration, efficient nitrogen and phosphorus removal, and effective biomass recycling (Xie *et al.*, 2018). Symbiotic interactions between microalgae, bacteria, and fungi have been used for wastewater treatment (Kabir *et al.*, 2022; Leng *et al.*, 2020; Leong & Chang, 2022; Zhang *et al.*, 2021). Some factors encouraging the UASB and HRAP integration are:

(i) The anaerobic treatment partially removes turbidity and suspended solids from the wastewater, which contributes to the light incidence in the water column and, consequently, microalgae growth and photosynthetic activity (Couto *et al.*, 2021; de Godos *et al.*, 2016).
(ii) The anaerobic effluent is rich in nutrients (NH_4^+ and PO_4^{3-}), which are more readily available and essential for microalgae growth. It is noteworthy that the main N form assimilated by microalgae is NH_4^+.
(iii) Bioremediation can be combined with simultaneous valuable bioproducts production, like pigments, biodiesel, bioCH_4, and biofertilizer (Leong & Chang, 2022).

Microalgae are involved in O_2 production, CO_2 consumption, and nutrient removal via photosynthesis. At the same time, bacteria are responsible for fixing and regenerating inorganic nutrients (NH_4^+, PO_4^{3-}, H_2S), consuming organic matter, and producing vitamins and siderophores (Lian *et al.*, 2018). Fungi and bacteria are involved in organic matter degradation in the anaerobic digestate, while microalgae can assimilate the CO_2 released during the degradation (Zhang *et al.*, 2022). This way, the produced biomass can remove N-NH_4^+ (volatilization or assimilation), PO_4^{3-} (precipitation or assimilation), and acetate along with metallic ions, for example calcium (Ca), magnesium (Mg), and iron (Fe) (Pacheco *et al.*, 2015). The biomass can also produce polypeptides, called chelating agents, capable of binding to heavy metals, for example mercury (Hg), cadmium (Cd), and lead (Pb) (Kabir *et al.*, 2022). Oliveira *et al.* (2021) observed that the presence of Cu and Zn, found in swine wastewater, altered the dynamics of HRAPs regarding nutrient removal, productivity, and biochemical composition of the biomass. Similarly, Oliveira *et al.* (2023) concluded that nutrient removal and biomass biochemical composition should be considered to combine the recovery of Zn and nutrients with the production of value-added biomass. Therefore, environmental parameters must also be considered, as they have effects on gene expression and can promote some biological pathways to the detriment of others, thus modifying the microbial structure and the inherent metabolism involved in the biotechnological process (Bose *et al.*, 2020; Lopatkin & Collins, 2020).

Vassalle *et al.* (2020b) investigated the combination of anaerobic (UASB) and aerobic (HRAP) treatment to treat municipal wastewater and reported that the HRAP was responsible for only 38% of COD removal, while the global mean removal efficiency of this variable was 72%. Still, according to the authors, HRAP was found efficient in removing estrogens (90–95%) and pharmaceuticals (64–70%).

Concerning nutrients, Oss *et al.* (2022), in a wastewater treatment plant (WWTP) composed of a UASB reactor followed by HRAP, produced activated carbon (C) from biomass and achieved removal rates for COD, N, and P similar to values already presented in the literature (Craggs *et al.*, 2012; Park & Craggs, 2010). Zkeri *et al.* (2021) compared two systems composed of a methanogenic moving-bed biofilm reactor (AnMBBR) followed by an aerobic MBBR (AeMBBR) and sequencing batch reactor (SBR) with *Chlorella sorokiniana*. The authors reported that the AnMBBR + AeMBBR combination removed COD, NH_4^+, total Kjeldahl nitrogen, and PO_4^{3-} by 93 (±4)%, 97 (±3)%, 99 (±1)%, and 49 (±15)%, respectively, while the AnMBBR + SBR combination removed COD, but only partially the other pollutants.

Table 8.1 summarizes the biomass production using anaerobic effluent as a culture medium. Under Brazilian environmental conditions, studies have reported productivities (based on volatile solids (VS) value) around 6.5 g/m²/day (Assis *et al.*, 2019; Assis *et al.*, 2017), operating with autochthonous

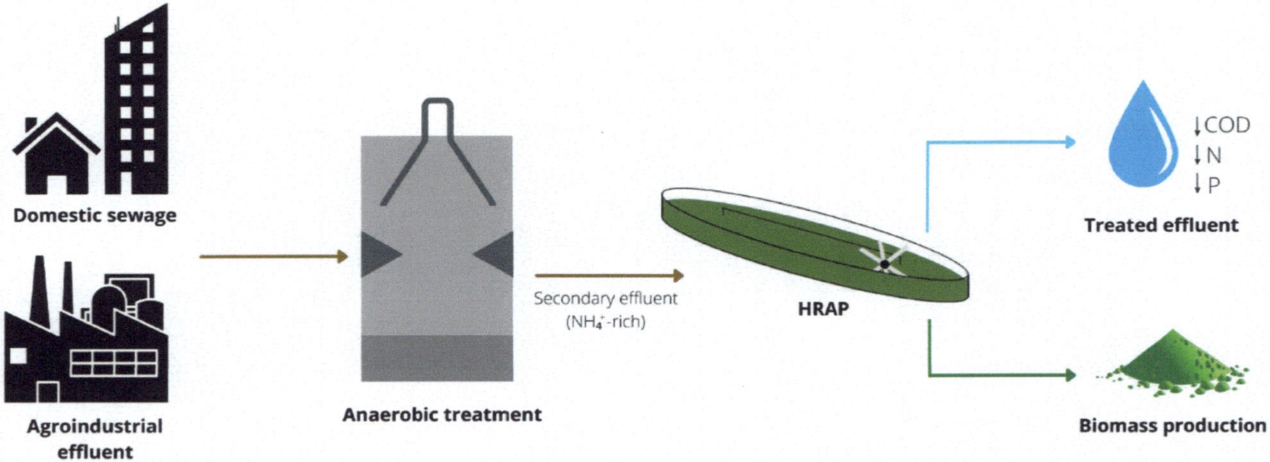

Figure 8.1 Schematic diagram of UASB and HRAP integration to increase pollutant removal and produce algal biomass. COD: chemical oxygen demand; N: nitrogen; P: phosphorus; HRAP: high-rate algal pond.

species adapted to the culture medium. This value can still be improved, and studies on biomass production optimization through operational strategies should be encouraged, considering the adversities of outdoor conditions and competition between microorganisms. Thus, a wide field of study can explore the treatability of both anaerobic systems and treatment system arrangements that allow maximum resource recovery.

8.2.2 Digestate

Organic waste AD produces a by-product named digestate. It contains many nutrients and other compounds that can cause undesired environmental impacts when discarded directly into the environment (Chen *et al.*, 2018). In this context, several alternatives have been investigated to value this nutrient-rich by-product. Recently, an emerging possibility is to couple microalgae cultivation with anaerobic digestate treatment (Barreiro-Vescovo *et al.*, 2020; Chen *et al.*, 2018; Patel *et al.*, 2021). Thus, digestate as a culture medium for producing microalgae biomass is an alternative to replace the demands for drinking water and fertilizers of conventional microalgae cultivation, reducing costs and environmental impacts (Al-Mallahi & Ishii, 2022). Given this, algal phycoremediation is a sustainable and efficient alternative to treat anaerobic digestate and allows simultaneous nutrient recycling (Leong & Chang, 2022).

Additionally, microalgae biomass is rich in lipids and proteins. Therefore, it may have several applications, such as biofuels, biofertilizers, and value-added products, such as biopolymers and pigments (Calijuri *et al.*, 2022). It creates an opportunity to develop the biorefinery concept and circular economy (Chen *et al.*, 2018). Microalgae cultivation using digestate has been studied in recent research. Among them, there is the AD of food (Barzee *et al.*, 2022; Patel *et al.*, 2021), animal (Lu *et al.*, 2022), and urban solid (Barreiro-Vescovo *et al.*, 2020) waste, as well as the combination of different residues (Chen *et al.*, 2018; Seelam *et al.*, 2022) (Figure 8.2).

Due to the remarkable ability of microalgae to adapt to extreme conditions and the possibility of nutrient recovery, microalgae cultivation in anaerobic digestate is a promising strategy. The digestate is rich in bioactive substances, such as monosaccharides, free amino acids, nucleic acids, and fulvic acid, stimulating microalgae development and providing greater tolerance to abiotic and biotic stress (Chong *et al.*, 2022). In addition, the processes that rule AD mineralize P and N into PO_4^{3-} and NH_4^+, respectively, which are the preferred forms assimilated by microalgae (Al-Mallahi & Ishii, 2022). Still, the volatile organic acids (VOA) in the anaerobic digestate are promising compounds for microalgae production (Patel *et al.*, 2021).

There are, however, some challenges in microalgae cultivation in anaerobic digestate, mainly concerning their physical and chemical characteristics. Excess of suspended solids, turbidity, NH_4^+, and metals in the digestate limit microalgae growth. Beyond that, a disbalanced nutrient proportion and presence of other competing organisms are other factors that can limit microalgae growth (Al-Mallahi & Ishii, 2022; Praveen *et al.*, 2018). Marcilhac *et al.* (2014) investigated the effect of light intensity and digestate color on nutrient removal and concluded that the initial optical density is inversely proportional to productivity and N assimilation. According to the authors, this fact is due to reduced light penetration and, consequently, reduced photosynthetic efficiency. To solve the problem of limiting light use due to suspended solids, Chen *et al.* (2018) proposed a membrane photobioreactor with a 0.1 μm pore size that resulted in removal efficiencies of 43.9% of NH_4^+ and 64.9% of PO_4^{3-}.

High NH_4^+ concentrations can also limit microalgae development, despite being the preferred assimilation form. Free ammonia (NH_3) is a toxic N form (Jiang *et al.*, 2021) that easily penetrates the cell membrane and accumulates in the cytoplasm, impairing photosynthetic processes (Uggetti *et al.*, 2014). Praveen *et al.* (2018) evaluated the effects of high N concentrations on growth (NH_4^+ between 20 and 120 mg/L) and concluded that microalgae growth was inhibited at concentrations exceeding 100 mg/L. However, the values can vary from 100 to 1,600 mg/L, depending on the microalgae species used and the cultivation conditions (Al-Mallahi & Ishii, 2022).

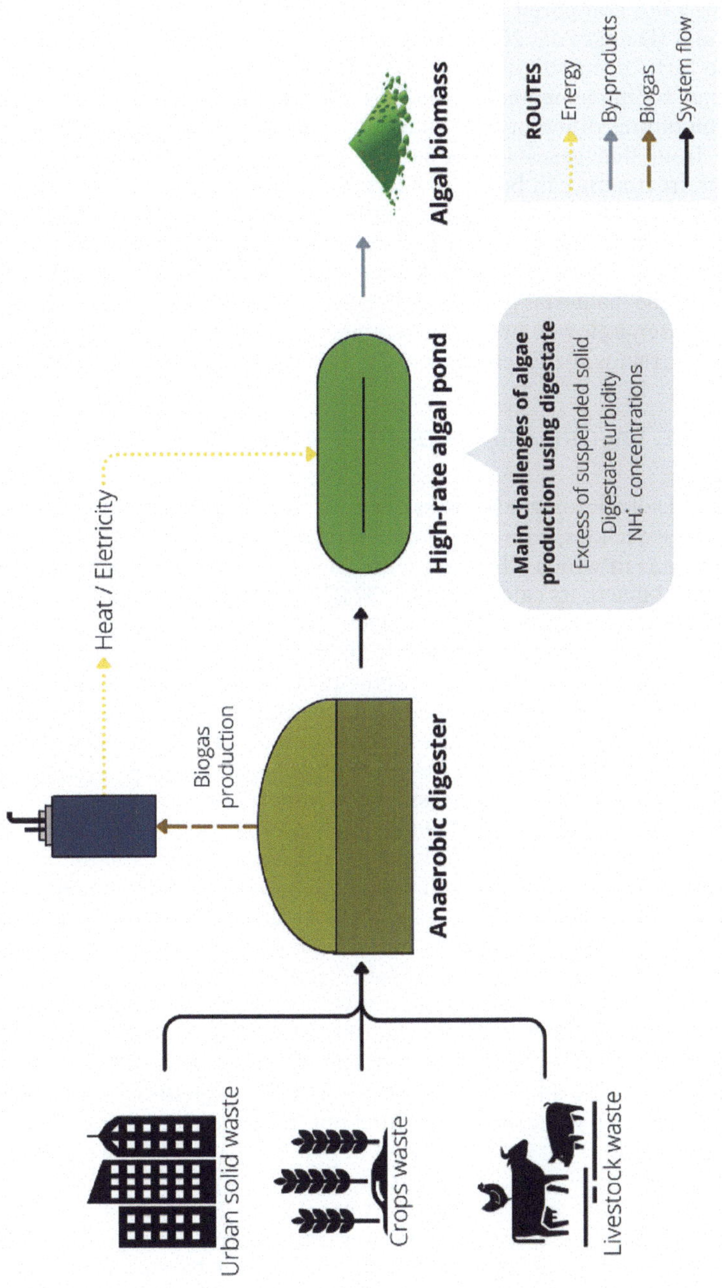

Figure 8.2 Schematic diagram of microalgae production using digestate from the anaerobic treatment of biomass.

Another limiting factor for using digestate as a culture medium is the nutrient proportion, given that an adequate C/N ratio is required for a synergistic microalgae and bacteria interaction (Fallahi *et al.*, 2021). Anaerobic digestate has a C/N ratio of 2:3 (Barzee *et al.*, 2022; Lu *et al.*, 2022). Therefore, it is considered a low ratio compared to the adequate C/N ratio for microalgae cultivation, which ranges between 6 and 8 (Dang *et al.*, 2022; Woertz *et al.*, 2009). The low C/N ratio of the anaerobic digestate is related to NH_4^+ accumulation and high pH value.

Some solutions can be implemented to improve the microalgae cultivation stage in anaerobic digestate, such as combining different raw materials during the AD stage (Chong *et al.*, 2022) or supplementing CO_2 from flue gases in the microalgae cultivation stage (Assis *et al.*, 2019). In addition, digestate pretreatment can be carried out. Pretreatment aims to facilitate the breakdown of complex organic compounds, which may reduce the suspended solids concentration, and mitigate possible toxicities due to high organic and inorganic matter concentrations, consequently reducing turbidity and promoting digestate sanitization (Chong *et al.*, 2022). For example, Praveen *et al.* (2018) investigated the microalgae–bacterial process performance through two stages: (1) digestate dilution with municipal wastewater, followed by (2) pretreatment in activated sludge, achieving COD, nitrate (NO_3^-), NH_4^+ and PO_4^{3-} removal efficiencies of 87%, 100%, 30% and 77%, respectively.

8.3 AD AS ENERGETIC VALORIZATION ROUTE OF ALGAE BIOMASS

8.3.1 AD of microalgae

Biogas production via AD of microalgae biomass obtained in wastewater treatment has been an energy recovery alternative since decades, with renewed research attention in recent years (Choudhary *et al.*, 2020). AD produces biogas in which CH_4 represents 55–70% of the composition, responsible for the process's energy potential due to its calorific value (37.27 MJ/m^3) (Ganesh Saratale *et al.*, 2018). Some biogas valorization routes are generating heat or electricity, liquefaction into methanol, compression into fuel for automobiles, and fuel gas (Zabed *et al.*, 2020).

Increasing AD performance, with greater methane (CH_4) production, depends on several factors, for example operational and environmental conditions, as well as substrate composition. The organic loading rate is a key factor for the AD efficiency among the operational factors. It prevents VOA accumulation or shortage, which influences the reactor pH (ideal range 6.6–7.4), a critical factor for balancing acidogenic and methanogenic processes. A volumetric organic load of 1.6–4.8 kg VS/m^3 is usually recommended for obtaining a high AD rate (Zabed *et al.*, 2020). Still, it may vary depending on the biomass and reactor types, the biomass biochemical composition, and the anaerobic microbial population. Other parameters are reflected in the production of VOA, including the C/N ratio, which, when low, causes high NH_4^+ concentrations in the digester, disturbing the microbial metabolism and consequently accumulating VOA. C/N ratios between 20 and 30 are considered adequate, and a C/N ratio equal to 25 usually gives better CH_4 yields (Zabed *et al.*, 2020).

Another variable that influences the AD quality is biomass moisture. A high solids content in the reactor decreases the available water, affecting the alkalinity availability, free NH_4^+, and VOA concentrations (Zabed *et al.*, 2020). Higher rates of CH_4 production have been reported at 60–80% moisture (Kwietniewska & Tys, 2014). Also, the reactor hydraulic retention time (HRT) should not be too short (~16 days). This HRT value avoids washing out methanogenic archaea, causing a low CH_4 bioconversion efficiency. Contrary, it should also not be overly long (>50 days) to prevent the depletion of substrates and nutrients. For pilot and commercial plants, the optimal HRT varies between 30 and 50 days, whereas on a laboratory scale, it ranges between 15 and 30 days (Zabed *et al.*, 2020).

Regarding operational conditions, low temperatures generate high accumulation of VOA, which is reflected in the pH, affecting the methanogenic archaea metabolism. In contrast, elevated temperatures increase NH_4^+ toxicity in addition to foaming and odor formation. The operating range of mesophilic AD is 30–40°C and for thermophilic it is 50–60°C, with 35°C and 55°C being the ideal temperature, respectively (Zabed *et al.*, 2020). Figure 8.3 presents a schematic design of an anaerobic reactor,

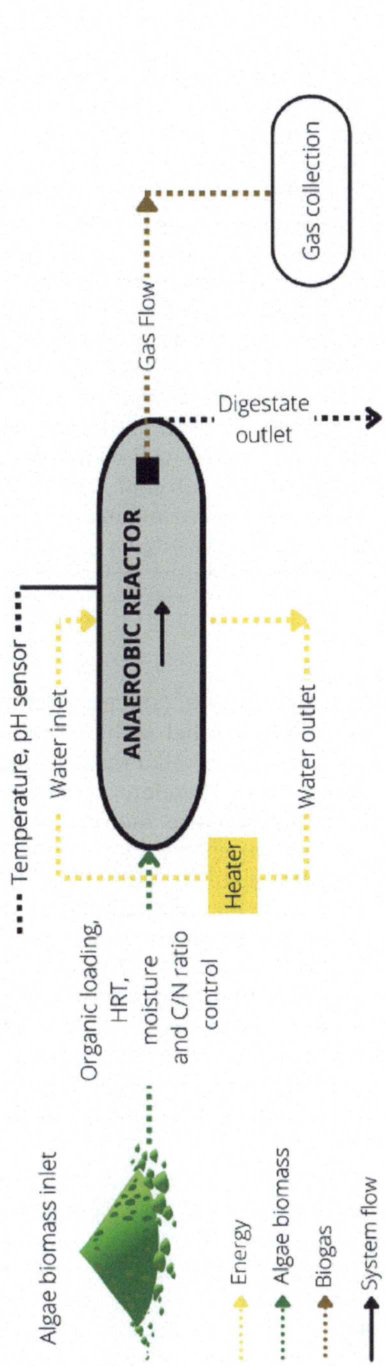

Figure 8.3 Schematic diagram of an anaerobic reactor with indications of the primary control parameters for enhancing biogas productivity of microalgae AD. HRT: hydraulic retention time; C/N ratio: carbon/nitrogen ratio.

highlighting the primary control parameters that can be utilized to enhance the biogas productivity of microalgae AD.

AD can be performed in different reactor types: fixed dome, floating drum, plastic, or textile reactors (Zabed *et al.*, 2020). As a rule, the reactor must create an oxygen-free environment. Furthermore, it must be protected against water, gas, and light leaks. In addition, it must contain protection mechanisms against corrosive chemicals and gases and avoid adverse weather conditions (Zabed *et al.*, 2020). The responses as a function of all these factors and parameters determine the CH_4 yield of algal biomass subjected to AD. Concerning the microalgae biomass produced in wastewater, several types of culture media and species have already been studied at bench scale (Choudhary *et al.*, 2020).

The main limitations of microalgae AD are: (1) low biomass biodegradability due to the microalgae cell walls resistance, causing low CH_4 potential (degradation extent) and low conversion rate (degradation speed); and (2) NH_4^+ inhibition risk due to the biomass low C/N ratio (as mentioned in Section 8.2.2). Moreover, NH_4^+ inhibition limits the maximum rate of organic discharge from the digesters and requires a longer HRT; therefore, a digester with a larger volume is required (Karuppiah & Ebenezer Azariah, 2019; Solé-Bundó *et al.*, 2019b). The respective solutions to these issues are: (a) pretreatments to disturb the microalgae cell wall and make its intracellular content more available (de Oliveira *et al.*, 2022; Yukesh Kannah *et al.*, 2021); and (b) anaerobic co-digestion (AcoD) to increase alkalinity, provide a balanced macro and micronutrients composition, stabilizing the process at high organic loading rates and increasing CH_4 yields (de la Lama-Calvente *et al.*, 2022; Veerabadhran *et al.*, 2021). Table 8.2 presents the CH_4 yield of microalgae AD, AcoD, and biomass pretreatments.

8.3.2 Pretreatment of microalgal biomass

Pretreatment methods can be divided into two groups: (a) energy-intensive, which are mechanical/physical (ultrasound, microwave, and milling), thermal or hydrothermal, and (b) energy-efficient, which are biological, enzymatic, chemical (acidic or alkaline) or combined (thermochemical) (de Oliveira *et al.*, 2022; Yukesh Kannah *et al.*, 2021). Therefore, strategies must be adopted for energy-intensive treatments to achieve better energy performance. Xiao *et al.* (2019a) proposed a hydrothermal pretreatment system for algal biomass using solar energy and obtained 348 mL CH_4/g VS. The CH_4 production was 57% higher than without pretreatment (221.70 mL CH_4/g VS). Biogas production with solar-powered hydrothermal pretreatment achieved a maximum exergy efficiency (40.85%) (Xiao *et al.*, 2019b). Biogas production with hydrothermal pretreatment with solar energy achieved a net energy ratio of 0.69, with emissions of –166.13 g $CO_{2\,eq}$ per kWh. Also, it achieved a leveled cost of 0.17 USD/m^3, representing a better performance than biogas without solar energy pretreatment (Xiao *et al.*, 2020). With a thermo-acid pretreatment, Barros *et al.* (2022) estimated that biogas production with microalgae biomass AD produced at the tertiary level would result in an energy surplus of 2.8% in the WWTP. Fu *et al.* (2023) estimated that with thermo-alkaline pretreatment, also for energy efficiency, there would be an energy surplus in the system, increasing the CH_4 production.

8.3.3 Anaerobic co-digestion

In WWTPs, an AcoD option exists between sludge and microalgae or microalgae with other residues. Damtie *et al.* (2020) obtained a 36% increase in CH_4 production when studying AcoD from biologically pre-treated algal biomass and primary sludge. Solé-Bundó *et al.* (2019b) obtained a 65% increase in CH_4 production in the AcoD of microalgae with primary sludge and a generation of 4.5 times the energy consumed, whereas the microalgae mono-digestion AD generated 2.7 times the energy consumed. Solé-Bundó *et al.* (2019a) achieved a 60% increase in CH_4 yield applying a thermal pre-treatment in the microalgae biomass and 15% after AcoD with WWTP residues (oil, grease, and fat). Zhang *et al.* (2020) added glycerol to the AcoD of microalgae and potato processing residues and found an increase of more than 50% in CH_4 production. Assemany *et al.* (2020) performed the

CH_4 Yield	Microalgae Species	Microalgae Culture Medium	AD Feedstock	Reference
348 L CH_4 per kg VS	*Chlorella pyrenoidosa*	Freshwater	Solar-driven hydrothermal pre-treated microalgae biomass	Xiao *et al.* (2019a)
221.70 L CH_4per kg VS			Raw microalgae biomass mono-digestion	
382 mL CH_4per g VS	*Chlorella* and *Desmodesmus*	Freshwater	Co-digestion of microalgae biomass and rice straw	Srivastava *et al.* (2022)
252 mL/g VS	Microalgal consortium	Wastewater	Thermo-acid hydrolysis pre-treated microalgae biomass	Barros *et al.* (2022)
308 mL/g VS	*Chlorella vulgaris*	Freshwater	Aerobic digestion as a pre-treatment for co-digestion of microalgae biomass and sludge	Damtie *et al.* (2020)
188 mL/g VS			Raw microalgae biomass mono-digestion	
0.33 m^3 CH_4 per kg VS	Microalgal consortium	Wastewater	Co-digestion of microalgae biomass and primary sludge	Solé-Bundó *et al.* (2019a)
0.20 m^3 CH_4 per kg VS			Raw microalgae biomass mono-digestion	
0.73 (±0.07) L CH_4 per g COD	*Chlorella vulgaris*	Freshwater	Raw microalgae biomass co-digested with 1% v/v glycerol	Zhang *et al.* (2020)
0.30 (±0.04) L CH_4 per g COD			Raw microalgae biomass mono-digestion	
207.35 mL CH_4 per g VS	*Chlorella* sp.	Freshwater	Co-digestion of algal residues after lipid extraction with mixed enzymes pretreatment and energy grass	Zhang *et al.* (2018)
128.75 mL CH_4 per g VS			Algal residue after lipid extraction from untreated microalgae mono-digestion	
128.80 mL CH_4 per g VS			Algal residue after lipid extraction from mixed enzymes pretreatment microalgae mono-digestion	
368.94 mL CH_4 per g VS	*Chlorella pyrenoidosa*	Wastewater	Co-digestion ratio of 1:2 of thermo-alkaline pretreated microalgae biomass and secondary sludge	Fu *et al.* (2023)
328.43 mL CH_4 per g VS			Co-digestion ratio of 1:1 of thermo-alkaline pretreated microalgae biomass and secondary sludge	
293.39 mL CH_4 per g VS			Co-digestion ratio of 2:1 of thermo-alkaline pretreated microalgae biomass and secondary sludge	
0.10 m^3 CH_4 per kg VS	Microalgal consortium	Wastewater	Co-digestion of municipal wastewater growth microalgae biomass and 10% olive mill wastewater	Assemany *et al.* (2020)
0.062 m^3 CH_4 per kg VS			Mono-digestion of municipal wastewater growth microalgae biomass	
0.13 m^3 CH_4 per kg VS			Co-digestion of brewing industry wastewater growth microalgae biomass and 10% olive mill wastewater	
0.16 m^3 CH_4 per kg VS			Mono-digestion of brewing industry wastewater growth microalgae biomass	

Note: AD: anaerobic digestion; VS: volatile solids; COD: chemical oxygen demand; v/v: volume/volume.

AcoD of microalgae biomass grown in brewing industry wastewater and olive mill wastewater. They obtained 61% more CH_4 compared to the microalgae biomass mono-digestion. As presented in Section 8.4, in WWTPs, microalgae can also capture CO_2 from biogas (Nguyen *et al.*, 2021), a form of biogas purification (Miyawaki *et al.*, 2021).

The generation of multiple by-products can be an approach to make microalgae biogas even more attractive. For example, Zhang *et al.* (2018), through the lipid extraction of *Chlorella* sp. with pretreatment by mixed enzymes (cellulase, xylanase, and pectinase), achieved 169% more energy in the combined biodiesel and CH_4 production than with biodiesel alone or with AcoD of the residual biomass and C-rich material (energy grasses). Srivastava *et al.* (2022) extracted lipids for biodiesel production and performed the AcoD of the remaining biomass with rice straw. They obtained a 382 mL CH_4/g VS yield, almost 50% higher than the control. Another opportunity is co-production of CH_4 and hydrogen gas (H_2) through AD in two stages. The acidogenic and methanogenic processes are carried out separately, allowing recovery of the H_2 generated in the first phase (Zabed *et al.*, 2020).

8.4 ALGAE CULTIVATION FOR BIOGAS UPGRADING

The biogas composition should be at least 95% CH_4 before feeding into the natural gas grid (Khan *et al.*, 2021). However, biogas is usually composed of 45–70% CH_4, 20–55% CO_2, and other gases, namely, nitrogen gas (N_2) (0–3%), O_2 (0–1%), water vapor (1–10%), H_2S (0–10,000 ppm), NH_3 (0–100 ppm), and traces of hydrocarbons, siloxanes, and chlorine (Bose *et al.*, 2019). These gases, except for the CH_4, decrease the heating value of the biogas or can produce environmental pollutants (Angelidaki *et al.*, 2018). Beyond that, they can corrode metal components of boilers, internal combustion engines, and gas pipelines (Khan *et al.*, 2021). Thus, the biogas can be cleaned by removing these compounds, and the heating value can be increased through a process named 'biogas upgrading'.

Many conventional biogas upgrading technologies can be used, such as pressure swing adsorption, chemical scrubbing, water scrubbing, organic solvent scrubbing, and membrane separation (Nguyen *et al.*, 2021). However, emerging biogas upgrading systems are being investigated as economic and environmental alternatives, such as adsorption by biochar, cryogenic upgrading, and biological upgrading. Among the biological upgrading systems, microalgae have attracted research interest (Miyawaki *et al.*, 2021; Thi Nguyen *et al.*, 2019; Toro-Huertas *et al.*, 2019; Xie *et al.*, 2023). It is noteworthy that selecting the appropriate technology for upgrading raw biogas depends on its final use, the economics involved, and the efficiency of the upgrading process (Khan *et al.*, 2021).

When using microalgae for biogas upgrading, CO_2 can be assimilated as a C source to produce chemical energy through photosynthesis (Thi Nguyen *et al.*, 2019). Microalgae can remove CO_2 from biogas using open or closed systems (Figure 8.4). The most common open system is the HRAP, also named raceway pond. It can be interconnected to an absorption bubble column (ABC). An ABC is fed with raw biogas, and the liquid containing microalgae produced in the HRAP is recirculated, allowing microalgae to capture the CO_2 from the biogas (Toro-Huertas *et al.*, 2019). Zabed *et al.* (2020) stated that although cultivation in open systems is techno-economically more convenient than in closed systems, open systems pose a higher risk of contamination with relatively lower biogas purification. The main drawback of the closed system is the higher energy requirements for light penetration and high capital costs. The potential CH_4 recovery by the photoautotrophic biogas upgrading process is 97%, with H_2S removal achieved simultaneously (Khan *et al.*, 2021).

Khan *et al.* (2021) reported that CO_2 solubility, mass transfer to microalgae, difficulty in biogas harvesting, and CH_4 solubility in microalgae media are the main challenges of open or closed systems. These limitations can be overcome by using indirect biogas upgrading systems (Figure 8.4). As Nguyen *et al.* (2021) stated, indirect methods can overcome the limitations of direct biogas upgrading. In this approach, CO_2 can be captured in a carbonate solution such as potassium carbonate (K_2CO_3).

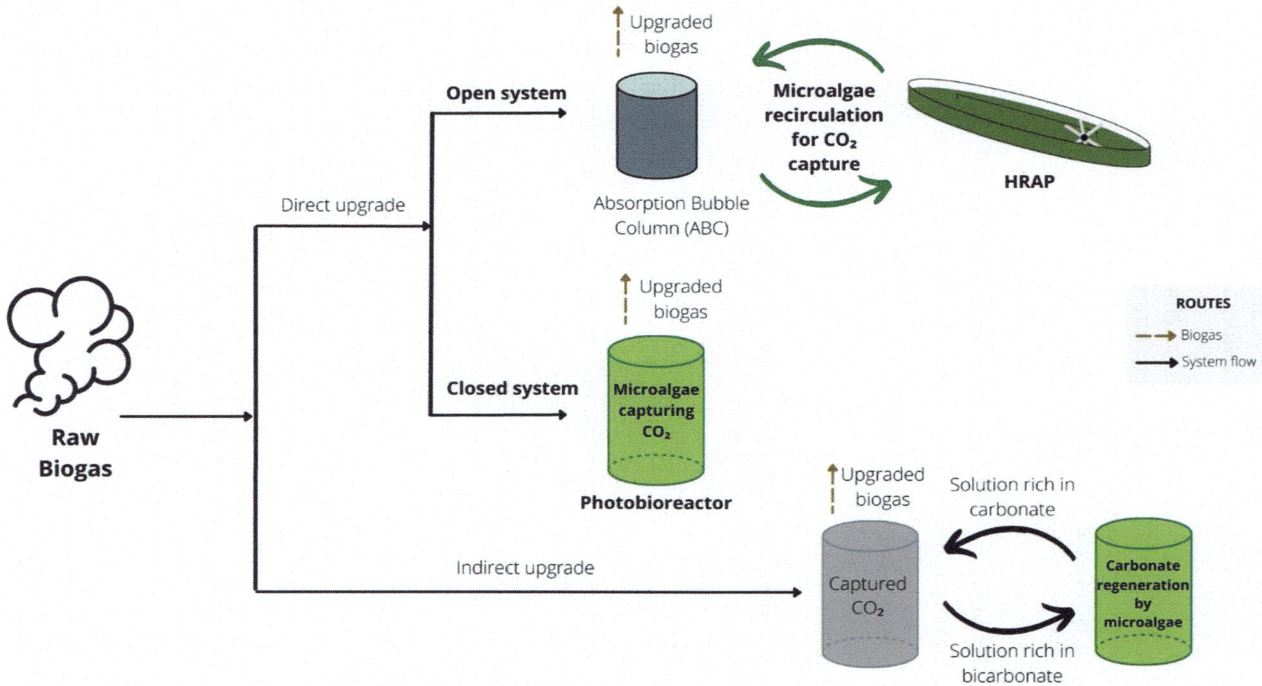

Figure 8.4 Biogas upgrading routes through algal technology. HRAP: high-rate algal ponds.

Notably, these systems are limited to only specific microalgal species that can tolerate an environment with strong ion concentrations and high alkali levels.

In a recent study removing H_2S and NH_4^+ by microalgae was also approached (Xie *et al.*, 2023). The authors could upgrade biogas while recovering N and P using microalgae treatment. They used *Chlorella vulgaris* in closed photobioreactors containing synthetic swine manure digestate. Different biogas-to-digestate liquid feed ratios were investigated to achieve a ratio that would maximize both CH_4 production and nutrient recovery. The authors achieved simultaneous biogas decarbonization and desulfurization with a 1:1 to 40:1 biogas-to-digestate ratio range, and nearly all CH_4 remained in the upgraded biogas. NH_4^+ was removed at higher biogas-to-digestate ratios. This finding demonstrates that the proposed system is suitable for treating high H_2S and NH_4^+ concentrations, both common contaminants from biomass processing units.

8.5 COUPLING TECHNOLOGIES FOR SUSTAINABLE BIOREFINERIES

Sustainable biomass conversion into a wide range of biobased products (food, feed, chemicals, and materials) and bioenergy (biofuels, power, and heat) is known as biorefinery (de Jong *et al.*, 2012). A biorefinery is usually associated with products with high environmental, social, and economic sustainability levels. According to Trivedi *et al.* (2015), a new biorefinery approach must integrate industry and the environment, improving resource use and minimizing the ecological footprint of the entire system.

Biorefineries integrate different processes into the same installation (physical, chemical, thermochemical, or biotechnological) to obtain a wide product range. There are many biomass types and possible combinations between platforms and end products, with the flexibility of a biorefinery being a key feature in incorporating new processes into existing facilities (Pascual *et al.*, 2015). As mentioned in Sections 8.2, 8.3, and 8.4, several studies have proposed coupling microalgae biotechnology with AD in many different ways. A biorefinery could be established if those multiple pathways are integrated (Figure 8.5). However, environmental, energy, and economic sustainability aspects must be better understood, requiring more effort in future research.

8.5.1 Biorefinery based on integrated microalgae and AD technologies

As stated in Section 8.2, UASB reactors are commonly used during wastewater treatment and can be integrated with microalgae cultivation for domestic sewage polishing. In this scenario, the microalgae biomass produced in the HRAP can be used as an anaerobic substrate in the UASB reactor, consisting of an AcoD between domestic wastewater and algal biomass. For a population of 20,000 inhabitants, this configuration provided an energy surplus between 0.15 and 0.32 KWh/m^3, and revenue between 10,321.89 and 21,822.60 USD/year, indicating the UASB reactor energy sustainability associated with HRAP (Gonçalves *et al.*, 2020). In addition, the energy production ranged from 70 to 180% more than consumed and could be applied in the WWTP or the neighboring community (Vassalle *et al.*, 2020b).

In the agroindustrial context, wastewater treatment based on microalgae tertiary treatment can also be interconnected to an anaerobic digester for bioenergy and biofertilizer production from sludge and microalgae AcoD. Avila *et al.* (2022) evaluating a circular bioeconomy model for nutrient and energy recovery from winery wastewater, highlighted the secondary sludge and algal biomass AcoD as a strategy to increase CH_4 yield and the importance of using other bioproducts from this route to reduce fertilizer costs. A virtual model proposed by Siqueira *et al.* (2022) to anaerobically treat vinasse integrated with microalgae biotechnology presented an electricity surplus of +14.49 MJ/m^3 of vinasse and a positive net energy ratio equal to 2, establishing a better integration of WWTPs and biorefineries.

8.5.2 Environmental impacts of integrated microalgae and AD technologies

Regarding environmental sustainability, the life-cycle assessment (LCA) is a powerful tool for measuring new processes, technology, or product impacts on the environment (Marangon *et al.*,

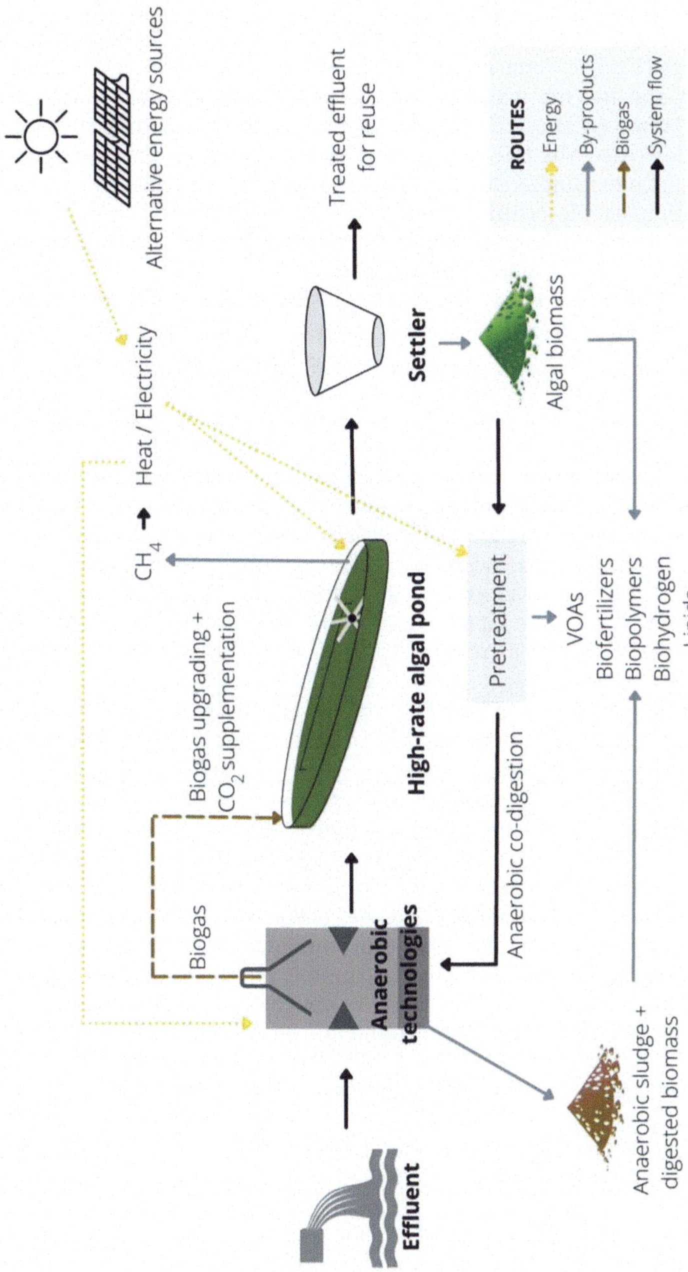

Figure 8.5 Sustainable biorefinery proposal by integrating microalgae and AD biotechnologies. VOA: volatile organic acids.

2022). Arashiro *et al.* (2018) performed an LCA of different systems, including a scenario with a HRAP followed by an energy recovery system through AcoD of the algal biomass with primary sludge. The authors emphasized the importance of AcoD in electric energy generation and the positive impact of this system on the environment. However, considering the eutrophication potential, it was the biggest polluter, mainly due to the lower nutrient removal efficiency. The lower energy consumption compared to the activated sludge scenario presented lower environmental impacts in climate change, ozone depletion, fresh and marine water eutrophication, photochemical oxidants formation, and fossil fuels depletion. However, the HRAP construction stage, demanding more material, harmed abiotic resource depletion. Also, the HRAP was responsible for greater environmental damage due to atmospheric emissions of nitrogenous compounds. Comparing all scenarios, the authors concluded that biomass valorization as biofertilizer instead of energy valorization via AD was the most economical alternative. Although presenting the most expensive operation, this option had a lower area requirement and a greater commercialization potential.

Arashiro *et al.* (2022) analyzed the integration of microalgae and AD in two different scenarios by comparing microalgae-based systems for wastewater treatment and bioproducts generation. The first was a HRAP followed by a closed photobioreactor treating domestic wastewater with biogas recovery after microalgae biomass AD. The second was characterized by an UASB reactor followed by a HRAP treating food industry wastewater with biogas recovery from the UASB reactor. Both scenarios included the recovery of other bioproducts, that is natural pigments and biofertilizer. The authors concluded that the wastewater type was the most decisive parameter in the LCA, as the second scenario presented lower environmental impacts in 8 of 10 categories. Compared to domestic wastewater, industrial wastewater resulted in lower air emissions due to lower NH_4^+ concentrations, higher biogas production, and lower heavy metal concentrations in the digestate.

Tua *et al.* (2021) investigated improvements in an existing municipal WWTP integrating a microalgal cultivation unit with the AD of the produced biomass. Microalgae were cultivated in the centrate from sludge dewatering and with CO_2 supplementation from flue gas of the combined heat and power unit. The biomass was separated in a settler and sent to AD for extra biogas production. Among the environmental indicators, the proposed system improved 7 of 15 indicators, mainly due to the electric energy generation. However, the system had negative environmental impacts, mainly due to nitrogenous compounds released into the environment, impacting particulate matter, terrestrial and marine acidification, and eutrophication categories. Another effect of the new proposed system was related to human toxicity, linked to residual biomass that can generate an environmental burden after co-incineration and subsequent disposal in landfills (carcinogenic toxicity). The non-carcinogenic toxicity was linked to the zinc (Zn) contribution to soil pollution when using biomass for agricultural purposes.

Alternatively, microalgae biotechnology can be used within the source-separated nutrient approach. Li *et al.* (2022) proposed a scenario that municipal wastewater and human urine were placed in different modules for microalgal cultivation coupled to struvite and biofuel production (heat, electricity, bio-oil, biogas, and biochar). The authors concluded that separating nutrients by urine precipitation was essential for the system's environmental sustainability, regardless of cost.

8.5.3 Insights for improving the sustainability performance of integrated microalgae and AD technologies

After considering different proposals from the literature, the main sustainability aspects of integrated microalgae and AD technologies can be highlighted (Figure 8.5). The beneficial use of digestate and residual biomass after AD as a nutrient source in microalgae cultivation and a valuable by-product is essential (Bussa *et al.*, 2020). Otherwise, it will be considered an emission to the environment, causing pollution. In that way, when valorizing the digestate as a biofertilizer, heavy metal recovery before soil application should be considered (Arashiro *et al.*, 2022). Emissions during microalgae cultivation are another point of interest, especially N emissions. Thus, pH control in the HRAP and CO_2 supplementation may be an alternative to minimize the negative impacts of ammonia volatilization (Tua *et al.*, 2021). The

CO_2 source for C supplementation during microalgae growth is also critical (Bussa *et al.*, 2020). In that way, the biogas upgrading through CO_2 bio-assimilation in the cultivation reactor may represent an environmental and economic benefit (see Section 8.4). Biogas production improvement is highly appreciated to increase the system energy yield. Besides AcoD, biomass pretreatment (see Section 8.3) before AD can be a good option to improve CH_4 production and the system's energy feasibility (Xiao *et al.*, 2020). In addition, renewable energy sources integration, such as solar energy, should be considered to reduce impacts related to electricity consumption (Arashiro *et al.*, 2022).

Lastly, AD will become an important technology for future biorefinery development. The process is already used as an auxiliary technology to recover waste streams. However, its use as a leading technology should be promoted. The challenge is to rethink existing biogas plants and expand their range of final products, going much further than selling electricity (Pascual *et al.*, 2015). For example, other bioproducts can be obtained: (1) biopolymers, bioalcohol, and medium-chain fatty acids through the VOAs platform and (2) biofertilizers, such as struvite and NH_4^+ salts via the digestate platform.

Rajendran and Murthy (2019) stated that acquisition of raw materials, plant operation aspects, and modernization costs are the major uncertainties regarding LCA and economic assessment for biogas production. Also, operational capacity and energy efficiency are the ones that most impact the system's economic performance (Aui *et al.*, 2019). Thus, proposals for biorefineries and technology integration and their sustainability will vary depending on local characteristics. Regional specificities must be taken into account to propose routes that are more favorable within each context. It is highly appreciated that regional aptitude (mainly in economic terms) is explored, considering the market cost, public acceptance, and by-products applicability, minimizing transport and logistic costs. So, there will not be an optimal biorefinery system applicable to any case, and the various local factors involved should be considered. For example, Bussa *et al.* (2020) concluded the high potential of integrated microalgal cultivation with AD in rural regions with cattle farming and in areas with a higher degree of urbanization where large municipal WWTPs were in operation. By doing a geospatial analysis, the authors stated that low potential areas require larger transportation distances for substrates or digestates, reducing the environmental benefits while increasing the economic burden.

8.6 CHALLENGES AND FUTURE PERSPECTIVES

The management of anaerobic digestate and microalgae cultivation is a sustainable strategy from environmental and economic points of view. In this way, proper treatment is provided for this nutrient-rich by-product with a high organic load, and, at the end of the treatment, a microalgae biomass is obtained with several applications that can guarantee the overall viability of the process. Coupling these two technologies on a large scale is a possibility that has already been studied. However, it needs further research to solve some limitations due to the presence of high suspended solids concentrations and ammonia toxicity, among other factors. The challenge, especially for high-strength wastewater treatment, is the need to dilute the anaerobic effluent so that the microalgae can withstand the organic load. Considering full-scale wastewater treatment, this would be disadvantageous due to the consumption of water and inputs and the need for larger units to hold the diluted effluent. Thus, future research can focus on strategies to overcome this bottleneck, for example, by combining two complementary effluents. In addition, there are still opportunities to evaluate the performance of microalgae technologies to treat micropollutants recently attracting attention, such as pharmaceuticals, endocrine disruptors and microplastics.

Regarding AD's technical limitations, pretreatment methods are recommended in further studies, especially those classified as energy efficient or associated with renewable energy sources. In addition, the co-production of CH_4, H_2, and biodiesel, together with other valuable by-products and the AcoD of microalgae and other biomass types deserve continuous efforts. Furthermore, these techniques can be performed concurrently and applied to WWTPs, making the microalgae AD energy recovery more attractive. Despite all the technological advances, upgrading biogas through microalgae needs further research studies to make it feasible on a larger scale. Finally, coupling AD and microalgae technologies could be an affordable way to encourage a biobased green circular economy model, able to improve microalgae biomass production, nutrient recovery, wastewater treatment, and biogas upgrading.

REFERENCES

Al-Mallahi J. and Ishii K. (2022). Attempts to alleviate inhibitory factors of anaerobic digestate for enhanced microalgae cultivation and nutrients removal: a review. *Journal of Environmental Management*, **304**, 114266, 10.1016/j.jenvman.2021.114266

Angelidaki I., Treu L., Tsapekos P., Luo G., Campanaro S., Wenzel H. and Kougias P. G. (2018). Biogas upgrading and utilization: current status and perspectives. *Biotechnology Advances*, **36**, 452–466, 10.1016/j.biotechadv.2018.01.011

Arashiro L. T., Montero N., Ferrer I., Acién F. G., Gómez C. and Garfí M. (2018). Life cycle assessment of high rate algal ponds for wastewater treatment and resource recovery. *Science of the Total Environment*, **622–623**, 1118–1130, 10.1016/j.scitotenv.2017.12.051

Arashiro L. T., Josa I., Ferrer I., van Hulle S. W. H., Rousseau D. P. L. and Garfí M. (2022). Life cycle assessment of microalgae systems for wastewater treatment and bioproducts recovery: natural pigments, biofertilizer and biogas. *Science of the Total Environment*, **847**, 157615, 10.1016/j.scitotenv.2022.157615

Assemany P., Marques I. d. P., Calijuri M. L., Lopes da Silva T. and Reis A. (2018). Energetic valorization of algal biomass in a hybrid anaerobic reactor. *Journal of Environmental Management*, **209**, 308–315, 10.1016/j.jenvman.2017.12.054

Assemany P., de Paula Marques I., Calijuri M. L. and Reis A. (2020). Complementarity of substrates in anaerobic digestion of wastewater grown algal biomass. *Waste Biomass Valorization*, **11**, 5759–5770, 10.1007/s12649-019-00875-8

Assis L. R., Calijuri M. L., do Couto E. d. A. and Assemany P. P. (2017). Microalgal biomass production and nutrients removal from domestic sewage in a hybrid high-rate pond with biofilm reactor. *Ecological Engineering*, **106**, 191–199, 10.1016/j.ecoleng.2017.05.040

Assis D., Lúcia M., Peixoto P., Saleme A., Paula A. de., Assis T. C. de., Calijuri M. L., Assemany P. P., Pereira A. S. A. de. P. and Martins M. A. (2019). Using atmospheric emissions as CO_2 source in the cultivation of microalgae: productivity and economic viability. *The Journal of Cleaner Production*, **215**, 1160–1169, 10.1016/j.jclepro.2019.01.093

Assis L. R. de., Calijuri M. L., Assemany P. P., Silva T. A. and Teixeira J. S. (2020). Innovative hybrid system for wastewater treatment: high-rate algal ponds for effluent treatment and biofilm reactor for biomass production and harvesting. *Journal of Environmental Management*, **274**, 111183, 10.1016/j.jenvman.2020.111183

Aui A., Li W. and Wright M. M. (2019). Techno-economic and life cycle analysis of a farm-scale anaerobic digestion plant in Iowa. *Waste Management*, **89**, 154–164, 10.1016/j.wasman.2019.04.013

Avila R., Justo Á, Carrero E., Crivillés E., Vicent T. and Blánquez P. (2022). Water resource recovery coupling microalgae wastewater treatment and sludge co-digestion for bio-wastes valorisation at industrial pilot-scale. *Bioresource Technology*, **343**, 126080, 10.1016/j.biortech.2021.126080

Barreiro-Vescovo S., Barbera E., Bertucco A. and Sforza E. (2020). Integration of microalgae cultivation in a biogas production process from organic municipal solid waste: from laboratory to pilot scale. *ChemEngineering*, **4**, 25, 10.3390/chemengineering4020025

Barros R., Raposo S., Morais E. G., Rodrigues B., Afonso V., Gonçalves P., Marques J., Cerqueira P. R., Varela J., Teixeira M. R. and Barreira L. (2022). Biogas production from microalgal biomass produced in the tertiary treatment of urban wastewater: assessment of seasonal variations. In *Energies*, **15**(15), MDPI AG, p. 5713, https://doi.org/10.3390/en15155713

Barzee T. J., Yothers C., Edalati A., Rude K., Chio A., el Mashad H. M., Franz A. and Zhang R. (2022). Pilot microalgae cultivation using food waste digestate with minimal resource inputs. *Bioresource Technology Reports*, **19**, 101200, 10.1016/j.biteb.2022.101200

Benett C. G. S., Buzetti S., Silva K. S., Bergamaschine A. F. and Fabricio J. A. (2008). Produtividade e composição bromatológica do capim-marandu a fontes e doses de nitrogênio. *Ciência e Agrotecnologia*, **32**, 1629–1636, 10.1590/S1413-70542008000500041

Bose A., Lin R., Rajendran K., O'Shea R., Xia A. and Murphy J. D. (2019). How to optimise photosynthetic biogas upgrading: a perspective on system design and microalgae selection. In *Biotechnology Advances*, **37**(8), Elsevier BV, p. 107444, https://doi.org/10.1016/j.biotechadv.2019.107444

Bose A., O'Shea R., Lin R. and Murphy J. D. (2020). A perspective on novel cascading algal biomethane biorefinery systems. *Bioresource Technology*, **304**, 123027, 10.1016/j.biortech.2020.123027

Bussa M., Zollfrank C. and Röder H. (2020). Life-cycle assessment and geospatial analysis of integrating microalgae cultivation into a regional economy. *The Journal of Cleaner Production*, **243**, 118630, 10.1016/j.jclepro.2019.118630

Calijuri M. L., Silva T. A., Magalhães I. B., Pereira A. S. A. d. P., Marangon B. B., Assis L. R. d. and Lorentz J. F. (2022). Bioproducts from microalgae biomass: technology, sustainability, challenges and opportunities. *Chemosphere*, **305**, 135508, 10.1016/j.chemosphere.2022.135508

Chen Y., Ho S.-H., Nagarajan D., Ren N. and Chang J.-S. (2018). Waste biorefineries – integrating anaerobic digestion and microalgae cultivation for bioenergy production. *Current Opinionin Biotechnology*, **50**, 101–110, 10.1016/j.copbio.2017.11.017

Chernicharo C. A. L. (2006). Post-treatment options for the anaerobic treatment of domestic wastewater. *Reviews in Environmental Science and Biotechnology*, **5**, 73–92, 10.1007/s11157-005-5683-5

Chernicharo C. A. de, Ribeiro T. B., Pegorini E. S., Possetti G. R. C., Miki M. K. and Souza S. N. de. (2018). Contribuição para o aprimoramento de projeto, construção e operação de reatores UASB aplicados ao tratamento de esgoto sanitário – Parte 1: Tópicos de Interesse. *Revista DAE*, **66**, 5–16, 10.4322/dae.2018.038

Chong C. C., Cheng Y. W., Ishak S., Lam M. K., Lim J. W., Tan I. S., Show P. L. and Lee K. T. (2022). Anaerobic digestate as a low-cost nutrient source for sustainable microalgae cultivation: a way forward through waste valorization approach. *Science of the Total Environment*, **803**, 150070, 10.1016/j.scitotenv.2021.150070

Choudhary P., Assemany P. P., Naaz F., Bhattacharya A., Castro J. d. S., Couto E. d. A. d. C., Calijuri M. L., Pant K. K. and Malik A. (2020). A review of biochemical and thermochemical energy conversion routes of wastewater grown algal biomass. *Science of the Total Environment*, **726**, 137961, 10.1016/j.scitotenv.2020.137961

CONAMA – Conselho Nacional do Meio Ambiente (2011). Resolução N° 430, De 13 De Maio De 2011 8.

Couto E., Calijuri M. L. and Assemany P. (2020). Biomass production in high rate ponds and hydrothermal liquefaction: wastewater treatment and bioenergy integration. *Science of the Total Environment*, **724**, 138104, 10.1016/j.scitotenv.2020.138104

Couto E., Calijuri M. L., Assemany P. and Cecon P. R. (2021). Evaluation of high rate ponds operational and design strategies for algal biomass production and domestic wastewater treatment. *Science of the Total Environment*, **791**, 148362, 10.1016/j.scitotenv.2021.148362

Craggs R., Sutherland D. and Campbell H. (2012). Hectare-scale demonstration of high rate algal ponds for enhanced wastewater treatment and biofuel production. *The Journal of Applied Phycology*, **24**, 329–337, 10.1007/s10811-012-9810-8

Damtie M. M., Shin J., Jang H. M. and Kim Y. M. (2020). Synergistic co-digestion of microalgae and primary sludge to enhance methane yield from temperature-phased anaerobic digestion. *Energies (Basel)*, **13**, 4547, 10.3390/en13174547

Dang B.-T., Nguyen T.-T., Ngo H. H., Pham M.-D.-T., Le L. T., Nguyen N.-K.-Q., Vo T.-D.-H., Varjani S., You S.-J., Lin K. A., Huynh K.-P.-H. and Bui X.-T. (2022). Influence of C/N ratios on treatment performance and biomass production during co-culture of microalgae and activated sludge. *Science of the Total Environment*, **837**, 155832, 10.1016/j.scitotenv.2022.155832

de Godos I., Arbib Z., Lara E. and Rogalla F. (2016). Evaluation of high rate algae ponds for treatment of anaerobically digested wastewater: effect of CO_2 addition and modification of dilution rate. *Bioresource Technology*, **220**, 253–261, 10.1016/j.biortech.2016.08.056

de Jong E., Higson A., Walsh P. and Wellisch M. (2012). Bio-based chemicals value added products from biorefineries.

de la Lama-Calvente D., Cubero J., José Fernández-Rodríguez M., Jiménez-Rodríguez A. and Borja R. (2022). Anaerobic co-digestion of microalgae and industrial wastes: a critical and bibliometric review, in: progress in microalgae research – a path for shaping sustainable futures. IntechOpen. https://doi.org/10.5772/intechopen.104378

de Oliveira M. C., Bassin I. D. and Cammarota M. C. (2022). Microalgae and cyanobacteria biomass pretreatment methods: a comparative analysis of chemical and thermochemical pretreatment methods aimed at methane production. *Fermentation*, **8**, 497, 10.3390/fermentation8100497

Fallahi A., Rezvani F., Asgharnejad H., Khorshidi Nazloo E., Hajinajaf N. and Higgins B. (2021). Interactions of microalgae-bacteria consortia for nutrient removal from wastewater: a review. *Chemosphere*, **272**, 129878, 10.1016/j.chemosphere.2021.129878

Franco-Morgado M., Tabaco-Angoa T., Ramírez-García M. A. and González-Sánchez A. (2021). Strategies for decreasing the O_2 content in the upgraded biogas purified via microalgae-based technology. *Journal of Environmental Management*, **279**, 111813, 10.1016/j.jenvman.2020.111813

Fu J., Yan B., Gui S., Fu Y. and Xia S. (2023). Anaerobic co-digestion of thermo-alkaline pretreated microalgae and sewage sludge: methane potential and microbial community. *Journal of Environmental Sciences*, **127**, 133–142, 10.1016/j.jes.2022.04.020

Ganesh Saratale R., Kumar G., Banu R., Xia A., Periyasamy S. and Dattatraya Saratale G. (2018). A critical review on anaerobic digestion of microalgae and macroalgae and co-digestion of biomass for enhanced methane generation. *Bioresource Technology*, **262**, 319–332, 10.1016/j.biortech.2018.03.030

Golueke C. G., Oswald W. J. and Gotaas H. B. (1957). Anaerobic digestion of algae. *Applied Microbiology*, **5**, 47–55, 10.1128/am.5.1.47-55.1957

Gonçalves R. F., Assis T. I., Maciel G. B., Borges R. M. and Cassini S. T. A. (2020). Co-digestion of municipal wastewater and microalgae biomass in an upflow anaerobic sludge blanket reactor. In *Algal Research*, **52**, Elsevier BV, p. 102117, https://doi.org/10.1016/j.algal.2020.102117

Greene J. P. (2019). Degradation and biodegradation standards for biodegradable food packaging materials, in: reference module in food science. Elsevier. https://doi.org/10.1016/B978-0-08-100596-5.22437-2

Jiang R., Qin L., Feng S., Huang D., Wang Z. and Zhu S. (2021). The joint effect of ammonium and pH on the growth of *Chlorella vulgaris* and ammonium removal in artificial liquid digestate. *Bioresource Technology*, **325**, 124690, 10.1016/j.biortech.2021.124690

Kabir S. bin, Khalekuzzaman M., Hossain N., Jamal M., Alam M. A. and Abomohra A. E.-F. (2022). Progress in biohythane production from microalgae-wastewater sludge co-digestion: an integrated biorefinery approach. *Biotechnology Advances*, **57**, 107933, 10.1016/j.biotechadv.2022.107933

Karuppiah T. and Ebenezer Azariah V. (2019). Biomass pretreatment for enhancement of biogas production, in: anaerobic digestion. IntechOpen. https://doi.org/10.5772/intechopen.82088

Khan A. A., Gaur R. Z., Tyagi V. K., Khursheed A., Lew B., Mehrotra I. and Kazmi A. A. (2011). Sustainable options of post treatment of UASB effluent treating sewage: a review. *Resources, Conservation and Recycling*, **55**, 1232–1251, 10.1016/j.resconrec.2011.05.017

Khan M. U., Lee J. T. E., Bashir M. A., Dissanayake P. D., Ok Y. S., Tong Y. W., Shariati M. A., Wu S. and Ahring B. K. (2021). Current status of biogas upgrading for direct biomethane use: a review. *Renewable and Sustainable Energy Reviews*, **149**, 111343, 10.1016/j.rser.2021.111343

Kimura S., Yamada T., Ban S., Koyama M. and Toda T. (2019). Nutrient removal from anaerobic digestion effluents of aquatic macrophytes with the green alga, *Chlorella sorokiniana*. *Biochemical Engineering Journal*, **142**, 170–177, 10.1016/j.bej.2018.12.001

Krishnamoorthy S., Manickam P. and Muthukaruppan V. (2019). Evaluation of distillery wastewater treatability in a customized photobioreactor using blue-green microalgae – laboratory and outdoor study. *Journal of Environmental Management*, **234**, 412–423 https://doi.org/10.1016/j.jenvman.2019.01.014

Kwietniewska E. and Tys J. (2014). Process characteristics, inhibition factors and methane yields of anaerobic digestion process, with particular focus on microalgal biomass fermentation. *Renewable and Sustainable Energy Reviews*, **34**, 491–500, 10.1016/j.rser.2014.03.041

Leng L., Zhang W., Leng S., Chen J., Yang L., Li H., Jiang S. and Huang H. (2020). Bioenergy recovery from wastewater produced by hydrothermal processing biomass: progress, challenges, and opportunities. *Science of The Total Environment*, **748**, 142383, 10.1016/j.scitotenv.2020.142383

Leong Y. K. and Chang J.-S. (2022). Integrated role of algae in the closed-loop circular economy of anaerobic digestion. *Bioresource Technology*, **360**, 127618, 10.1016/j.biortech.2022.127618

Li S., Li X. and Ho S.-H. (2022). Microalgae as a solution of Third World energy crisis for biofuels production from wastewater toward carbon neutrality: an updated review. *Chemosphere*, **291**, 132863, 10.1016/j.chemosphere.2021.132863

Lian J., Wijffels R. H., Smidt H. and Sipkema D. (2018). The effect of the algal microbiome on industrial production of microalgae. *Microbial Biotechnology*, **11**, 806–818, 10.1111/1751-7915.13296

Lopatkin A. J. and Collins J. J. (2020). Predictive biology: modelling, understanding and harnessing microbial complexity. *Nature Reviews Microbiology*, **18**, 507–520, 10.1038/s41579-020-0372-5

Lu R., Yan H., Liu Y., Wang Y., Cui X., Wu X., Yu Z., Ruan R. and Zhang Q. (2022). Enhancement of nutrients recovery and cell metabolism in piggery anaerobic digestate by the co-cultivation of indigenous microalgae and bacteria. *The Journal of Cleaner Production*, **375**, 134193, 10.1016/j.jclepro.2022.134193

Magalhães I. B., Ferreira J., de Siqueira Castro J., de Assis L. R. and Calijuri M. L. (2022). Agro-industrial wastewater-grown microalgae: a techno-environmental assessment of open and closed systems. Science of the Total Environment, 155282, 10.1016/j.scitotenv.2022.155282

Marangon B. B., Castro J. S., Assemany P. P., Couto E. A. and Calijuri M. L. (2022). Environmental performance of microalgae hydrothermal liquefaction: life cycle assessment and improvement insights for a sustainable renewable diesel. *Renewable and Sustainable Energy Reviews*, **155**, 111910, 10.1016/j.rser.2021.111910

Marcilhac C., Sialve B., Pourcher A.-M., Ziebal C., Bernet N. and Béline F. (2014). Digestate color and light intensity affect nutrient removal and competition phenomena in a microalgal-bacterial ecosystem. *Water Research*, **64**, 278–287, 10.1016/j.watres.2014.07.012

Miyawaki B., Mariano A. B., Vargas J. V. C., Balmant W., Defrancheschi A. C., Corrêa D. O., Santos B., Selesu N. F. H., Ordonez J. C. and Kava V. M. (2021). Microalgae derived biomass and bioenergy production enhancement through biogas purification and wastewater treatment. *Renewable Energy*, **163**, 1153–1165, 10.1016/j.renene.2020.09.045

Morais N. W. S. and Santos A. B. dos. (2019). Análise dos padrões de lançamento de efluentes em corpos hídricos e de reúso de águas residuárias de diversos estados do Brasil. *Revista DAE*, **67**, 40–55, 10.4322/dae.2019.004

Nguyen L. N., Kumar J., Vu M. T., Mohammed J. A. H., Pathak N., Commault A. S., Sutherland D., Zdarta J., Tyagi V. K. and Nghiem L. D. (2021). Biomethane production from anaerobic co-digestion at wastewater treatment plants: a critical review on development and innovations in biogas upgrading techniques. *Science of the Total Environment*, **765**, 142753, 10.1016/j.scitotenv.2020.142753

Oliveira A. P. de S., Assemany P., Ribeiro Júnior J. I., Covell L., Nunes-Nesi A. and Calijuri M. L. (2021). Swine wastewater treatment in high rate algal ponds: effects of Cu and Zn on nutrient removal, productivity and biomass composition. *Journal of Environmental Management*, **299**, 113668, 10.1016/j.jenvman.2021.113668

Oliveira A. P. de S., Assemany P., Covell L., Tavares G. P. and Calijuri M. L. (2023). Microalgae-based wastewater treatment for micropollutant removal in swine effluent: high-rate algal ponds performance under different zinc concentrations. *Algal Research*, **69**, 102930, 10.1016/j.algal.2022.102930

Oss R. N., Gonçalves R. F., Cassini S. T., Junior M. Â. S., Cipriano D. F. and de Freitas J. C. C. (2022). Single step production of activated carbon from microalgae cultivated with urban wastewater. *Algal Research*, **64**, 102669, 10.1016/j.algal.2022.102669

Pacheco M. M., Hoeltz M., Moraes M. S. A. and Schneider R. C. S. (2015). Microalgae: cultivation techniques and wastewater phycoremediation. *The Journal of Environmental Science and Health, Part A, Toxic/Hazardous Substances and Environmental Engineering*, **50**, 585–601, 10.1080/10934529.2015.994951

Park J. B. K. and Craggs R. J. (2010). Wastewater treatment and algal production in high rate algal ponds with carbon dioxide addition. *Water Science and Technology*, **61**, 633–639, 10.2166/wst.2010.951

Pascual A., Regueiro L. and Silvestre G. (2015). Evolución de las plantas de biogas agroindustrial hacia nuevos modelos basados en el concepto de biorefinería. *Retema: Revista Técnica de Medio Ambiente*, **185**, 34–40.

Patel A., Mahboubi A., Horváth I. S., Taherzadeh M. J., Rova U., Christakopoulos P. and Matsakas L. (2021). Volatile fatty acids (VFAs) generated by anaerobic digestion serve as feedstock for freshwater and marine oleaginous microorganisms to produce biodiesel and added-value compounds. *Frontiers in Microbiology*, **12**, 10.3389/fmicb.2021.614612

Praveen P., Guo Y., Kang H., Lefebvre C. and Loh K.-C. (2018). Enhancing microalgae cultivation in anaerobic digestate through nitrification. *Chemical Engineering Journal*, **354**, 905–912, 10.1016/j.cej.2018.08.099

Rajendran K. and Murthy G. S. (2019). Techno-economic and life cycle assessments of anaerobic digestion – a review. *Biocatalysis and Agricultural Biotechnology*, **20**, 101207, 10.1016/j.bcab.2019.101207

Santiago A. F., Calijuri M. L., Assemany P. P., Calijuri M. do C. and Reis A. J. D. dos (2013). Algal biomass production and wastewater treatment in high rate algal ponds receiving disinfected effluent. *Environmental Technology*, **34**, 1877–1885, 10.1080/09593330.2013.812670

Seelam J. S., Fernandes de Souza M., Chaerle P., Willems B., Michels E., Vyverman W. and Meers E. (2022). Maximizing nutrient recycling from digestate for production of protein-rich microalgae for animal feed application. *Chemosphere*, **290**, 133180, 10.1016/j.chemosphere.2021.133180

Sikarwar V. S., Pohořelý M., Meers E., Skoblia S., Moško J. and Jeremiáš M. (2021). Potential of coupling anaerobic digestion with thermochemical technologies for waste valorization. *Fuel*, **294**, 120533, 10.1016/j.fuel.2021.120533

Siqueira J. C., Braga M. Q., Ázara M. S., Garcia K. J., Alencar S. N. M., Ramos T. S., Siniscalchi L. A. B., Assemany P. P. and Ensinas A. V. (2022). Recovery of vinasse with combined microalgae cultivation in a conceptual energy-efficient industrial plant: analysis of related process considerations. *Renewable and Sustainable Energy Reviews*, **155**, 111904, 10.1016/j.rser.2021.111904

Solé-Bundó M., Garfí M., Matamoros V. and Ferrer I. (2019a). Co-digestion of microalgae and primary sludge: effect on biogas production and microcontaminants removal. *Science of the Total Environment*, **660**, 974–981, 10.1016/j.scitotenv.2019.01.011

Solé-Bundó M., Passos F., Romero-Güiza M. S., Ferrer I. and Astals S. (2019b). Co-digestion strategies to enhance microalgae anaerobic digestion: a review. *Renewable and Sustainable Energy Reviews*, **112**, 471–482, 10.1016/j.rser.2019.05.036

Srivastava G., Kumar V., Tiwari R., Patil R., Kalamdhad A. and Goud V. (2022). Anaerobic co-digestion of defatted microalgae residue and rice straw as an emerging trend for waste utilization and sustainable biorefinery development. *Biomass Conversion and Biorefinery*, **12**, 1193–1202, 10.1007/s13399-020-00736-8

Sutherland D. L., Turnbull M. H. and Craggs R. J. (2017). Environmental drivers that influence microalgal species in fullscale wastewater treatment high rate algal ponds. *Water Research*, **124**, 504–512, 10.1016/j.watres.2017.08.012

Thi Nguyen M.-L., Lin C.-Y. and Lay C.-H. (2019). Microalgae cultivation using biogas and digestate carbon sources. *Biomass Bioenergy*, **122**, 426–432, 10.1016/j.biombioe.2019.01.050

Toro-Huertas E. I., Franco-Morgado M., de los Cobos Vasconcelos D. and González-Sánchez A. (2019). Photorespiration in an outdoor alkaline open-photobioreactor used for biogas upgrading. *Science of the Total Environment*, **667**, 613–621, 10.1016/j.scitotenv.2019.02.374

Trivedi J., Aila M., Bangwal D. P., Kaul S. and Garg M. O. (2015). Algae based biorefinery – How to make sense? *Renewable and Sustainable Energy Reviews*, **47**, 295–307, 10.1016/j.rser.2015.03.052

Tua C., Ficara E., Mezzanotte V. and Rigamonti L. (2021). Integration of a side-stream microalgae process into a municipal wastewater treatment plant: a life cycle analysis. *Journal of Environmental Management*, **279**, 111605, 10.1016/j.jenvman.2020.111605

Uggetti E., Sialve B., Latrille E. and Steyer J.-P. (2014). Anaerobic digestate as substrate for microalgae culture: the role of ammonium concentration on the microalgae productivity. *Bioresource Technology*, **152**, 437–443, 10.1016/j.biortech.2013.11.036

Vassalle L., Díez-Montero R., Machado A. T. R., Moreira C., Ferrer I., Mota C. R. and Passos F. (2020a). Upflow anaerobic sludge blanket in microalgae-based sewage treatment: co-digestion for improving biogas production. *Bioresource Technology*, **300**, 122677, 10.1016/j.biortech.2019.122677

Vassalle L., García-Galán M. J., Aquino S. F., Afonso R. J. d. C. F., Ferrer I., Passos F. and R Mota C. (2020b). Can high rate algal ponds be used as post-treatment of UASB reactors to remove micropollutants? *Chemosphere*, **248**, 125969, 10.1016/j.chemosphere.2020.125969

Veerabadhran M., Natesan S., MubarakAli D., Xu S. and Yang F. (2021). Using different cultivation strategies and methods for the production of microalgal biomass as a raw material for the generation of bioproducts. *Chemosphere*, **285**, 131436, 10.1016/j.chemosphere.2021.131436

Villar-Navarro E., Baena-Nogueras R. M., Paniw M., Perales J. A. and Lara-Martín P. A. (2018). Removal of pharmaceuticals in urban wastewater: high rate algae pond (HRAP) based technologies as an alternative to activated sludge based processes. *Water Research*, **139**, 19–29, 10.1016/j.watres.2018.03.072

Woertz I., Fulton L. and Lundquist T. (2009). Nutrient Removal & Greenhouse Gas Abatement with CO_2 Supplemented Algal High Rate Ponds. WEFTEC Annual Conference, Orlando, Florida.

Xiao C., Liao Q., Fu Q., Huang Y., Chen H., Zhang H., Xia A., Zhu X., Reungsang A. and Liu Z. (2019a). A solar-driven continuous hydrothermal pretreatment system for biomethane production from microalgae biomass. *Applied Energy*, **236**, 1011–1018, 10.1016/j.apenergy.2018.12.014

Xiao C., Liao Q., Fu Q., Huang Y., Xia A., Shen W., Chen H. and Zhu X. (2019b). Exergy analyses of biogas production from microalgae biomass via anaerobic digestion. *Bioresource Technology*, **289**, 121709, 10.1016/j.biortech.2019.121709

Xiao C., Fu Q., Liao Q., Huang Y., Xia A., Chen H. and Zhu X. (2020). Life cycle and economic assessments of biogas production from microalgae biomass with hydrothermal pretreatment via anaerobic digestion. *Renewable Energy*, **151**, 70–78, 10.1016/j.renene.2019.10.145

Xie B., Gong W., Tian Y., Qu F., Luo Y., Du X., Tang X., Xu D., Lin D., Li G. and Liang H. (2018). Biodiesel production with the simultaneous removal of nitrogen, phosphorus and COD in microalgal-bacterial communities for the treatment of anaerobic digestion effluent in photobioreactors. *Chemical Engineering Journal*, **350**, 1092–1102, 10.1016/j.cej.2018.06.032

Xie T., Herbert C., Zitomer D., Kimbell L., Stafford M. and Venkiteshwaran K. (2023). Biogas conditioning and digestate recycling by microalgae: acclimation of *Chlorella vulgaris* to H_2S-containing biogas and high

NH_4-N digestate and effect of biogas: digestate ratio. *Chemical Engineering Journal*, **453**, 139788, 10.1016/j.cej.2022.139788

Yukesh Kannah R., Kavitha S., Parthiba Karthikeyan O., Rene E. R., Kumar G. and Rajesh Banu J. (2021). A review on anaerobic digestion of energy and cost effective microalgae pretreatment for biogas production. *Bioresource Technology*, **332**, 125055, 10.1016/j.biortech.2021.125055

Zabed H. M., Akter S., Yun J., Zhang G., Zhang Y. and Qi X. (2020). Biogas from microalgae: technologies, challenges and opportunities. *Renewable and Sustainable Energy Reviews*, **117**, 109503, 10.1016/j.rser.2019.109503

Zhang Y., Kang X., Wang Z., Kong X., Li L., Sun Y., Zhu S., Feng S., Luo X. and Lv P. (2018). Enhancement of the energy yield from microalgae via enzymatic pretreatment and anaerobic co-digestion. *Energy*, **164**, 400–407, 10.1016/j.energy.2018.08.124

Zhang Y., Caldwell G. S., Blythe P. T., Zealand A. M., Li S., Edwards S., Xing J., Goodman P., Whitworth P. and Sallis P. J. (2020). Co-digestion of microalgae with potato processing waste and glycerol: effect of glycerol addition on methane production and the microbial community. *RSC Advances*, **10**, 37391–37408, 10.1039/D0RA07840A

Zhang C., Li S. and Ho S.-H. (2021). Converting nitrogen and phosphorus wastewater into bioenergy using microalgae-bacteria consortia: a critical review. *Bioresource Technology*, **342**, 126056, 10.1016/j.biortech.2021.126056

Zhang H., Xu B., Zhao C., Liu J., Zhao Y., Sun S. and Wei J. (2022). Simultaneous biogas upgrading and biogas slurry treatment by different microalgae-based technologies under various strigolactone analog (GR24) concentrations. *Bioresource Technology*, **351**, 127033, 10.1016/j.biortech.2022.127033

Zhen X., Luo M., Li Z., Lin Z., Zhang Y., Feng L. and Kang J. (2022). The sustainable utilization of anaerobic digestion effluents treating in suspended filler algae assisted systems. *Sustainable Energy Technologies and Assessments*, **53**, 102354, 10.1016/j.seta.2022.102354

Zicari S., Zhang R. and Kaffka S. (2019). Sugar beet, in: Integrated processing technologies for food and agricultural by-products. Elsevier, pp. 331–351. https://doi.org/10.1016/B978-0-12-814138-0.00013-7

Zkeri E., Iliopoulou A., Katsara A., Korda A., Aloupi M., Gatidou G., Fountoulakis M. S. and Stasinakis A. S. (2021). Comparing the use of a two-stage MBBR system with a methanogenic MBBR coupled with a microalgae reactor for medium-strength dairy wastewater treatment. *Bioresource Technology*, **323**, 124629, 10.1016/j.biortech.2020.124629

doi: 10.2166/9781789063547_0171

Chapter 9

Algae for wastewater treatment and biofuel production

S. B. Velasquez-Orta

School of Engineering, Department of Chemical Engineering, Merz Court, Newcastle University, Newcastle upon Tyne, NE1 7RU, UK
Corresponding author: sharon.velasquez-orta@newcastle.ac.uk

ABSTRACT

Biofuels can be used for the provision of electricity, heating, and transport. Interest in biofuels has been sparked by their suitability to decrease carbon emissions and fossil fuel dependency without major modifications to our existing energy infrastructure. Microalgae grown in wastewater are a suitable feedstock to produce two of the most utilized types of biofuels: bioethanol and biodiesel. Biodiesel is obtained as fatty acid methyl esters from microalgae via a chemical reaction known as transesterification. Bioethanol is produced from biomass by microbial fermentation. So far, microalgae growing in wastewater has been characterized for containing a mixture of lipids, carbohydrates, and proteins. Hence, another area of interest is the use of wastewater-derived microalgae for the sequential production of bioethanol, biodiesel, and protein compounds. A biorefinery concept emerges for the generation of multiple co-products from the wastewater-derived microalgae that can maximize the use of unit operations and the valorization of microalgal biomass. In this chapter, concepts for biodiesel and bioethanol production are evaluated and a biocircular economy prospected.

Keywords: bioethanol, biodiesel, sewage, pre-treatment, fermentation, transesterification, microalgae

9.1 INTRODUCTION

Microalgae for wastewater treatment have been studied for more than 70 years. Initial observations of microalgae species were in facultative lagoons. Even if microalgae were able to subsist in lagoon treatment systems, cultivations presented challenges such as: the requirement of a large surface area (Lavoie & de la Noüe, 1985); the needed maintenance of introduced microalgae, as per replacement and succession of species (Gantar *et al.*, 1991); and the difficulty to harvest diluted algal biomass (Tredici *et al.*, 1992). Current research in microalgal growth in wastewater aims to address these challenges, and solutions have gradually been elucidated such as the growth of microalgae using high-rate algal ponds or bioreactors. The main advantage of using microalgae emerges as currently nitrogen and phosphorus are not completely removed in wastewater treatment plants and microalgae are species that can remove them once most of the carbon has been depleted. Additionally, microalgae can co-exist with other microorganisms such as bacteria and yeast allowing the treatment of wastewater with high carbon loads (Hernández-García *et al.*, 2019).

Microalgae can assimilate inorganic and organic compounds from wastewater and at the same time accumulate biomolecules of interest, for example lipids, carbohydrates, or proteins, under unfavorable environmental conditions. The generated algal biomass can be used to produce different bioproduct such as biofertilizers, biohydrogen, biodiesel, bioethanol, bio-oil; or biomaterials like biofertilizers, biopolymers or biofilms (Jebali *et al.*, 2018; Salama *et al.*, 2017). In addition, biomass can be thermally processed to produce heat and electricity (Gouveia *et al.*, 2016; Romero Villegas *et al.*, 2017). For this reason, microalgae are considered a biomass source with good bioenergetic potential.

In the bioenergy context, microalgae are classified as a third-generation feedstock offering several advantages with respect to first- and second-generation terrestrial plants. Microalgae characteristics supersede terrestrial plants, as per the high growth rate and productivity; dual photosynthetic or heterotrophic growth; high microbial carbon dioxide fixation under autotrophic growth, no dependence on fertile soil, and no compromise in food production for human consumption. Microalgae growth in wastewater fits well with its later use in bioenergy products such as biodiesel and bioethanol. The quality of the produced biodiesel and bioethanol using microalgae grown in wastewater is comparable to other biomass feedstocks or microalgae grown in synthetic medium. Additionally, the risks of using biodiesel and bioethanol are lower than their fossil fuel-derived competitors.

This chapter provides an insight in the development of wastewater-grown microalgae for biodiesel and bioethanol production. Our laboratory has been researching the optimization of microalgae growth and their transformation to energy products through process understanding, integration and intensification (Velasquez-Orta *et al.*, 2022). The conversion of microalgae into biofuels requires the selection of operations that are economically viable and environmentally friendly. Biodiesel and bioethanol are the primary biofuel products globally produced. Previous communications have reviewed the conversion of microalgae to biodiesel or bioethanol, but few take as a basis the growth of microalgae in wastewater. Processing will play a significant role in the economic feasibility of biofuel production given their low-cost commodity.

9.2 CHARACTERIZATION OF MICROALGAE GROWN IN WASTEWATER FOR BIOFUEL PRODUCTION

Microalgae growth using wastewater has gained vast attention in the last three decades. According to Science Direct publication numbers, there were six times more publications in 2022, than a decade ago. Most studies utilize a consortium of microalgae for inoculation in wastewater. Commonly inoculated strains are *Chlorella* sp., cyanobacteria, *Desmodesmus* sp. or *Scenedesmus* sp. as they have positively prevailed in wastewater. Given the non-sterile nature of wastewater, cultivations end-up being a mixture of microbial strains where usually a desired microalgae dominates a consortium. As a result, the biochemical composition of microalgal biomass segregates into different fractions rather than a high fraction of a specific compound. The fractions can be generally divided into proteins, carbohydrates, and lipids. Carbohydrates are of interest for bioethanol production, whilst neutral lipids can be transformed into biodiesel. Velasquez-Orta *et al.* (2014) reported a total of 0.3 mg lipids/mg of biomass from mixed microalgae cultures growing in a wastewater treatment lagoon. Oliveira *et al.* (2018) inoculated *Scenedesmus* sp. in wastewater, after 16 days of growth, obtaining a biomass composition of 0.2–0.3 mg lipids/mg of dry biomass, 0.2–0.3 mg carbohydrates/mg of dry biomass and 0.4 mg protein/mg of dry biomass. Hernández-García *et al.* (2019) indicated 0.4 mg lipids/mg of biomass and 0.5 mg carbohydrates/mg of biomass after microalgal cultivation under nutrient limitation conditions. As can be seen, the composition of microalgae cultivations in wastewater varies and should be monitored in wastewater treatment systems. Algae cultivations produce usable fractions for both bioethanol and biodiesel production.

9.3 BIODIESEL PRODUCTION FROM MICROALGAE GROWN IN WASTEWATER

9.3.1 Biodiesel production process

Biodiesel is defined as a mixture that contains at least 96.5%, by weight, of fatty acid methyl esters (FAME), in accordance with the EN 14214:2003 standard. FAME is derived from the conversion of the neutral lipid fraction of microalgae, known as triglycerides (TAG). The transesterification of microalgal lipids involves a chemical reaction that converts the extracted lipids into FAME and glycerol as shown in Figure 9.1. Apart from the chemical reaction, a series of unit operations are required prior and after, to obtain biodiesel from microalgae. The starting stages involve microalgae cultivation, harvesting, cell disruption and drying. Figure 9.2 provides an overview of the refining stages needed after the transesterification reaction to ensure that the FAME mixture is classified as biodiesel. Figures 9.1 and 9.2 show the production of glycerol as a by-product from the transesterification reaction. Glycerol production amounts to around 15% of the total volume and can be refined to chemical, edible or cosmetic applications. Glycerol biodegradability enables its application as feedstock for biological transformation. Elahinik *et al.* (2022) proposed the use of glycerol effluents emanating from biodiesel and epoxy resin industrial plants. The glycerol-rich wastewater was used to obtain propionate via aerobic granular sludge fermentation. Glycerol has also been converted to bioelectricity using stackable microbial fuel cells (Zhao *et al.*, 2017). Some considerations on the use of glycerol should be its price volatility as per its by-product nature and regulations on the trading of glycerol emanating from a waste fraction.

The global share of biodiesel production was 23% in 2009 and rose to 37% in 2020 (BP, 2021). Production of biodiesel from microalgae grown in wastewater has been shown to be feasible since our first publication (Komolafe *et al.*, 2014). Combining these two systems provides the benefit of wastewater bioremediation with fuel biorefining. Once microalgal biomass is available, the amount of unit operations needed for biodiesel production can be reduced through intensification (Velasquez-Orta *et al.*, 2022). For example, *in situ* transesterification can potentially combine the stages involving total drying, cell disruption, lipid extraction, and transesterification. Hence, *in situ* transesterification offers a one-step approach for cell disruption, lipid extraction, and conversion. The chemical transesterification reaction requires significantly higher amounts of alcohol (×100 times the usual

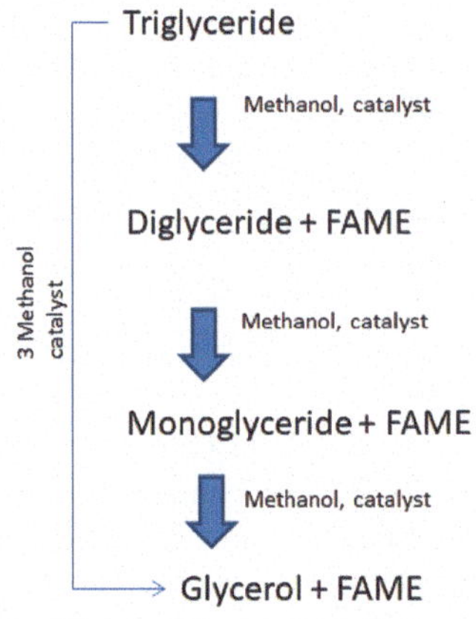

Figure 9.1 Overview of transesterification reaction to convert microalgae lipids into FAMEs.

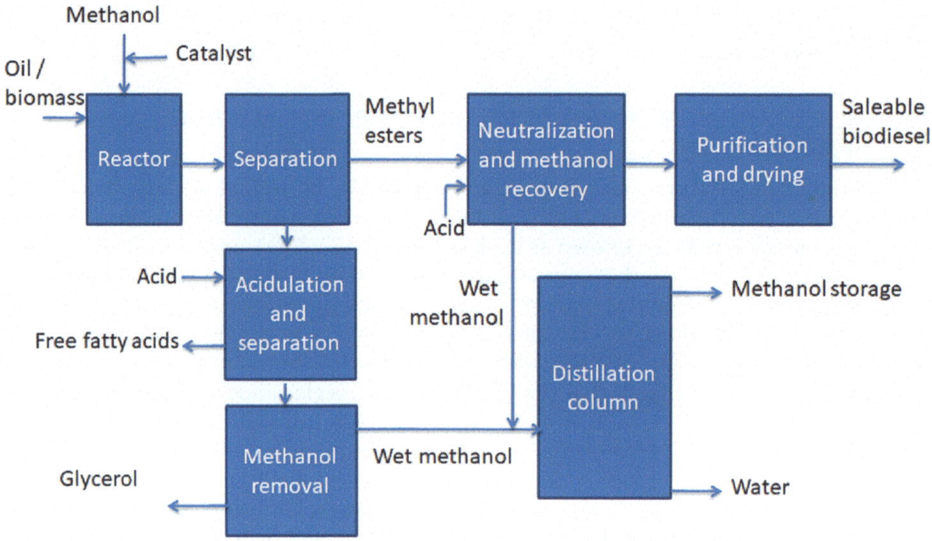

Figure 9.2 Overview of unit operations following the transesterification reaction for biodiesel production.

value), as per its dual act as both solvent lipid extraction and reactant. However, the alcohol can be recovered and reutilized. In contrast, the conventional route requires organic solvent mixtures for lipid extraction (e.g., chloroform, methanol, ethanol, hexane, or mixtures) which can be difficult to recover.

9.3.2 Types of microalgae grown in wastewater for biodiesel production

Different types of microalgae have been studied for FAME harvesting based on their lipid production. The conventional FAME microalgal fingerprint is showcased in Figure 9.3, generated from information taken from Komolafe *et al.* (2014), demonstrated the conversion of *Desmodesmus* sp. grown in wastewater into FAME via *in situ* transesterification. The highest recovery was reported as 77.6 (±2.3) wt% of FAME at a reaction time of 75 min, equivalent to 0.2 mg/mg of microalgae biomass, using a catalyst/lipid (NaOH) molar ratio of 0.15:1 and a methanol/lipid molar ratio of 600:1. Vasistha *et al.* (2023) obtained approximately 0.3 mg FAMEs per mg of *Coelastrella* sp. KJ-04 grown in distillery wastewater.

Figure 9.3 shows the common fingerprint of microalgae-derived biodiesel, the highest fractions correspond to oleic (C18:19c) and ɣ-linoleic (C18:3n6) methyl esters (20–27%), followed by steric (C18:0) and palmitoleic methyl esters (C16:1n9t). These C16 and C18 carbon chains make-up 60% of the overall FAMEs. Vasistha *et al.* (2023) also reported carbon chains C16–C18 with no more than 2 degrees of unsaturation (16–18 < 3), which seems to be a deterministic factor in FAMEs obtained from green microalgae.

9.4 BIOETHANOL PRODUCTION FROM MICROALGAE GROWN IN WASTEWATER

9.4.1 Bioethanol production process

Bioethanol is a type of biofuel with the formula: C_2H_5OH produced from fermentation of plant material with high sugar/carbohydrate content. Its overall fermentation reaction is provided in Figure 9.4 and can be simplified into two main reaction mechanisms. Reaction 1 showcases the conversion of glucose to ethanol. Reaction 2 demonstrates the glucose consumed for microbial (yeast) growth. Bioethanol is currently the highest globally consumed biofuel having an 82% share of all biofuels commercialized. Its production has attracted extensive biomass studies, lately including the use of lignocellulosic

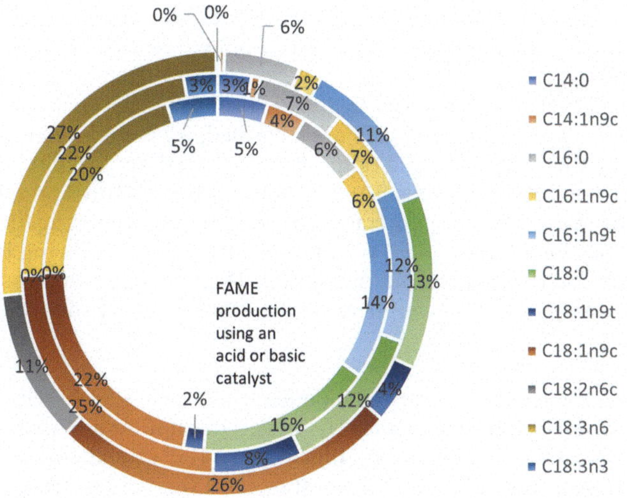

Figure 9.3 Fingerprint of FAMEs contained in microalgal biodiesel production. Percentages by weight of the types of methyl esters found. Inner circle corresponds to the lipid extraction of algae crude oil, middle circle showcases FAME production using an acid catalyst, outer circle shows FAME composition using an alkaline catalyst. (Source: Biodiesel composition obtained from Komolafe *et al.*, 2014).

biomass and fermentations via synthetic developed strains. The major producers of bioethanol are in America. The United States and Brazil had an annual increase from 34.4 to 59.7 billion litres between 2019 and 2020. The demand for bioethanol continues to grow and it is expected to increase by 9.7% in 2026. Bioethanol is a building block in the production of other chemicals and solvents. These products include drugs, plastics, lacquers, polishes, plasticizers, and cosmetics. Hence, ethanol is an essential commodity and organic chemical needed in large volumes for consumer products and industry.

Microalgae including *Chlorella*, *Dunaliella*, *Chlamydomonas*, *Scenedesmus*, and *Spirulina* have a carbohydrate content up to about 50% (w/w), which make them good candidates for bioethanol production (Chen *et al.*, 2013; Dragone *et al.*, 2011). Cultivation strategies, such as nutrient starvation, can help promote the accumulation of energy-rich compounds: carbohydrates and lipids (Hernández-García *et al.*, 2019). Most studies in the literature report bioethanol production from microalgae cultivated in synthetic medium, however, microalgae *Scenedesmus obliquus* was shown to be able to grow in wastewater more than 70 years ago (Gotaas *et al.*, 1954). *Scenedesmus* sp. has been one of the most studied species because of its ability to remove a high percentage of phosphorus (85–99%) and nitrogen (88–99%) as well as its microalgal biomass productivity between 0.073 and 0.15 g/L/d

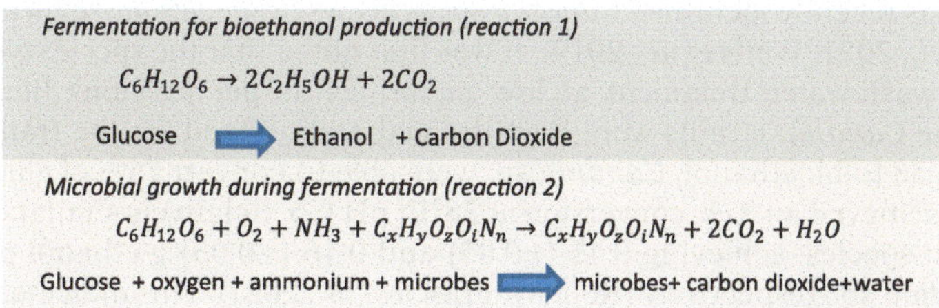

Figure 9.4 Overview of the main fermentation and growth reactions during microalgae sugars conversion to bioethanol.

(Ji *et al.*, 2015; Ruiz *et al.*, 2014; Zhang *et al.*, 2014). Hernández-García *et al.* (2019) observed that cultures of *Desmodesmus* sp. accumulated up to 41% carbohydrate by weight (and 20%w of lipid) after nutrient-limiting conditions.

Microalgal biomass is first harvested and hydrolysed to obtain fermentable sugars, which can be transformed into ethanol. Once simple sugars are obtained, a conventional fermentation process is conducted, continued by the separation and refining of the bioethanol produced (Figure 9.5). Usually, batch fermentations are conducted between 8 and 12 hours. The fermented products are then separated using a centrifuge. The wine/beer (liquid fraction) output from centrifugation is then distilled to achieve a mixture of 95% bioethanol and 5% water (Figure 9.5a). Following this, ethanol is further refined using processes such as azeotropic distillation (Figure 9.5b) or pressure swing adsorption (Figure 9.5c). These two last processes can increase the ethanol purity to 99.6% (w/w).

9.4.2 Hydrolysis

Hydrolysis or saccharification of the harvested biomass is a crucial step to release fermentable sugars. Miranda *et al.* (2012) compared different methods of cell disruption and extraction of sugar from *S. obliquus*, including physical (homogenization, sonication at 120°C temperature) and physicochemical (acid or alkaline hydrolysis), concluding that the best method was acid hydrolysis using sulphuric acid. Acid hydrolysis provides a high efficiency in converting cellulosic materials into fermentable sugars (Harun & Danquah, 2011; Phwan *et al.*, 2018). Figure 9.6 showcases the hydrolysis reaction of starch to produce simpler carbohydrate molecules. Romero-Frasca *et al.* (2021) conducted an acid hydrolysis of microalgae using 0.1M sulphuric acid, a temperature of 85–90°C, and constant stirring for 120 min. The reaction was then neutralized using a 5 M sodium hydroxide solution.

9.4.3 Fermentation

The hydrolysate obtained is then fermented into bioethanol as shown in Figure 9.4. Fermentation releases carbon dioxide which can be recovered and incorporated into the system for microalgae cultivation. Bioethanol has a high-octane number and high heat of vaporization. Hence it is an adequate gasoline replacement or blend in concentrations between 10 and 80% (v/v) following minor engine modifications (e.g. the intake minifold needs to be redesigned as per bioethanol's high heat of vaporization).

Initial studies mainly reported yields on the bioethanol production from microalgal biomass grown in synthetic media. Ho *et al.* (2013) obtained bioethanol from the acid hydrolysate of *S. obliquus* (51.8%, carbohydrate content) using *Zymomonas mobilis* for the fermentation process. After 4 h of fermentation, an ethanol concentration of 8.6 g/L was obtained with a yield of 0.22 g of ethanol/g of biomass. In this study, acid hydrolysis (2% H_2SO_4) was used to saccharify the wet biomass of microalgae, achieving a glucose yield of 96–98% and a transformation to ethanol of 99.8%, respectively. Reyimu and Özçimen (2017) reported bioethanol with yields of 0.04 g of ethanol/g biomass using *Tetraselmis suecica* cultivated in treated municipal wastewater. On the contrary, Tighiri and Erkurt (2016) reported an ethanol yield of 0.05 (g ethanol/g biomass), using biomass from a mixed culture of microalgae, also cultivated in wastewater.

Our laboratory has recently identified *Candida* sp. as growing species during wastewater treatment (Romero-Frasca *et al.*, 2021; Walls *et al.*, 2019). It was first noted that the species were able to produce bioethanol during wastewater treatment at low quantities as per previous literature (Reyimu & Özçimen, 2017). The *Candida* strains were then isolated and utilized for the transformation of acid hydrolysed microalgae to bioethanol. *Candida* sp. were able to convert 75% of glucose to bioethanol, whilst *S. cerevisiae* achieved an 87% conversion at 28°C, pH 6.5. Relatively similar ethanol yields were determined for both species, achieving 0.45 (±0.05) and 0.46 (±0.05) g ethanol per g glucose for *S. cerevisiae* and *Candida* sp., respectively (Romero-Frasca *et al.*, 2021). This indicated that the wild-type species of *Candida* have the potential to conduct fermentations using wastewater as growth medium. Additionally it also demonstrated acid hydrolysis as a viable method for producing bioethanol from microalgae, without significant inhibition in alcoholic fermentation due to possible toxic compounds.

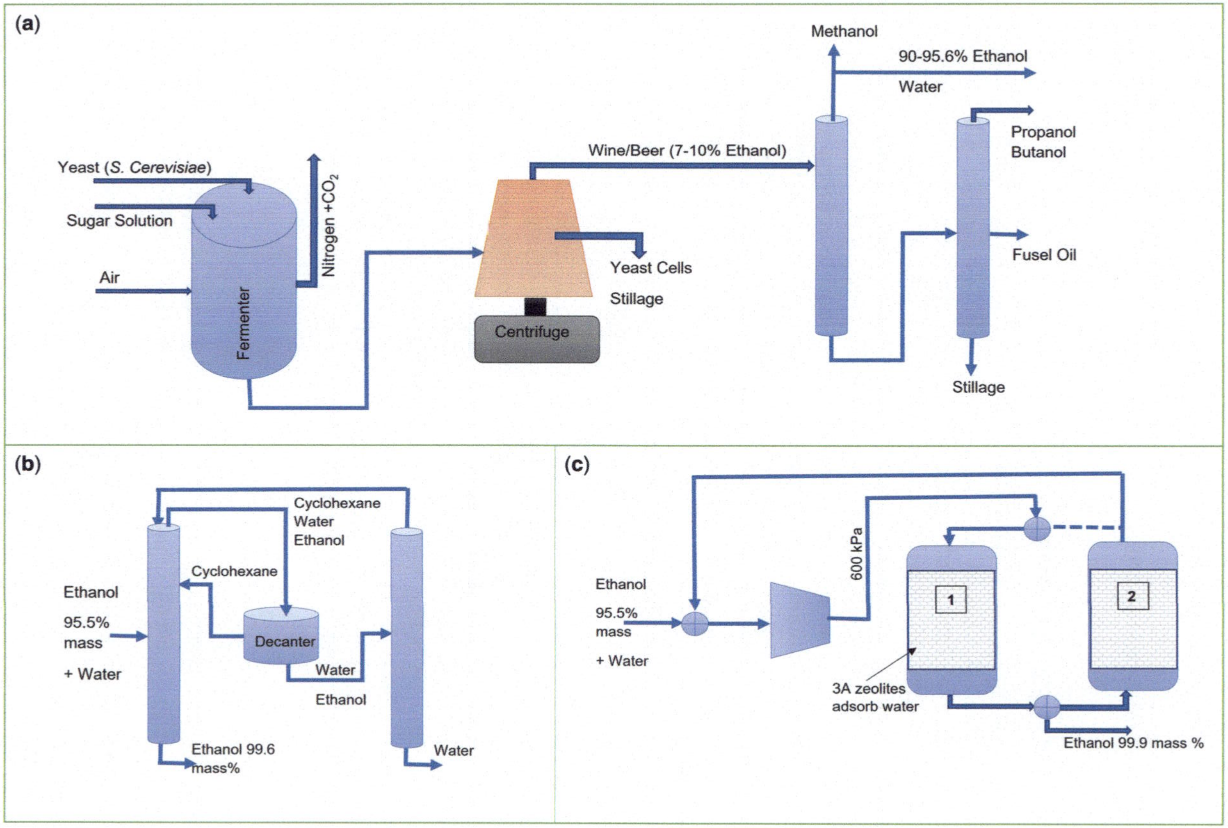

Figure 9.5 Overview of unit operations following the fermentation reaction for bioethanol production. (a) First separation and distillation unit operations to achieve 95% bioethanol, (b) azeotropic distillation and (c) pressure swing adsorption where bioethanol is dehydrated. The system consists of two identical adsorption columns (1,2) that switch on/off in sequential mode. This is done to maintain the pressure from one column to another and save energy.

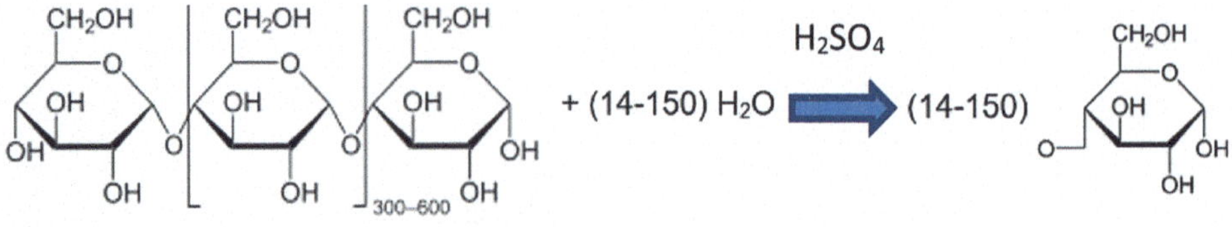

Figure 9.6 Example of an acid hydrolysis reaction.

9.5 CONCLUSIONS AND PERSPECTIVES

A circular bioeconomy framework can be established by using microalgae for the dual purpose of wastewater treatment and biofuel recovery. One of the main initial detriments of biofuel production from microalgae is the high microbial cultivation costs using synthetic medium, hence by using wastewater this cost is alleviated. However, there are still other challenges that need to be resolved on microalgal biofuel production systems. For example, currently biomass pre-treatment and refining require energy-intensive unit operations. Hence, costs could be reduced through investigating other process integration routes similar to this dual bioremediation and biorefining approach. Another approach is the process intensification of microalgae growth and processing. In this last one, *in situ* transesterification has shown advantages over separate extraction and transesterification processes. Further research should elucidate other mechanisms for integration and intensification. One example is the concept of a biorefinery system where high-value compounds are produced using similar unit operations to obtain biofuels.

The production of biodiesel and bioethanol from microalgae grown in wastewater has been shown at the laboratory- and pilot scales. Wild microalgae can contain up to 40%w lipids and 50%w carbohydrates after nutrient limitation conditions. Microalgal biodiesel is produced via a transesterification reaction. The reaction will only convert neutral lipids to FAME, with reported conversions between 80 and 99% using either alkaline or acid catalysts. However, a biodiesel production route involves a series of refining processes, apart from the main reaction, that require special consideration. Microalgal biodiesel have a specific fingerprint with carbon chains of C16 and C18. Regarding bioethanol production, the complex structure of microalgae biomass needs pre-treatment via lysis and hydrolysis before fermentation. In hydrolysis, using an acid has been shown a straight forward mechanism, however, costs and associated risks at a large scale hinder its industrial economic use. It is interesting that bioethanol was also found to be produced during wastewater treatment, giving the possibility for future explorations using a dual fermentation and wastewater treatment process.

REFERENCES

BP (2021). Statistical Review of World Energy, 70th Edition. https://www.bp.com/content/dam/bp/business-sites/en/global/corporate/pdfs/ energy-economics/statistical-review/bp-stats-review-2021-full-report.pdf, [Accessed May 05, 2023].

Chen C. Y., Zhao X. Q., Yen H. W., Ho S. H., Cheng C. L., Lee D. J., Bai F. W. and Chang J. S. (2013). Microalgae-based carbohydrates for biofuel production. *Biochemical Engineering Journal*, **78**, 1–10, https://doi.org/10.1016/j.bej.2013.03.006

Dragone G., Fernandes B. D., Abreu A. P., Vicente A. A. and Teixeira J. A. (2011). Nutrient limitation as a strategy for increasing starch accumulation in microalgae. *Applied Energy*, **88**, 3331–3335, https://doi.org/10.1016/j.apenergy.2011.03.012

Elahinik A., Haarsma M., Abbas B., Pabst M., Xevgenos D., van Loosdrecht M. C. M. and Pronk M. (2022). Glycerol conversion by aerobic granular sludge. *Water Research*, **227**, 119340, https://doi.org/10.1016/J.WATRES.2022.119340

Gantar M., Obreht Z. and Dalmacija B. (1991). Nutrient removal and algal succession during the growth of *Spirulina platensis* and *Scenedesmus quadricauda* on swine wastewater. *Bioresource Technology*, **36**, 167–171, https://doi.org/10.1016/0960-8524(91)90175-J

Gotaas H. B., Oswald W. J. and Ludwig H. F. (1954). Photosynthetic reclamation of organic wastes. *Science Monitor*, **79**, 368–378.

Gouveia L., Graça S., Sousa C., Ambrosano L., Ribeiro B., Botrel E. P., Neto P. C., Ferreira A. F. and Silva C. M. (2016). Microalgae biomass production using wastewater: treatment and costs: scale-up considerations. *Algal Research*, **16**, 167–176, https://doi.org/10.1016/J.ALGAL.2016.03.010

Harun R. and Danquah M. K. (2011). Influence of acid pre-treatment on microalgal biomass for bioethanol production. *Process Biochemistry*, **46**, 304–309, https://doi.org/10.1016/J.PROCBIO.2010.08.027

Hernández-García A., Velásquez-Orta S. B., Novelo E., Yáñez-Noguez I., Monje-Ramírez I. and Orta Ledesma M. T. (2019). Wastewater-leachate treatment by microalgae: biomass, carbohydrate and lipid production. *Ecotoxicology and Environmental Safety*, **174**, 435–444, https://doi.org/10.1016/j.ecoenv.2019.02.052

Ho S. H., Huang S. W., Chen C. Y., Hasunuma T., Kondo A. and Chang J. S. (2013). Bioethanol production using carbohydrate-rich microalgae biomass as feedstock. *Bioresource Technology*, **135**, 191–198, https://doi.org/10.1016/J.BIORTECH.2012.10.015

Jebali A., Acién F. G., Rodriguez Barradas E., Olguín E. J., Sayadi S. and Molina Grima E. (2018). Pilot-scale outdoor production of *Scenedesmus* sp. in raceways using flue gases and centrate from anaerobic digestion as the sole culture medium. *Bioresource Technology*, **262**, 1–8, https://doi.org/10.1016/J.BIORTECH.2018.04.057

Ji M. K., Yun H. S., Park S., Lee H., Park Y. T., Bae S., Ham J. and Choi J. (2015). Effect of food wastewater on biomass production by a green microalga *Scenedesmus obliquus* for bioenergy generation. *Bioresource Technology*, **179**, 624–628, https://doi.org/10.1016/J.BIORTECH.2014.12.053

Komolafe O., Velasquez Orta S. B., Monje-Ramirez I., Noguez I. Y., Harvey A. P. and Orta Ledesma M. T. (2014). Biodiesel production from indigenous microalgae grown in wastewater. *Bioresource Technology*, **154**, 297–304, https://doi.org/10.1016/J.BIORTECH.2013.12.048

Lavoie A. and de la Noüe J. (1985). Hyperconcentrated cultures of *Scenedesmus obliquus*: a new approach for wastewater biological tertiary treatment? *Water Research*, **19**, 1437–1442, https://doi.org/10.1016/0043-1354(85)90311-2

Miranda J. R., Passarinho P. C. and Gouveia L. (2012). Pre-treatment optimization of *Scenedesmus obliquus* microalga for bioethanol production. *Bioresource Technology*, **104**, 342–348, https://doi.org/10.1016/J.BIORTECH.2011.10.059

Oliveira G. A., Carissimi E., Monje-Ramírez I., Velasquez-Orta S. B., Rodrigues R. T. and Ledesma M. T. O. (2018). Comparison between coagulation-flocculation and ozone-flotation for *Scenedesmus* microalgal biomolecule recovery and nutrient removal from wastewater in a high-rate algal pond. *Bioresource Technology*, **259**. pp. 334–342, https://doi.org/10.1016/j.biortech.2018.03.072

Phwan C. K., Ong H. C., Chen W. H., Ling T. C., Ng E. P. and Show P. L. (2018). Overview: comparison of pretreatment technologies and fermentation processes of bioethanol from microalgae. *Energy Conversion and Management*, **173**, 81–94, https://doi.org/10.1016/J.ENCONMAN.2018.07.054

Reyimu Z. and Özçimen D. (2017). Batch cultivation of marine microalgae *Nannochloropsis oculata* and *Tetraselmis suecica* in treated municipal wastewater toward bioethanol production. *Journal of Cleaner Production*, **150**, 40–46, https://doi.org/10.1016/J.JCLEPRO.2017.02.189

Romero-Frasca E., Velasquez-Orta S. B., Escobar-Sánchez V., Tinoco-Valencia R. and Orta Ledesma M. T. (2021). Bioprospecting of wild type ethanologenic yeast for ethanol fuel production from wastewater-grown microalgae. *Biotechnology for Biofuels*, **14**, 93, https://doi.org/10.1186/s13068-021-01925-x

Romero Villegas G. I., Fiamengo M., Acién Fernández F. G. and Molina Grima E. (2017). Outdoor production of microalgae biomass at pilot-scale in seawater using centrate as the nutrient source. *Algal Research*, **25**, 538–548, https://doi.org/10.1016/J.ALGAL.2017.06.016

Ruiz J., Arbib Z., Álvarez-Díaz P. D., Garrido-Pérez C., Barragán J. and Perales J. A. (2014). Influence of light presence and biomass concentration on nutrient kinetic removal from urban wastewater by *Scenedesmus obliquus*. *Journal of Biotechnology*, **178**, 32–37, https://doi.org/10.1016/J.JBIOTEC.2014.03.001

Salama E. S., Kurade M. B., Abou-Shanab R. A. I., El-Dalatony M. M., Yang I. S., Min B. and Jeon B. H. (2017). Recent progress in microalgal biomass production coupled with wastewater treatment for biofuel generation. *Renewable and Sustainable Energy Reviews*, **79**, 1189–1211, https://doi.org/10.1016/J.RSER.2017.05.091

Tighiri O. H. and Erkurt E. A. (2016). Integrated microalgae cultivation, nutrient removal and product utilization bio-processes. *Desalination and Water Treatment*, **57**(47), 22329–22340, https://doi.org/10.1080/19443994.2015.1128987

Tredici M. R., Margheri M. C., Zittelli G. C., Biagiolini S., Capolino E. and Natali M. (1992). Nitrogen and phosphorus reclamation from municipal wastewater through an artificial food-chain system. *Bioresource Technology*, **42**, 247–253, https://doi.org/10.1016/0960-8524(92)90028-V

Vasistha S., Balakrishnan D., Manivannan A. and Rai M. P. (2023). Microalgae on distillery wastewater treatment for improved biodiesel production and cellulose nanofiber synthesis: A sustainable biorefinery approach. *Chemosphere*, **315**, 137666, https://doi.org/10.1016/J.CHEMOSPHERE.2022.137666

Velasquez-Orta S. B., Garcia-Estrada R., Monje-Ramirez I., Harvey A. and Orta Ledesma M. T. (2014). Microalgae harvesting using ozoflotation: effect on lipid and FAME recoveries. *Biomass and Bioenergy*, **70**, 356–363, https://doi.org/10.1016/J.BIOMBIOE.2014.08.022

Velasquez-Orta S. B., Mohiuddin O., Orta Ledesma T. and Harvey A. P. (2022). Process intensification of microalgal biofuel production. 3rd Generation Biofuels: Disruptive Technologies to Enable Commercial Production 269–290, https://doi.org/10.1016/B978-0-323-90971-6.00030-9

Walls L. E., Velasquez-Orta S. B., Romero-Frasca E., Leary P., Yáñez Noguez I. and Orta Ledesma M. T. (2019). Non-sterile heterotrophic cultivation of native wastewater yeast and microalgae for integrated municipal wastewater treatment and bioethanol production. *Biochemical Engineering Journal*, **151**, 107319, https://doi.org/10.1016/J.BEJ.2019.107319

Zhang C., Zhang Y., Zhuang B. and Zhou X. (2014). Strategic enhancement of algal biomass, nutrient uptake and lipid through statistical optimization of nutrient supplementation in coupling *Scenedesmus obliquus*-like microalgae cultivation and municipal wastewater treatment. *Bioresource Technology*, **171**, 71–79, https://doi.org/10.1016/J.BIORTECH.2014.07.060

Zhao N., Angelidaki I. and Zhang Y. (2017). Electricity generation and microbial community in response to short-term changes in stack connection of self-stacked submersible microbial fuel cell powered by glycerol. *Water Research*, **109**, 367–374, https://doi.org/10.1016/J.WATRES.2016.11.064

Part 4

Algal Biotechnology

doi: 10.2166/9781789063547_0183

Chapter 10

Advanced value-added bioproducts from microalgae

Pallavi Saxena[1], Pritikrishna Majhi[2], Ruchi Pithania[3], Sarieh Tarigholizadeh[1], Harish[4] and Tatiana Minkina[1]*

[1]*Academy of Biology and Biotechnology, Southern Federal University, Rostov-on-Don, Russia*
[2]*DBT-BIRAC E-YUVA Centre, GIETU, Gunupur, Odisha, India 765022*
[3]*Division of Biological Sciences, University of California, San Diego, USA 92093*
[4]*Department of Botany, Mohanlal Sukhadia University, Udaipur, Rajasthan, India 313001*
**Corresponding author: 19octpallavi89@gmail.com*

ABSTRACT

Currently, the world is facing major issues of degradation of land by natural as well as anthropogenic activities such as desertification, salinization, industrialization, pollution and population growth. The limited resources and the expanding global population require alternative resources to meet the demands in the future. Microalgae are contemplated as a favorable resource for high-value products, including carotenoids, phycobilin, astaxanthin, docosahexaenoic acid, eicosahexaenoic acid and omega-3/6 polyunsaturated fatty acids. Although the use of algae is not new, the idea of developing high value-added products concerning sustainability, economic viability, nutrition enrichment and environmental friendliness is attracting researchers to explore more about the potential of microalgal flora. Microalgae not only thrive under extremophilic conditions but also do not compete with plants for land resources. Having a short generation period, diverse biochemical composition, low-cost nutritional needs and fixation of CO_2 are also significant reasons to promote their products. Also, the biorefinery concept and sustainable cultivation possibilities can substantially add to enabling sustainable production of high-value biomolecules, while proposing opportunities for increasing sustainable food and fuel supplies. However, a few challenges like inadequate domestic demand, constant maintenance of ideal conditions for cultivation and food regulations still need to be overcome.

Keywords: bio-stimulant, microalgae, polyunsaturated fatty acids, carotenoids, pigments

10.1 INTRODUCTION

The worldwide population is expected to reach nearly 10 billion by the end of 2050. The rising population will certainly increase the demand for food, beverages, supplements, pharmaceuticals and personal care products (Rahman 2020). Now the world is looking for alternatives to fulfil the demand and sustainability criteria for the future and among the alternatives microalgae can be considered as a promising resource

(Caporgno & Mathys, 2018). The estimated microalgae-based industry was US$3.4 billion in 2020 which is expected to become USD 4.6 billion in 2027 (Zhuang *et al.*, 2022). To develop the microalgae industry to their full ability, further research is necessary in terms of improving product yields and lowering overall costs. The utilization of microalgae for industrial purposes is nevertheless facing issues like low yield, energy consumption, maintenance of the cultures and products. The ongoing technological advancements will surely pave the way for these issues, for instance, selection of heterotrophic strains are in preference to limit the impact on natural conditions. As these strains grow rapidly using organic compounds, they knock out the limitation of environmental conditions and help in utilizing waste resources like lignocellulosic materials. However, with the advancement of genetic and metabolic engineering, the progress of culture and screening procedures as well as incorporation of nanotechnology, microalgae will become the most reliable resource of products and applications (Harun *et al.*, 2011).

Microalgae having universal presence and thriving survival under extreme conditions could be one of the major advantages for the industry. Microalgae are abundantly present organisms: 0.2–0.8 million species are present within the algae kingdom, yet underexplored in different sectors. Although their use is not new to mankind, still their utilization for the maximum extent with sustainability brings a new horizon for industrial research. Further, microalgae are an efficient fixer of atmospheric CO_2, which could substantially lead to the decrease of the greenhouse effect and will empower environmental health (Liyanaarachchi *et al.*, 2021; Mironiuk & Chojnacka, 2018). Therefore, development of microalgal-based products by industries not only provides benefits to the human health but it also supports the improvement of environmental health. Moreover, nutritional needs of the microalgae are limited, under the presence of sunlight rapid generation of microalgae can be easily achieved, which could be cost-effective from the industrial point of view. In addition, microalgae will produce high biomass concentrations in comparison to the terrestrial plants without engaging hectares of land (Russell *et al.*, 2022). Also, algae being a primitive plant ensure easy extraction and purification of the bioactive metabolites/compounds for further application in comparison to the complex procedures required in higher plants (Mironiuk & Chojnacka, 2018). Microalgae-based products besides being organic, possess the nutrition values higher than the usual supplements and are more potent to human health (Korzeniowska *et al.*, 2018). Also, the use of microalgae-based pigments for cosmeceuticals and nutraceuticals is attracting a lot more attention (Saxena *et al.*, 2020; Zhuang *et al.*, 2022). This chapter is an overview of the recent research conducted on microalgae for the detection, extraction and commercialization of their biomolecules in various sectors, including their presence in the market and concerning challenges of industries.

10.2 MARKET VALUE OF ALGAE-BASED HIGH-VALUE COMPOUNDS

The market value assessment of algal products is based on their nutritional composition, formulation, level of purity and usage (Vieira, 2016). Also, it is important to understand the regulatory law framework, technical and economic aspects and risk management for the development of microalgae-based market products. The crucial challenges with the market are expensive operational cost, requirement of infrastructure and maintenance, optimization of commercial scale harvesting quantity and optimization of market financial affairs regarding microalgae-based products. Besides these difficulties, it is estimated that microalgae-based product markets will reach up to US$ 53.43 billion by 2026 (https://www.credenceresearch.com/report/algae-productsmarket)

According to Khattar *et al.* (2009), the global microalgae-based product astaxanthin market was assessed around US$555.4 million in 2016. Microalgae is the natural resource for this pigment which is widely in use for nutraceuticals and aquaculture industry due to its antioxidant properties and fortification. Its market value is way higher than its synthetic version in the market (Li *et al.*, 2011; Pérez-López *et al.*, 2014). Also due to its potential application in neuro and cardio-related diseases, their market values have been influenced remarkably (Wu *et al.*, 2015). For 130 metric tons of annual production currently more than $200 million have been utilized. The average market prices are estimated to be between 1000 and 2000 USD per kg depending upon the purity level. Due to the high cost till now only 1% of the market is covered by astaxanthin produced by microalgae (Shah *et al.*, 2016). Similarly, beta-carotene another biomolecule extracted from algae was having a 3.5% compound annual growth rate and was US$224 million in 2018 (Transparency Market Research, 2018) and the largest shareholder is Europe (Market Watch, 2018). For multiple applications in personal care, food and pharmaceuticals have raised its demands, also in the Asia-Pacific regions.

Furthermore, with the raising consciousness about health in mankind, industries are witnessing accelerated demand of Omega-3 (Market Research Future, 2019). Omega-3 is an essential fatty acid which is not produced in the human body. The market value of microalgae-based omega-3 is increasing by 13.5% per year. Currently its market is expanding in US, Europe and Asia-Pacific due to numerous health benefits. Likewise, the market size of eicosapentaenoic acid (EPA) and docosahexaenoic acid (DHA) are 300 million and 1.5 billion USD, respectively, and the price is 0.2–0.5 USD/g and 18–22 USD/g, respectively. Over 75% of the manufacture volume of microalgae was used in the health food marketplace as nutritional enhancements (Chacon-Lee and Gonzalez-Marino, 2010). The algae-based valued food additives and ingredients, for example DHA, represent a rising market. Martek's (now DSM) algae-derived DHA is found in 99% of all baby foods in the USA (Eckelberry, 2011).

High-value products that are extracted from microalgae thus improve the economics in a biorefinery approach and have market scope and opportunities (Figure 10.1). However, it needs to be

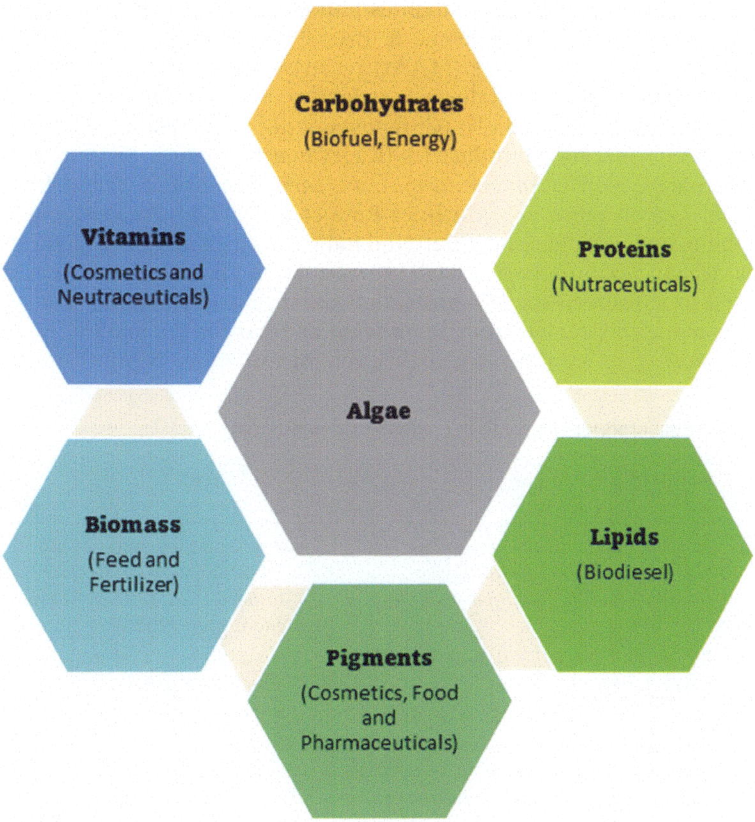

Figure 10.1 Various application of algae.

understood whether it is market driven or technology driven. Production economics such as the cost effectiveness of the food needs to be offset against the market opportunities and high-value products from microalgae: the technology also needs to be robust and reliable for its market flow.

10.3 HIGH-VALUE PRODUCTS USED IN DIFFERENT SECTORS

10.3.1 Cosmetics

Cosmetics are the products used globally by people to protect their skin and this industry is growing very fast due to modern lifestyle. The daily used cosmetics contain different synthetic chemicals which may cause adverse side effects on continuous exposure, due to these reasons, there is need to replace synthetic chemicals with environmentally sustainable products (Ariede *et al.*, 2017). Cosmeceutics is a nonofficial term and defined as cosmetic products with biologically active ingredients having medical or drug-like properties (Kligman, 2000; Martin & Glaser, 2011). Nowadays, natural ingredients from algae gained tremendous attention as an alternative for safe and high-quality products (Table 10.1). Microalgae contain natural pigments/metabolites having biological activities such as antioxidants, anti-bacterial, anti-aging, anti-inflammatory, anticancer and antiviral which makes them useful organisms in cosmetics industry for skin care products, anti-aging creams, sun protection products, thereby increasing the growth of this market (Fernando *et al.*, 2016; Talero *et al.*, 2015; Thomas & Kim, 2013; Wang *et al.*, 2017). There are hundreds of bioactive metabolites predicted from cyanobacteria and thousands more are predicted from eukaryotic microalgae.

Algae produce various secondary metabolites/antioxidants during their adaptation to stress and for survival in harsh conditions (Sansone & Brunet, 2019; Wang *et al.*, 2015). The secondary metabolite mycosporine-like amino acids (MAAs) received much attention these days due to their various applications in daily use materials such as fabrics, plastics, varnishes and paints to protect them against ultra-violet radiations (UVR) (Chrapusta *et al.*, 2017). Hence, MAAs are promising in many cosmetical and pharmaceutical industries (Kageyama & Waditee-Sirisattha, 2019). MAAs are present in some microalgae such as *Anabaena* sp., *C. vulgaris*, *D. salina*, *Eutreptiella* sp., *Scenedesmus* sp., *S. platensis* and *Leptolyngbya* sp. (Martínez-Ruiz *et al.*, 2022a). Another pigment, scytonemin which is present in the mucilaginous sheath around numerous marine cyanobacterial cells producing extracellular polysaccharides imparts yellowish-brown color to the cells (Martínez-Ruiz *et al.*, 2022a). Scytonemin is used in sunscreen for UV protection because it absorbs the light spectrum around 315–400 nm and is mainly extracted from *Scytonema* and *Nostoc* sp. majorly *N. punctiforme* (Sen & Mallick, 2022). *Nostoc* sp. can survive high levels of radiation. Natural antioxidants, such as

Table 10.1 Major microalgal products involved in cosmetic preparations.

Cosmeceutical Compound	Name of Microalgae	References
Scytonemin	*Nostoc punctiforme*, *Scytonema* sp., *Nostoc commune*, *Calothrix* sp., *Lyngbya* sp., *Leptolynbya mycodia*,	Stolz and Obermayer (2005), Nowruzi *et al.* (2020), Rastogi *et al.* (2020), Santiesteban-Romero *et al.* (2022), Sheibani and Naeimpoor (2023)
Sporopollenin	*Dunaliella salina*, *Chlorella fusca*, *Scenedesmus* sp.	Priyadarshani and Rath (2012), Pallela (2014), He *et al.* (2016), Gupta *et al.* (2023)
Mycosporine	*Isochrysis* sp., *Chlorella minutissima*, *Dunaliella tertiolecta*, *Chlorella sorokiniana*, *Thalassiosira weissflogii*, *Lyngbya purpurem*, *Oscilatoria* sp.	Stolz and Obermayer (2005), Kim and Chojnacka (2015), Chandra *et al.* (2020), Geraldes *et al.* (2020), Rosic (2021), Zaytseva *et al.* (2021)
	Dunaliella tertiolecta, *Chlorella sorokiniana*,	Tossavainen *et al.* (2019), Fawcett *et al.* (2022)

astaxanthin, carotenoids and lycopene protect the skin from oxidative stress and damage caused due to the production of free radicals through exposure of ultra-violet (UV) and further prevents skin aging (Gao *et al.*, 2021; Mourelle *et al.*, 2017). Lutein, a compound from different microalgae majorly *C. protothecoides, prevents skin damage caused by ultraviolet-C (UVC) (*Saha *et al.*, 2020). *Dunaliella tertiolecta* and *Tetraselmis suecica* produce high concentrations of α-tocopherol and vitamin E, which are widely used in cosmetic formulations (Arora & Philippidis, 2023). *Dunaliella salina* and *Spirulina platensis* sp. are rich in β-carotene and *Porphyridium* is rich in sulphated polysaccharides which prevent the formation of reactive oxygen species (ROS), inhibit lipid peroxidation and ultimately prevent oxidative damage to skin cells and produce hyaluronic acid, a glycosaminoglycan which helps in skin hydration (Gupta *et al.*, 2023). The secondary metabolites of brown algae, *Macrocystis pyrifera* (i.e. phlorotannins) and *Turbinaria conoides* (i.e. laminarin, fucoidan and alginate) have antioxidant activity, thus preventing the formation of free radicals and protect skin from aging (Peng *et al.*, 2011). ß-1,3-Glucan polysaccharide, rich in *Chlorella* sp. and *Skeletonema* diatom, as well as *Porphyridium* and *Nostoc flegelliforme*, acts as a free-radical collector and active immunostimulator which makes them potential candidates for preventing aging (Hamed, 2016; Shao *et al.*, 2013). The main carotenoids present in microalgae are β-carotene, lycopene, astaxanthin, zeaxanthin, violaxanthin and lutein and the most common microalgae commercially used for pigment production are *Dunaliella salina, Haematococcus pluvialis, Chlorella* sp., *Scenedesmus* sp., *Muriellopsis* sp., *Spirulina* sp. and *Porphyridium* sp. (Sathasivam & Ki, 2018).

The natural pigments present in microalgae and cyanobacteria are chlorophylls, carotenoids (carotenes and xanthophylls) and phycobilins and used in cosmetics such as in lipstick, eye shadows and eyeliners as a natural colorant (Begum *et al.*, 2016; Morocho-Jácome *et al.*, 2020). For example, ß-carotenes from *Dunaliella salina*; astaxanthin from *Haematococcus pluvialis* (red color), phycocyanobilins (blue pigment) from *Spirulina* and phycoerythrobilins from rhodophyte *Porphyridium* are natural dyes (Hamed, 2016). Other pigments such as chlorophyll are easily extracted and used in deodorants, due to their ability to mask odors, as well as in toothpastes and hygiene products (Mourelle *et al.*, 2018). Canthaxanthin pigment from *Nannochloropsis* sp. is commercially used in tanning pills (Rebelo *et al.*, 2020). Different type of lipids such as triacylglycerides, waxes, fatty acids, ceramides, glycerophospholipids, sterols, hydrogenated, esterified and oxidized lipids are used in cosmetics as dermatological delivery and moisturizing agents (De Luca *et al.*, 2021). The extracts from algae *Arthrospira platensis, H. pluvialis* and *T. suecica proteins* and polysaccharides are used in gels as thickeners and moisturizers (Martínez-Ruiz *et al.*, 2022b). Various marine strains secrete extracellular polysaccharides which act as physical barriers protecting the cells from external stimuli. Color and fragrance are two important characteristics for cosmetic products. The coloring is mostly done through pigments and essential oils provide aroma. Some commercially available products produced by *D. salina*, its extract known as blue retinol, help in growth and proliferation of skin cells (Mourelle *et al.*, 2017). Another product, silidine from the purple-red alga *Porphyridium cruentum* improves the skin texture and decreases redness. GoldenChlorella and AlgaPür Algae Oils from exopolysaccharides are beneficial for skin and hair treatments. Some companies are using extracts from algae. Recently, lipid extract from *Phaeodactylum tricornutum* is used as an anti-aging agent because it stimulates cell detoxification from oxidized proteins through proteasome, thus preventing aging by inhibiting the accumulation of harmful proteins (Vasilopoulou *et al.*, 2021). *Chlorella vulgaris* extract is also used for collagen repair and supporting tissue regeneration, thus reducing wrinkle (Ariede *et al.*, 2017; Wang *et al.*, 2015). A protein-rich extract from *Arthrospira* repairs the aging, tightens the skin and prevents stria formation (Bilal *et al.*, 2017).

10.3.2 Pharmaceuticals

The naturally derived valuable compounds from algae are gaining attention due to their useful biomedical properties such as anticancer, antidiabetic, antiviral and antimicrobial compounds. These compounds can be primary and/or secondary metabolites and used as a sustainable and cheap source for various pharmaceutical products such as antibodies, recombined proteins, vaccines and drug delivery in the pharmaceutical sector (Table 10.2). The high-value compounds from microalgae and cyanobacteria are screened for anti-diabetic properties having specific enzymes with catalytic activities (Abo-Shady *et al.*, 2023). Some examples of enzymes are α-amylase, α-glucosidase, *N*-acetyl-glucosaminidase, aldose reductase, hexokinase, glucose-6-phosphatase, dipeptidyl peptidase IV, glucose transporter 4 and glycogen synthase kinase-3β from *Chlorella* sp. *Nitzschia laevis, Isochrysis galbana*, *Chaetoceros furcellatus*, *Skeletonema marinoi* and *Porosira glacidis* species (Lauritano & Ianora, 2016; Mutanda *et al.*, 2020).

Pharmaceutical industries showed much interest in lipids such as polyunsaturated fatty acids (PUFAs), phytosterols and carotenoids and used in prevention and treatment of cardiovascular

Table 10.2 Multiple applications of pigments extracted from different microalgal strains.

Name of the Pigment	Name of the Alga	Applications	References
Astaxanthin	*Chlorella zofingiensis, Haematococcus pluvialis, Monoraphidium Chlorococcum* sp., *Scenedesmus* sp., *Chlamydomonas nivalis, Nannochloropsis* sp., *Chlamydocapsa* sp., *Chlorella vulgaris, Eremosphaera viridis, Neochloris wimmeri and Coelastrella striolata, Chromochloris zofngiensis*	Nutritional food, cosmetics, Aquaculture, poultry and food	Borowitzka (2013), Allen *et al.* (2018), Mao *et al.* (2020), Perozeni *et al.* (2020), Zhang *et al.* (2021), Ritu *et al.* (2023)
Canthaxanthin	*Chlorella* sp. *Asterarcys quadricellulare, Coelastrum* sp. *Tetraspora* sp. *Coelastrella* sp., *Chlorococcal* sp.	Aquaculture, poultry and food	Nasrabadi and Razavi (2010), Singh *et al.* (2019), Rebelo *et al.* (2020), Maswanna *et al.* (2022), Janchot *et al.* (2019), Corato *et al.* (2022)
β-Carotene	*Dunaliella salina. Tetradesmus obliquus, Scenedesmus sp. Chlamydomonas reinhardtii*	Biomedical Research	Borowitzka (2013), Tran *et al.* (2019), Singh *et al.* (2020), Harvey and Ben-Amotz (2020), Goswami *et al.* (2022)
Zeaxanthin	*Chlorella ellipsoidea, Dunaliella salina, Synechococcus* sp., *Synechocystis* sp., *Rhodosorus* sp., *Chromochloris zofingiensis, Nannochloropsis oculata*	Antioxidant, food pigment	Koo *et al.* (2012), Bourdon *et al.* (2021), Chen *et al.* (2022), Victor and Camarena-Bernard (2023)
Lutein	*Scenedesmus* sp., *Muriellopsis* sp., *Chlorella sorokiniana, Scenedesmus almeriensis*	Antioxidant	Fernández-Sevilla *et al.* (2010), Xie *et al.* (2019), Molino *et al.* (2019), Patel *et al.* (2022)
Echinenone	*Botryococcus braunii*	Antioxidant	Borowitzka, 2013; Indrayani *et al.* 2022
Phycoerythrin	*Spirulina* sp. *Rhodomonas* sp., *Porphyridium purpureum,*	Pharmaceuticals	Allen *et al.* (2018), Sosa-Hernández *et al.* (2019), Rodas-Zuluaga *et al.* (2021), Ji *et al.* (2022), Derbel *et al.* (2022)

diseases and blood coagulation (Xia *et al.*, 2021). Most studied lipids from microalgae are PUFAs and its derivatives such as DHA, EPA, α-linolenic acid, arachidonic acid (ARA) and docosapentaenoic acid are used for the treatment of diabetes, inflammatory bowel disorders, skin and respiratory disorders (Khan *et al.*, 2015). DHA and EPA also act as anti-inflammatory agents and used to reduce hypertension, stroke and arthritis. Additionally, DHA compounds are also used for the proper function and development of the nervous system (Jha *et al.*, 2017). ARA and DHA are essential for the development of eyes and brain in infant and are used as fortifications to infant formula (Mimouni *et al.*, 2012). Some examples of microalgae producing lipids are *Phaeodactylum tricornutum* producing EPA, *Nitzschia conspicua* producing arachidonic acid and *Schizochytrium* sp. accumulating DHA, EPA and palmitic acid (Ramos-Vega *et al.*, 2018; Xia *et al.*, 2021). To produce high amounts of PUFA, different extraction methods and systems need to be evaluated.

Phytosterols inhibit cholesterol absorption in the intestine and thus reduce cholesterol in humans (Le Goff *et al.*, 2019). Carotenoids from *Chlorella ellipsoidea* and *Chlorella vulgaris* have anti-proliferative effect on a human colon carcinoma cell line thus promoting cell death particularly in colon cancer (Cha *et al.*, 2008). β-Carotene from *Dunaliella salina* has good anti-inflammatory and immunomodulatory effects (Hyrslova *et al.*, 2022).

Cyanobacteria (*Spirulina*) and microalage such as *Chlorella vulgaris*, *Scenedesmus quadricauda* and *Dunaliella* sp. produce sulphated polysaccharides which have a wide range of bioactivities such as antiviral, anticancer and anti-inflammatory (de Morais *et al.*, 2015; Kiran & Venkata Mohan, 2021). The polysaccharides prevent the attachment of viruses with the target molecule. Microalga *Gyrodinium impudium* produce p-KG103, which inhibits the growth of tumor cells by stimulating cytokine production (Guo *et al.*, 2017). The polysaccharide from *Chlorella pyrenoidosa* showed promising improvement in hyperlipidemia disorder in rats (Wan *et al.*, 2020). Various polysaccharides from microalgae, such as alginate, carrageenan, laminarin and fucoidan, are used in drug delivery via nanoparticles (Yang *et al.*, 2022). The secondary metabolites fucoxanthin, sargaquinoic acid, sargahydroquinoic acid and sargaquinal produced by *Sargassum heterophyllum* show anti-malarial properties (Mutanda *et al.*, 2020). The polyphenols, phycobiliproteins and vitamins have antioxidant properties which help in preventing the oxidative damage caused by free radicals inhibit the growth of cancer cells and help to fight against various diseases such as chronic disorders, cardiovascular diseases and inflammations (Coulombier *et al.*, 2021).

Production of recombinant proteins using algal expression systems is also gaining consideration as it has many advantages like rapid growth rate, post-translational modifications like mammalian cells, cost effective and easy scale-up for bioprocessing and purification. *C. reinhardtii* was used to express malaria antigens from *Plasmodium falciparum* (Shamriz & Ofoghi, 2019). A chimeric gene having hemagglutinin-neuraminidase and fusion epitopes of Newcastle disease virus was successfully expressed in *C. reinhardtii* through an agrobacterium-mediated genetic transformation system (Shahriari *et al.*, 2019). *Phaeodactylum tricornutum* diatom and *D. salina* microalgae were successfully engineered to produce human IgG antibodies against the Hepatitis B virus surface protein (Geng *et al.*, 2003, Vanier *et al.*, 2015). Still some hurdles need to be addressed such as safety evaluation of transgenic strains, downstream processing, purification, cost and clinical trials.

10.3.3 Food supplements

10.3.3.1 Protein content of algae

An increase in the global population and their high demand for meat and dairy products has created pressure on protein supply. To meet those needs, it is highly essential to find out alternative protein sources. In this context, available protein in microalgae offers an excellent nutritional substitute by delivering all essential amino acids (Hariskos and Posten, 2014; Bhagia *et al.*, 2022). Protein is an integral component of the structure and metabolism of microalgal cells. Many microalgae contain very high amounts of protein, ranging from 42% to 70% (Barkia *et al.*, 2019; Milovanovic *et al.*, 2012; Plaza *et al.*, 2009). Microalgae can produce 2.5–7.5 tons of proteins annually per hectare (Bleakley & Hayes,

2007). *Chlorella* sp. has more than 70% of protein content, which has been commercialized recently (Eilam *et al.*, 2023). Similarly, *Spirulina* sp. and *Arthrospira* sp. are two well-known protein-rich microalgal strains. However, some other cyanobacteria like *Lyngbya majuscule*, *Nostoc* sp., *Anabaena* sp. and *Porphyridium* sp. are observed due to the production of microcolin-A (immunosuppressive agent), cyanovirin (antiviral agent against HIV) and enzyme superoxide dismutase (antioxidant), respectively (Arya and Gupta, 2001). *Microcystis aeruginosa* produces amino acids like serine, glycine, proline and valine. A carbonic enzyme anhydrase is produced by *Isochrysis galbana* that converts carbon dioxide into carbonic acid and bicarbonate (Khan *et al.*, 2018).

10.3.3.2 Single-cell protein

Single-cell protein (SCP) refers to a conventional or substituent for a protein found either from pure or mixed cultures of microalgae (also extracted from fungi, bacteria, or yeast) mostly used for animal and human consumption. These macromolecules with various chemical structures lead to various important functions (morphological, technological and physiological). Those protein components can be used as individual protein concentrates and can be integrated into processed food products. In this regard, microalgae are considered one of the most reliable protein sources and most of the algal domain is involved in the food sector. They also possess higher protein contents than conventional plant and animal protein sources. For example, according to Moorhead *et al.* (2011), *Chlorella* sp. is considered a human diet and used in mariculture due to the presence of immune-active substances, for example, 3-glucan β-1. Similarly, the protein content in *Spirulina platens* is 65% which is 36%, 37%, 22%, 24%, 26% and 24% greater than dried skimmed milk, chicken, soy flour, beef, fish and peanuts, respectively. Some other microalgal strains like *Aphanimezonon* sp., *Nostoc* sp., *Dunaliella* sp., *Porphyridium* sp., *Arthrospira* sp., *Scenedesmus* sp., *Anabaena* sp. and *Tetraselmis* sp. are involved in SCP production.

10.3.3.3 Carbohydrates

Algal cells are an important food source as they have an excellent content of carbohydrates. They may be monosaccharides, oligosaccharides, or polysaccharides that perform structural and metabolic activities. They can attach to proteins or lipids as glycoproteins or glycolipids (Arad & Levy-Ontman, 2010). Moreover, the microalgae can also generate carbohydrates such as glucose and starch through photosynthesis which are the basic constituents of the cell wall. Some species have a high carbohydrate content (Barkia *et al.*, 2019; Harun *et al.*, 2011), that is *Spirogyra* sp. and *Porphyridium cruentum* contain 35–65% and 40–60% carbohydrate, respectively. Microalgal polysaccharides, another form of carbohydrates, play a vital role in manufacturing pharmaceuticals such as antiviral, antibacterial, antioxidant and anticancer compounds. The microalgal polysaccharides are also involved in synthesizing cosmetics, nutritional components, anti-herpes drugs and pharmacological compounds in the business market. These are produced in different forms depending on the microalgae species. More specifically, several cyanophytes synthesize glycogen, some accumulate semi-amylopectin (Nakamura *et al.*, 2005) and various species of chlorophyta can synthesize starch in the shape of 2 glucose polymers, namely amylose and amylopectin (Busi *et al.*, 2014). Similarly, diatoms are well known for synthesizing floridian starch and chrysolaminarin (Gugi *et al.*, 2015). The microalgal polysaccharides benefit the cosmetic industry, acting as hygroscopic agents and antioxidants for topical creams and lotions (Gujar *et al.*, 2019).

10.3.3.4 Lipids

Microalgal lipids have attracted much attention for their commercialization due to biodiesel production. Moreover, poly-unsaturated fatty acids, such as omega fatty acids, have noticeably high trade values in infant formulations and nutraceuticals (Qu *et al.*, 2013). This provides structural support to plasma membranes and acts as energy reservoirs. The lipid percentage of microalgae is a major portion of neutral (acylglycerols, carotenoids and free fatty acids) and polar (phospholipids and galactolipids) lipids. Most microalgae are well-off in polar lipids in their exponential growth phase and commonly pile triacylglycerols in their stationary phase under unfavorable conditions (Rodolfi *et al.*, 2009). Fatty acids in microalgae are normally categorized as unsaturated and saturated fatty acids. These saturated–unsaturated fatty acids are mostly associated with neutral and polar lipids.

The lipid content of algal biomass ranges from 20% to 50% of dry cell weight (w/w). The production of different lipids depends upon the types of microalgal species and different cultivation conditions like salinity, temperature, growth phase and availability of nutrients, light intensity and pH (Guschina & Harwood, 2006). It was also reported that the lipid content increases considerably by limiting the nitrogen amount during their stationary phase. Microalgal lipids are given the most attention as healthy food supplements and vegan alternatives to fish oil and can be utilized as a foundation for industrial chemicals like cosmetic industry waxes and polymer lubricants (Mendes *et al.*, 2003).

10.3.3.5 Vitamins

Microalgae are recommended for their high content of different vitamins. They produce a wide variety of cost-effective and commercially important products. Vitamins from microalgae are easily available and their production is highly dependent upon nitrogen availability in the biomass and culture medium (Bonnet *et al.*, 2010). The different vitamin composition of microalgae is observed mostly during both the log and stationary phases of growth (Chew *et al.*, 2017). The microalga *Dunaliella salina* is well known for synthesizing pyridoxine, vitamins E and A (tocopherols), nicotinic acid, biotin, thiamine and riboflavin (Santhosh *et al.*, 2016). Another microalga, *Haslea osteria*, is rich in vitamin E. High quantities of vitamin A, E, C and β-carotene are synthesized by *Porphyridium cruentum* (Sheih *et al.*, 2009). The algal vitamins are highly nutritious for animals and humans (Borowitzka, 2013). The diatom *Navicula* sp. releases a blue-colored pigment called marennine, which is rich in tocopherols (Gastineau *et al.*, 2018).

10.3.3.6 Minerals

Microalgae are also well known for the accumulation of trace metals; however, few reports are available on the mineral content of microalgae. Minerals in the microalgal biomass include phosphorus, zinc, fluorine, potassium, iron, calcium, magnesium, sodium, sulphur, chlorine, manganese, copper, iodine, cobalt, selenium and molybdenum (Alsenani *et al.*, 2015). They are present either in elemental form or incorporated as compound forms in microalgal biomass and carry out several important functions. According to Tokusoglu and Üunal (2003), a significant number of elements like zinc, manganese, phosphorus, iron, magnesium, potassium, calcium and sodium are present in *Chlorella stigmatophora*, *C. vulgaris*, *Isochrisis galbana*, *D. tertiolecta*, *Tetraselmis suecica* and *S. platensis* (Tokuşoglu & Üunal, 2003).

10.3.4 Agricultural products

Algal extracts can be applied in agriculture in the form of bio-stimulants, biofertilizers, or bioregulators of plant growth. Plant growth regulators can alter cell division, root and shoot elongation, initiation of flowering and other metabolic functions, whereas fertilizers are the supplements needed for normal growth of the plant (Figure 10.2). Microalgae can be exploited as natural soil conditioners and biofertilizers to significantly improve the soil characteristics. Recent studies indicate that algae contain several phytohormones as well as high amounts of micronutrients and macronutrients that are essential for plant growth, health and development with better growth and crop yield (Guo *et al.*, 2020; Renuka *et al.*, 2018). Moreover, microalgae can also be utilized to improve soil health and to reduce erosion by crust formation; to treat wastewater for irrigation by removing agrochemical, fertilizers and pesticides as well as for metal removal and nitrogen recovery (Castro *et al.*, 2020).

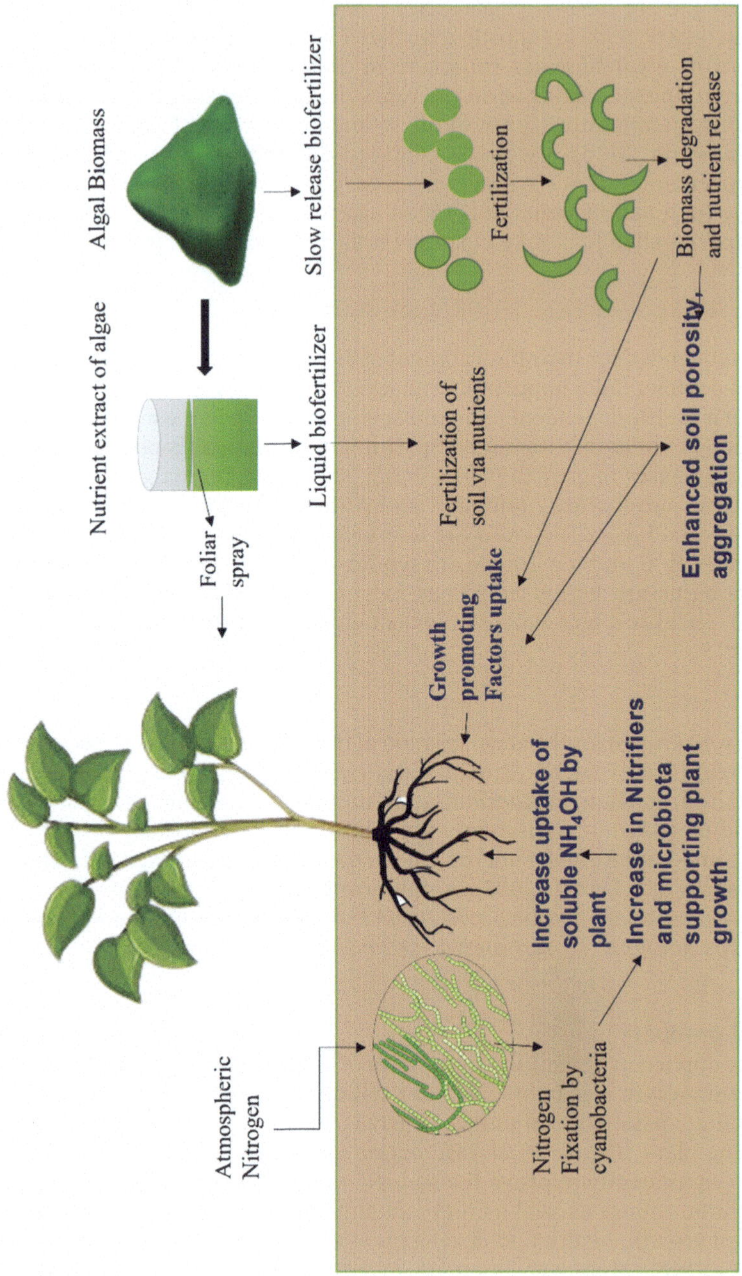

Figure 10.2 Use and role of algae products for enhancing plant growth and development.

10.3.4.1 Biofertilizer/biostimulants

Nitrogen, phosphorus, potassium, carbon and some trace elements are essential for plants for better growth, development and productivity and their deficiency can be corrected by applying ample biofertilizers. Because long-term and excessive usage of chemical or synthetic fertilizers leads to deposition of heavy metals in the cultivation land and eventually cause ecosystem imbalance (Ritika & Utpal, 2014). Biofertilizers comprise of living or dormant microbes alone or in combination, which improve the chemical and biological traits of soils, refurbish the soil fertility and enhance plant growth (Ammar *et al.*, 2022). The leftover crude of defatted or residual biomass after the extraction of value-added products can be used as biofertilizers and hence, reduce the production cost (Guo *et al.*, 2020). Microalgal extract and their biomass (such as *Spirulina* sp., *Chlorella* sp. and Cyanobacteria) itself can be used as suitable biofertilizer source (Balasubramaniam *et al.*, 2021). They are being regarded as the major organic matter sources in the agro-ecosystem as they can produce several polysaccharides, plant growth hormones, antibacterial chemicals and other metabolites required for plant growth (Guo *et al.*, 2020; Ronga *et al.*, 2019). Algae are capable of photosynthesis and carbon dioxide sequestration; hence, they can add enough organic carbon to the soil. Similarly, cyanobacteria contain heterocysts in their cell, thus enabling atmospheric nitrogen fixation. In this regard, some mutant strains of cyanobacteria can also be employed to enhance their resistance towards harsh, extreme, or unfavorable conditions or their efficiency for stimulating growth of different plants. A biofertilizer obtained from blue green algae (BGA) in the brand name 'Algalization' is commercialized having great economic viability in paddy cultivation. This helps to fix nitrogen under anaerobic conditions and deposits about 25– 30 kg N/ha/season which enhances the crop yield by 10–15% (Mehta *et al.*, 2018).

According to Bilal *et al.* (2017), it is worth noting that microalgae could supply a set of plant-protecting biological substances which can enhance germination percentage, stem and leaf growth and flowering. They can also be used as plant biostimulants for seed germination (Stirk & van Staden, 2020). A few algal extracts are available in the market as commercial biofertilizers for plants in the name of Acadian (Canada), Seamac Ultra Plus Liquid and Turfcomplex (UK), Ekologik R (Chile), Maxicrop (UK), Kelpak 66 (South Africa), Seasol (Australia), Göemill (France), Algamino Plant (Poland), SeaCrop16 (USA) and Actiwave R (Italy).

10.3.4.2 Plant growth-promoting substances/hormones

After extraction of high-value products, some of the nutrients remain in the processed/residual biomass that can be used as biofertilizer for the growth of plants. Some algal extracts can be obtained by the extraction in water simply by boiling, autoclaving and homogenization and it has great application in modern agriculture. They can be used to promote health, growth and crop yield of many cereals and vegetable plants due to the availability of numerous biological components in them. Those extracts can be applied on both soil and plants as well as used as hydroponic solutions or foliar applications (Ali *et al.*, 2022).

Algae are also considered as rich sources of plant growth promoting hormones or substances, that is, cytokinins, auxins, gibberellins, abscisic acid, ethylene, betaine and polyamines (Ammar *et al.*, 2022). Extracts from some specific algal strains can be used effectively and commercially as growth stimulants and as amendments in agricultural crops and crop production systems (Ronga *et al.*, 2019). It has been documented that few microalgae are a rich source of phytosterols belonging to the steroid group having specific biological activities (Fernandes & Cordeiro, 2021; Luo *et al.*, 2015). A study has been done by Plaza *et al.* (2018) regarding the phytohormone content in *Scenedesmus* sp. and *Arthrospira* sp., where they have found that *Scenedesmus* sp. showed higher concentrations of phytohormones as compared to *Arthrospira* sp. Another study has shown the impacts of *Aulosira fertilissima* on the growth of rice (*Oryza sativa* L.) and reported the occurrence of root-promoting hormones (auxins, cytokinins and gibberellic acid) that induced increased growth of rice seedlings (Ronga *et al.*, 2019). Another study showed the enhanced effects of *Chlorella thermophilla* biomass on rice plant (*Oryza sativa* L.) seedlings, grown on chromium-enriched soil (Majhi & Samantaray, 2021).

Similarly, a herbicide-resistant mutated strain of *Anabaena variabilis* was designed that showed plant growth promoting activity in rice plants (Singh & Datta, 2007).

Cyanobacterial strains like *Tolypothrix* sp., *Anabaena* sp., *Aulosira* sp. and *Nostoc* sp. can maintain soil fertility, physico-chemical properties and fix atmospheric nitrogen with some positive effects on plants and soil (Song *et al.*, 2005). Symbiotically, *Azolla* and *Anabaena* sp. provide various growth-promoting components such as indole-3-acetic acid, 3-methyl indole, indole-3-propionic acid and vitamin B_{12} and it adds about 60 kg/ha of nitrogen to the soil. It has been reported that dry microalgal biomass possesses around 1% phosphorus and 7% nitrogen (Wijffels & Barbosa, 2010). Moreover, the pyrolysis of algal biomass results in the formation of biochar, which can be a promising source of agricultural biofertilizer, bioenergy production and CO_2 sequestration (Mona *et al.*, 2021).

10.3.4.3 Biopesticides

Pesticides include insecticides, herbicides and fungicides and are mostly applied in agriculture to control pests and pathogens to get high crop yields (Abu-Ghosh *et al.*, 2021). Vigorous application of synthetic pesticides leads to several environmental problems, ground water contamination toxicity to humans and animals and induce harmful transformation on non-target pests (Rani *et al.*, 2020; Yadav & Sharma, 2019). Biopesticides are well known for their activity against plant pathogens which typically possess antioxidant, antimicrobial, antiviral and antifungal properties as well as encourage crop development (Gonçalves, 2021). In this regard, some strains of green algae and cyanobacteria are regarded as the most effective biocontrol agents against fungal pathogens and several soil-borne diseases and can increase the defence mechanisms in plants (Renuka *et al.*, 2018). Chlorellins, from *Chlorella* sp., is an algal isolated bioactive compound having pesticidal effects against pathogenic bacteria (Gupta *et al.*, 2013). Cryptophycin 1 from *Nostoc* sp. (ATCC 53789) is another biocontrol agent found to be most active against fungi and yeasts due to antimitotic and antiproliferative activities (Abu-Ghosh *et al.*, 2021).

10.3.5 Construction sector

The total energy consumption by the building sector is about 40% and it annually contributes up to 30% of the global GHGs emissions. Furthermore, it is also expected that GHG emissions from buildings become double over the next 20 years. Therefore, the vindication of GHG emissions from buildings is one of the utmost requirements of every national climate change strategy which needs holistic approaches to recover building energy performance (Elrayies, 2018; UNEP, 2009). The application of algae in architecture has so many benefits in terms of reducing carbon dioxide emissions, energy saving, oxygen generation, biofuel production, wastewater treatment at micro and macro level using building facades and creating urban space (Ilvitskaya & Chistyakova, 2020). Implementation of algae can reduce the overuse of agricultural land and transportation cost.

New techniques are now introduced to design unique photobioreactors to convert natural resources into energy. Holistic urban spaces, building façade and individual small architectural buildings, integrated with vertical flat panel, helical tubes and tube panel photobioreactors are the major contributions of microalgae-based photobioreactor systems (Ilvitskaya & Lobkova, 2018; Pruvost *et al.*, 2016). The most famous photobioreactor integrated building blocks are Process Zero Concept Building (Los Angeles, California, USA), B.I.Q House (Hamburg, Germany) and Photo. Synth.Etica (Dublin, Ireland). Similarly, the formation of integral urban spaces involves Alga Energety City (Istanbul, Turkey), Carbon T.A.P. (Tunnel Algae Park) (Philadelphia, USA) and Culture Urbaine (Geneva, Switzerland). Moreover, Urban Algae Canopy, EcoLogicStudio, living beings (by Jacob Dunias and Ethan Frier) and Street lamp (by Pierre Callech) are the most known small microalgal architectural forms. Another unique architectural and spatial construction is the Algae Dome culturing *Spirulina* sp. inside it, a four-meter-high bioreactor presented at the CHART art fair in Copenhagen (Denmark). There are some major factors that should be given importance during this kind of construction (Figure 10.3). They are day light performance, potential visibility, capital cost, environmental viability, thermal performance and acoustical performance (Elrayies, 2018).

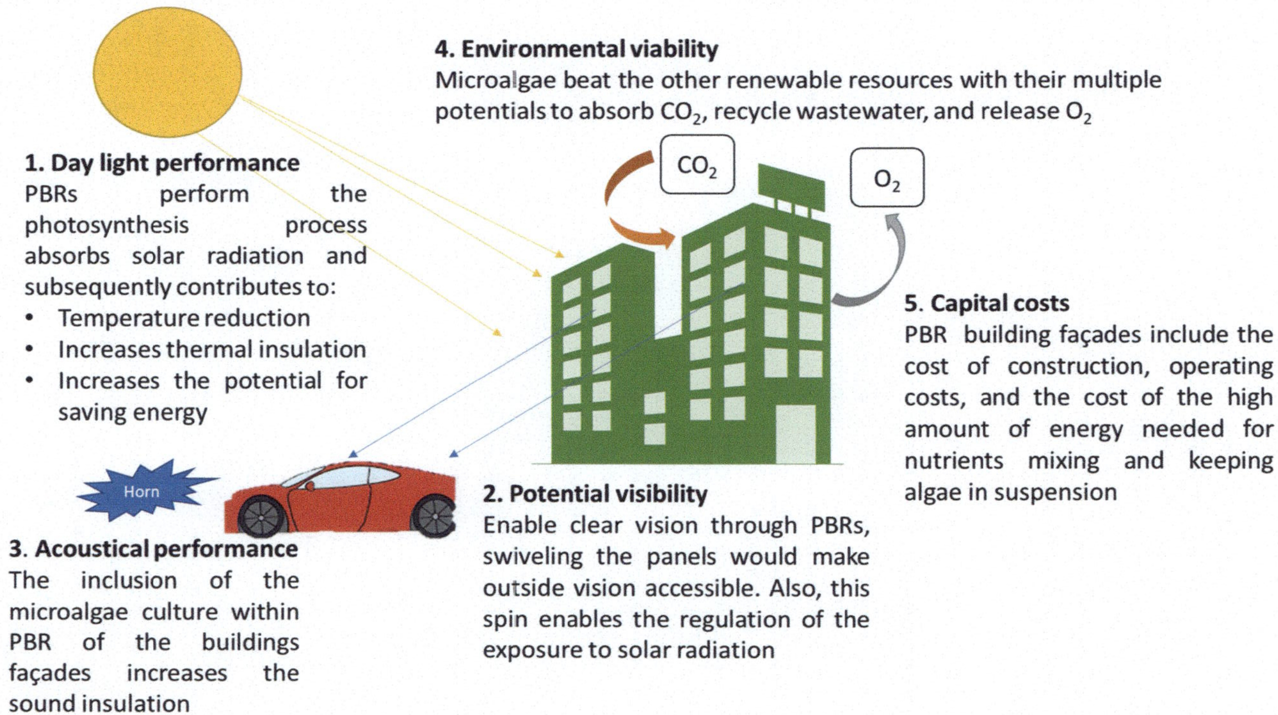

Figure 10.3 Potential factors taken into consideration while using algal photobioreactors as construction material.

10.4 CONSTRAINTS OF ALGAL BIOMASS PRODUCTION AND APPLICATION

Microalgae are a potential source of fuel, fodder and food yield (Table 10.3). The nutrient media required for cultivation and the energy-demanding methods of harvesting microalgal culture have a high recurring rate which is the major obstacle to the improvement of algal technologies. From the biotechnological point of view, microalgae need significant investigation. Hence, extensive innovation is required in various sectors of algal biotechnologies. This lacuna leads to the failure of algal research, although more than a thousand algal species are available worldwide. Among the 10 000 identified species, only a few are investigated to date for their potential chemical composition and very few are cultivated on an industrial scale. Similarly, genetic modification of microalgae along with their cultivation mode is another important goal in the field of algal research. The past few decades have accepted the use of microalgal biomass and their biomolecules in the improvement of many innovative food and other commercial products.

Table 10.3 Global production of nutraceutical products from different algal strains.

Microalgae Genus	Main Producers	Products
Spirulina sp.	Myanmar *Spirulina* Factory (Myanmar) https://www.spirulinasource.com/slideshows/myanmar-spirulina/	Tablets, pasta, chips and liquid extract
	Cyanotech Corp. (USA) https://www.cyanotech.com/	Tablets, beverages, powders, extracts
	Earthrise Nutritionals (USA) https://www.earthrise.com/	Tablets, powders, extracts
	Pondicherry Spirulina Farms (India) http://www.pyfarms.com/	Powder, capsules
Chlorella sp.	Hainan Simai Pharmacy Co. (China) https://www.chinafirm.biz/company-simai-pharmaceutical-haikou-35358	Powders, extracts
	Taiwan *Chlorella* Manufacturing Co. (Taiwan) https://www.taiwanchlorella.com/	Nectar, tablets, powders, noodles
	Algomed, Klotze (Germany) https://www.algomed.de/en/homepage/	Powders
Haematococcus pluvalis	Parry Nutraceuticals Ltd. (India) https://www.parrynutraceuticals.com/	Soft gel, oleoresins and beadlets
	Britannia Health Products Ltd. (U.K.) https://www.britannia-pharm.co.uk/	Capsule
	Nutrex Hawaii (USA) https://www.nutrex-hawaii.com/	Soft gel
Dunaliella bardawil	AquaCarotene Ltd. (Australia) http://www.aquacarotene.com/	Whole-dried biomass
	Betatene® (Australia) https://www.apfoodonline.com/industry/betatene-australias-own-natural-beta-carotene/	Tablet, soft gel, powders, capsule
	Cyanotech Corporation (USA) https://www.cyanotech.com/	Capsule, soft gel, oil
Aphanizomenon flos-aquae	Vision (USA)	Powder, capsules, crystals
	Blue Green Foods (USA) https://bluegreenfoods.com/	Capsules, crystals

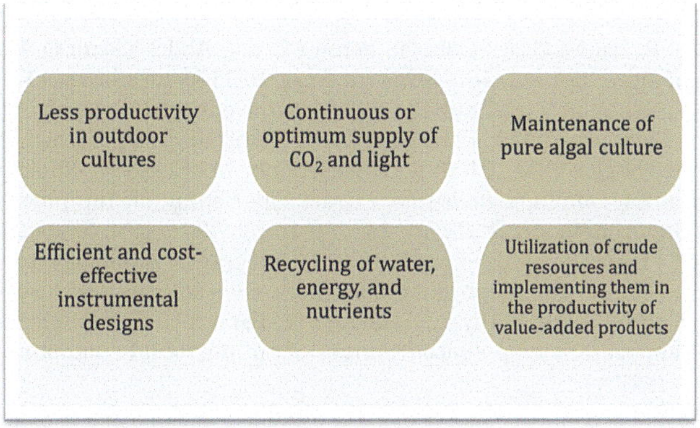

Figure 10.4 Major challenges in growing microalgae at large scale.

Although the microalgae-based product market is steadily expanding now, it is still not profitable as its substantial growth is troubled by the manufacturing techniques employed. Cost-effective production and optimized recovery operation are the two major challenges of this time (Figure 10.4). For example, the harvested biomass's wet slurry contains nearly 75% water which needs to be extracted using electrical or mechanical energy. Then, the dried biomass undergoes an extraction process to harvest the desired products. To date, there is not a single extraction technique that is commercially feasible. Moreover, the yield and nature of the desired product are also influenced by cell disruption techniques, which further require specific optimization steps. The biorefinery concept will be economically beneficial only when the extracts of biomass and the biomass itself can be utilized to produce commercially attractive value-added products.

10.5 CONCLUSION

Microalgae are sustainable and precious resources that have an ideal role in biofuel production, wastewater treatment and applications in agriculture, nutrition, pharmacy and the construction sector. They have high productivity properties and can expand even in wastelands. The growing population has created an opportunity for finding more suitable sustaining solutions. Changing the way of life needs development to create alternative options to provide both nutritional and health security in an eco-friendly and economical manner. Though microalgae were used many years ago to nourish their culture and harvest, they are growing rapidly now. If this continues to expand, then only the revolutionary changes in the pharmaceutical, cosmetics, energy and food industries can be performed in the coming years. The unique chemical contents in microalgae provide various functional ingredients, leading to the synthesis of high-value-added products. Additionally, there are some algal toxins, heavy metals and undefined compounds present in algal biomass which need profound research regarding them to address their deleterious effects on their consumption. Hence, this chapter provides specific data based on the available microalgal resources, which should further focus on the economical and nutritional improvement of microalgal applications.

ACKNOWLEDGMENT

The study was supported by the Ministry of Science and Higher Education of the Russian Federation, agreement no. 075-15-2022-1122.

REFERENCES

Abo-Shady A. M., Gheda S. F., Ismail G. A., Cotas J., Pereira L. and Abdel-Karim O. H. (2023). Antioxidant and antidiabetic activity of algae. *Life*, **13**, 460, https://doi.org/10.3390/life13020460

Abu-Ghosh S., Dubinsky Z., Verdelho V. and Iluz D. (2021). Unconventional high-value products from microalgae: a review. *Bioresource Technology*, **329**, 124895, https://doi.org/10.1016/j.biortech.2021.124895

Ali J., Jan I., Ullah H., Ahmed N., Alam M., Ullah R., El-Sharnouby M., Kesba H., Shukry M., Sayed S. and Nawaz T. (2022). Influence of *Ascophyllum nodosum* extract foliar spray on the physiological and biochemical attributes of okra under drought stress. *Plants*, **11**, 790, https://doi.org/10.3390/plants11060790

Allen J., Unlu S., Demirel Y., Black P. and Riekhof W. (2018). Integration of biology, ecology and engineering for sustainable algal-based biofuel and bioproduct biorefinery. *Bioresources and Bioprocessing*, **5**(1), 1–28.

Alsenani F., Tupally K. R., Chua E. T., Eltanahy E., Alsufyani H., Parekh H. S. and Schenk P. M. (2020). Evaluation of microalgae and cyanobacteria as potential sources of antimicrobial compounds. *Saudi Pharmaceutical Journal*, **28**(12), 1834–1841.

Ammar E. E., Aioub A. A., Elesawy A. E., Karkour A. M., Mouhamed M. S., Amer A. A. and El-Shershaby N. A. (2022). Algae as bio-fertilizers: between current situation and future prospective. *Saudi Journal of Biological Sciences*, **29**(5), 3083–3096.

Arad S. M. and Levy-Ontman O. (2010). Red microalgal cell-wall polysaccharides: biotechnological aspects. *Current Opinion in Biotechnology*, **21**(3), 358–364.

Ariede M. B., Candido T. M., Jacome A. L. M., Velasco M. V. R., de Carvalho J. C. M. and Baby A. R. (2017). Cosmetic attributes of algae – a review. *Algal Research*, **25**, 483–487, https://doi.org/10.1016/j.algal.2017.05.019

Arora N. and Philippidis G. P. (2023). The prospects of algae-derived vitamins and their precursors for sustainable cosmeceuticals. *Processes*, **11**, 587, https://doi.org/10.3390/pr11020587

Arya V. and Gupta V. K. (2011). A review on marine immunomodulators. *Int J Pharm Life Sci*, **2**(5), 751–758.

Balasubramaniam V., Gunasegavan R. D. N., Mustar S., Lee J. C. and Mohd Noh M. F. (2021). Isolation of industrial important bioactive compounds from microalgae. *Molecules*, **26**(4), 943. https://doi.org/10.3390/MOLECULES26040943

Barkia I., Saari N. and Manning S. R. (2019). Microalgae for high-value products towards human health and nutrition. *Marine Drugs*, **17**(5), 304.

Begum H., Yusoff F. M. D., Banerjee S., Khatoon H. and Shariff M. (2016). Availability and utilization of pigments from microalgae. *Critical Reviews in Food Science and Nutrition*, **56**, 2209–2222, https://doi.org/10.1080/10408398.2013.764841

Bhagia S., Bornani K., Ozcan S. and Ragauskas A. J. (2021). Terephthalic Acid Copolyesters Containing Tetramethylcyclobutanediol for High-Performance Plastics. *ChemistryOpen*, **10**(8), 830–841.

Bilal M., Rasheed T., Ahmed I. and Iqbal M. N. H. (2017). High-value compounds from microalgae with industrial exploitability – a review. *Frontiers in Bioscience Scholar*, **9**, 319–342.

Bleakley S. and Hayes M. (2017). Algal proteins: extraction, application and challenges concerning production. *Foods*, **6**(5), 33.

Bonnet S., Webb E. A., Panzeca C., Karl D. M., Capone D. G. and Wilhelmy S. A. S. (2010). Vitamin B12 excretion by cultures of the marine cyanobacteria Crocosphaera and Synechococcus. *Limnology and Oceanography*, **55**(5), 1959–1964.

Borowitzka M. A. (2013). High-value products from microalgae – their development and commercialisation. *Journal of Applied Phycology*, **25**, 743–756, https://doi.org/10.1007/s10811-013-9983-9

Bourdon L., Jensen A. A., Kavanagh J. M. and McClure D. D. (2021). Microalgal production of zeaxanthin. *Algal Research*, **55**, 102266, https://doi.org/10.1016/j.algal.2021.102266

Busi M. V., Barchiesi J., Martín M. and Gomez-Casati D. F. (2014). Starch metabolism in green algae. *Starch-Stärke*, **66**(1-2), 28–40.

Caporgno M. P. and Mathys A. (2018). Trends in microalgae incorporation into innovative food products with potential health benefits. *Frontiers in Nutrition*, **5**, 58.

Cha K. H., Koo S. Y. and Lee D. U. (2008). Antiproliferative effects of carotenoids extracted from *Chlorella ellipsoidea* and *Chlorella vulgaris* on human colon cancer cells. *Journal of Agricultural and Food Chemistry*, **56**, 10521–10526, https://doi.org/10.1021/jf802111x

Chacón-Lee T. L. and González-Mariño G. E. (2010). Microalgae for "healthy" foods-possibilities and challenges. *Comprehensive Reviews in Food Science and Food Safety*, **9**(6), 655–675.

Chandra R., Pons-Faudoa F. P., Saldívar R. P. and Rittmann B. E. (2020). Effect of ultra-violet exposure on production of mycosporine-like amino acids and lipids by *Lyngbya purpurem*. *Biomass and Bioenergy*, **134**, 105475, https://doi.org/10.1016/j.biombioe.2020.105475

Chen Q., Chen Y., Xiao L., Li Y., Zou S. and Han D. (2022). Co-production of lutein, zeaxanthin and β-carotene by utilization of a mutant of the green alga *Chromochloris zofingiensis*. *Algal Research*, **68**, 102882, https://doi.org/10.1016/j.algal.2022.102882

Chew K. W., Yap J. Y., Show P. L., Suan N. H., Juan J. C., Ling T. C., Lee D. J. and Chang J. S. (2017). Microalgae biorefinery: high value products perspectives. *Bioresource Technology*, **229**, 53–62.

Chrapusta E., Kaminski A., Duchnik K., Bober B., Adamski M. and Bialczyk J. (2017). Mycosporine-like amino acids: potential health and beauty ingredients. *Marine Drugs*, **15**, 326, https://doi.org/10.3390/md15100326

Corato A., Le T. T., Baurain D., Jacques P., Remacle C. and Franck F. (2022). A fast-growing oleaginous strain of *Coelastrella* capable of astaxanthin and canthaxanthin accumulation in phototrophy and heterotrophy. *Life*, **12**, 334, https://doi.org/10.3390/life12030334

Coulombier N., Jauffrais T. and Lebouvier N. (2021). Antioxidant compounds from microalgae: a review. *Marine Drugs*, **19**, 549, https://doi.org/10.3390/md19100549

De Luca M., Pappalardo I., Limongi A. R., Viviano E., Radice R. P., Todisco S., Martelli G., Infantino V. and Vassallo A. (2021). Lipids from microalgae for cosmetic applications. *Cosmetics*, **8**, 52, https://doi.org/10.3390/cosmetics8020052

de Morais M. G., da Vaz B. S., de Morais E. G. and Costa J. A. V. (2015). Biologically active metabolites synthesized by microalgae. BioMed Research International, **2015**, 835761.

de Siqueira Castro J., Calijuri M. L., Mattiello E. M., Ribeiro V. J. and Assemany P. P. (2020). Algal biomass from wastewater: soil phosphorus bioavailability and plants productivity. *Science of the Total Environment*, **711**, 135088.

Derbel H., Elleuch J., Tounsi L., Nicolo M. S., Rizzo M. G., Michaud P. and Abdelkafi S. (2022). Improvement of biomass and phycoerythrin production by a strain of *Rhodomonas* sp. isolated from the Tunisian coast of Sidi Mansour. *Biomolecules*, **12**, 885, https://doi.org/10.3390/biom12070885

Eckelberry R. (2011). Algae – food or chemical grade. Algae Industry Magazine. http://www.algaeindustrymagazine.com/algae-business-algae-food-or-chemical-grade/

Eilam Y., Khattib H., Pintel N. and Avni D. (2023). Microalgae – sustainable source for alternative proteins and functional ingredients promoting gut and liver health. *Global Challenges*, **7**, 2200177, https://doi.org/10.1002/gch2.202200177

Elrayies G. M. (2018). Microalgae: prospects for greener future buildings. *Renewable and Sustainable Energy Reviews*, **81**, 1175–1191, https://doi.org/10.1016/j.rser.2017.08.032

Fawcett C. A., Senhorinho G. N., Laamanen C. A. and Scott J. A. (2022). Microalgae as an alternative to oil crops for edible oils and animal feed. *Algal Research*, **64**, 102663, https://doi.org/10.1016/j.algal.2022.102663

Fernandes T. and Cordeiro N. (2021). Microalgae as sustainable biofactories to produce high-value lipids: biodiversity, exploitation and biotechnological applications. *Marine Drugs*, **19**(10), 573, https://doi.org/10.3390/md19100573

Fernández-Sevilla J. M., Fernández F. A. and Grima E. M. (2010). Biotechnological production of lutein and its applications. *Applied Microbiology and Biotechnology*, **86**, 27–40, https://doi.org/10.1007/s00253-009-2420-y

Fernando I. P. S., Nah J. W. and Jeon Y. J. (2016). Potential anti-inflammatory natural products from marine algae. *Environmental Toxicology and Pharmacology*, **48**, 22–30, https://doi.org/10.1016/j.etap.2016.09.023

Gao X., Jing X., Liu X. and Lindblad P. (2021). Biotechnological production of the sunscreen pigment scytonemin in cyanobacteria: progress and strategy. *Marine Drugs*, **19**, 129, https://doi.org/10.3390/md19030129

Gastineau R., Prasetiya F. S., Falaise C., Cognie B., Decottignies P., Morançais M., Méléder V., Davidovich N., Turcotte F., Tremblay R., Pasetto P., Dittmer J., Bardeau J.-F., Pouvreau J.-B. and Mouget J. L. (2018). Marennine-like pigments: blue diatom or green oyster cult. *Blue Biotechnology: Production and Use of Marine Molecules*, **1**, 529–551, https://doi.org/10.1002/9783527801718.ch16

Geng D., Wang Y., Wang P., Li W. and Sun Y. (2003). Stable expression of hepatitis B surface antigen gene in Dunaliella salina (Chlorophyta). *Journal of Applied Phycology*, **15**, 451–456.

Geraldes V., Jacinavicius F. R., Genuário D. B. and Pinto E. (2020). Identification and distribution of mycosporine-like amino acids in Brazilian cyanobacteria using ultrahigh-performance liquid chromatography with diode array detection coupled to quadrupole time-of-flight mass spectrometry. *Rapid Communications in Mass Spectrometry*, **34**, e8634, https://doi.org/10.1002/rcm.8634

Gonçalves A. L. (2021). The use of microalgae and cyanobacteria in the improvement of agricultural practices: a review on their biofertilising, biostimulating and biopesticide roles. *Applied Sciences*, **11**, 871, https://doi.org/10.3390/app11020871

Goswami R. K., Agrawal K. and Verma P. (2022). Microalgae *Dunaliella* as biofuel feedstock and β-carotene production: an influential step towards environmental sustainability. *Energy Conversion and Management: X*, **13**, 100154, https://doi.org/10.1016/j.ecmx.2021.100154

Guo Q., Shao Q., Xu W., Rui L., Sumi R., Eguchi F. and Li Z. (2017). Immunomodulatory and anti-IBDV activities of the polysaccharide AEX from *Coccomyxa gloeobotrydiformis*. *Marine Drugs*, **15**, 36, https://doi.org/10.3390/md15020036

Guo S., Wang P., Wang X., Zou M., Liu C. and Hao J. (2020). Microalgae as Biofertilizer in Modern Agriculture. In: Microalgae Biotechnology for Food, Health and High Value Products, M. Alam, J. L. Xu, and Z. Wang (eds.), Springer, Singapore. https://doi.org/10.1007/978-981-15-0169-2_12

Gügi B., Le Costaouec T., Burel C., Lerouge P., Helbert W. and Bardor M. (2015). Diatom-specific oligosaccharide and polysaccharide structures help to unravel biosynthetic capabilities in diatoms. *Marine Drugs*, **13**(9), 5993–6018.

Gujar A., Cui H., Ji C., Kubar S. and Li R. (2019). Development, production and market value of microalgae products. *Applied Microbiology Open Access*, **5**(2), 162.

Gupta V., Ratha S. K., Sood A., Chaudhary V. and Prasanna R. (2013). New insights into the biodiversity and applications of cyanobacteria (blue-green algae) – prospects and challenges. *Algal Research*, **2**, 79–97, https://doi.org/10.1016/j.algal.2013.01.006

Guschina I. A. and Harwood J. L. (2006). Lipids and lipid metabolism in eukaryotic algae. *Progress in Lipid Research*, **45**(2), 160–186.

Gupta A., Singh A. P., Singh V. K., Singh P. R., Jaiswal J., Kumari N., Upadhye V., Singh S. C. and Sinha R. P. (2023). Natural sun-screening compounds and DNA-repair enzymes: photoprotection and photoaging. *Catalysts*, **13**, 745, https://doi.org/10.3390/catal13040745

Hamed I. (2016). The evolution and versatility of microalgal biotechnology: a review. *Comprehensive Reviews in Food Science and Food Safety*, **15**, 1104–1123, https://doi.org/10.1111/1541-4337.12227

Hariskos I. and Posten C. (2014). Biorefinery of microalgae-opportunities and constraints for different production scenarios. *Biotechnology Journal*, **9**(6), 739–752.

Harun R., Jason W. S. Y., Cherrington T. and Danquah M. K. (2011). Exploring alkaline pre-treatment of microalgal biomass for bioethanol production. *Applied Energy*, **88**(10), 3464–3467.

Harvey P. J. and Ben-Amotz A. (2020). Towards a sustainable *Dunaliella salina* microalgal biorefinery for 9-cis β-carotene production. *Algal Research*, **50**, 102002, https://doi.org/10.1016/j.algal.2020.102002

He X., Dai J. and Wu Q. (2016). Identification of sporopollenin as the outer layer of cell wall in microalga *Chlorella protothecoides*. *Frontiers in Microbiology*, **7**, 1047. https://medium.com/space10/how-algae-could-help-solve-some-of-the-worlds-biggest-problems-1fa7774a16b1

Hyrslova I., Krausova G., Mrvikova I., Stankova B., Branyik T., Malinska H., Huttl M., Kana A. and Doskocil I. (2022). Functional properties of *Dunaliella salina* and its positive effect on probiotics. *Marine Drugs*, **20**, 781, https://doi.org/10.3390/md20120781

Ilvitskaya S. V. and Chistyakova A. G. (2020). Microalgae in architecture as an energy source. In: IOP Conference Series: Materials Science and Engineering **944**(1), 012010. IOP Publishing.

Ilvitskaya S. V., Alexeev Y. V. and Anufriev A. A. (2020). The recessed buildings as a perspective direction of 'Green' architecture. In IOP Conference Series: Materials Science and Engineering, Vol. 753, No. 4, p. 042008. IOP Publishing.

Indrayani I., Egeland E. S., Moheimani N. R. and Borowitzka M. A. (2022). Carotenoid production of *Botryococcus braunii* CCAP 807/2 under different growth conditions. *Journal of Applied Phycology*, **34**, 1177–1188, https://doi.org/10.1007/s10811-022-02682-6

Janchot K., Rauytanapanit M., Honda M., Hibino T., Sirisattha S., Praneenararat T. and Waditee-Sirisattha R. (2019). Effects of potassium chloride-induced stress on the carotenoids canthaxanthin, astaxanthin and lipid accumulations in the green *Chlorococcal* microalga strain TISTR 9500. *Journal of Eukaryotic Microbiology*, **66**, 778–787, https://doi.org/10.1111/jeu.12726

Jha D., Jain V., Sharma B., Kant A. and Garlapati V. K. (2017). Microalgae-based pharmaceuticals and nutraceuticals: an emerging field with immense market potential. *ChemBioEng Reviews*, **4**, 257–272, https://doi.org/10.1002/cben.201600023

Ji L., Liu Y., Luo J. and Fan J. (2022). Freeze-thaw-assisted aqueous two-phase system as a green and low-cost option for analytical grade B-phycoerythrin production from unicellular microalgae *Porphyridium purpureum*. *Algal Research*, **67**, 102831, https://doi.org/10.1016/j.algal.2022.102831

Kageyama H. and Waditee-Sirisattha R. (2019). Antioxidative, anti-inflammatory and anti-aging properties of mycosporine-like amino acids: molecular and cellular mechanisms in the protection of skin-aging. *Marine Drugs*, **17**, 222, https://doi.org/10.3390/md17040222

Khan M., Rahman M. M., Zaman S., Jahangir T. A. and Razu M. H. (2015). Omega-3 polyunsaturated fatty acids from algae. In: Recent Advances in Microalgal Biotechnology, J. L. Liu, Z. Sun and H. Gerken (eds.), OMICS Group eBooks, Oster City, CA, USA, pp. 1–17.

Khan M. I., Shin J. H. and Kim J. D. (2018). The promising future of microalgae: current status, challenges and optimization of a sustainable and renewable industry for biofuels, feed and other products. *Microbial Cell Factories*, **17**(1), 1–21.

Khattar J. I. S., Singh D. P. and Kaur G. (2009). Algal Biology and Biotechnology. I.K. International, New Delhi.

Kim S. K. and Chojnacka K. (eds.). (2015). Marine algae extracts: processes, products, and applications. John Wiley & Sons.

Kiran B. R. and Venkata Mohan S. (2021). Microalgal cell biofactory – therapeutic, nutraceutical and functional food applications. *Plants (Basel)*, **10**, 836, https://doi.org/10.3390/plants10050836

Kligman D. (2000). Cosmeceuticals. *Dermatologic Clinics*, **18**, 609–615, https://doi.org/10.1016/S0733-8635(05)70211-4

Koo S., Cha K., Song D. G., Chung D. and Pan C. H. (2012). Optimization of pressurized liquid extraction of zeaxanthin from *Chlorella ellipsoidea*. *Journal of Applied Phycology*, **24**, 725–730, https://doi.org/10.1007/s10811-011-9691-2

Korzeniowska K., Górka B., Lipok J. and Wieczorek P. P. (2018). Algae and their extracts in medical treatment. In: Algae Biomass: Characteristics and Applications. Developments in Applied Phycology, K. Chojnacka, P. Wieczorek, G. Schroeder and I. Michalak (eds.), Vol 8, Springer, pp. 73–87, https://doi.org/10.1007/978-3-319-74703-3_7

Lauritano C. and Ianora A. (2016). Marine organisms with anti-diabetes properties. *Marine Drugs*, **14**, 220, https://doi.org/10.3390/md14120220

Le Goff M., Le Ferrec E., Mayer C., Mimouni V., Lagadic-Gossmann D., Schoefs B. and Ulmann L. (2019). Microalgal carotenoids and phytosterols regulate biochemical mechanisms involved in human health and disease prevention. *Biochimie*, **167**, 106–118, https://doi.org/10.1016/j.biochi.2019.09.012

Li J., Zhu D., Niu J., Shen S. and Wang G. (2011). An economic assessment of astaxanthin production by large scale cultivation of *Haematococcus pluvialis*. *Biotechnology Advances*, **29**, 568–574, https://doi.org/10.1016/j.biotechadv.2011.04.001

Liyanaarachchi V. C., Premaratne M., Ariyadasa T. U., Nimarshana P. H. V. and Malik A. (2021). Two-stage cultivation of microalgae for production of high-value compounds and biofuels: a review. *Algal Research*, **57**, 102353, https://doi.org/10.1016/j.algal.2021.102353

Luo X., Su P. and Zhang W. (2015). Advances in microalgae-derived phytosterols for functional food and pharmaceutical applications. *Marine Drugs*, **13**, 4231–4254, https://doi.org/10.3390/md13074231

Majhi P. and Samantaray S. M. (2021). Bio-reduction of hexavalent chromium by an indigenous green alga and its impact on the germination of rice seed in chromium enriched environment. *Bioremediation Journal*, **25**, 128–147, https://doi.org/10.1080/10889868.2020.1867048

Mao X., Zhang Y., Wang X. and Liu J. (2020). Novel insights into salinity-induced lipogenesis and carotenogenesis in the oleaginous astaxanthin-producing alga *Chromochloris zofingiensis*: a multi-omics study. *Biotechnology for Biofuels*, **13**, 1–, https://doi.org/10.1186/s13068-019-1642-1

Market Research Future (2019). Omega-3 PUFA market scope 2019, global industry analysis, key players, size, share, growth, trends and forecast to 2023. https://www.marketwatch.com/press-release/omega-3-pufa-market-scope-2019-global-industry-analysis-key-players-sizeshare-growth-trends-and-forecast-to-2023-2019-02-22

Market Watch (2018). Astaxanthin market challenges, key players, industry segments, development, opportunities, forecast report 2021. https://www.marketwatch.com/press-release/astaxanthin-market-challenges-key-players-industry-segments-development-opportunitiesforecast-report-2021-2018-05-23

Martin K. I. and Glaser D. A. (2011). Cosmeceuticals: the new medicine of beauty. *Missouri Medicine*, **108**, 60–63.

Martínez-Ruiz M., Martínez-González C. A., Kim D.-H., Santiesteban-Romero B., Reyes-Pardo H., Villaseñor-Zepeda K. R., Meléndez-Sánchez E. R., Ramírez-Gamboa D., Díaz-Zamorano A. L., Sosa-Hernández J. E.,

Coronado-Apodaca K. G., Gámez-Méndez A. M., Iqbal H. M. N. and Parra-Saldivar R. (2022a). Microalgae bioactive compounds to topical applications products –a review. *Molecules*, **27**, 3512, https://doi.org/10.3390/molecules27113512

Martínez-Ruiz M., Martínez-González C. A., Kim D. H., Santiesteban-Romero B., Reyes-Pardo H., Villaseñor-Zepeda K. R., Meléndez-Sánchez E. R., Ramírez-Gamboa D., Díaz-Zamorano Maswanna T. and Maneeruttanarungroj C. (2022b). Identification of major carotenoids from green alga *Tetraspora* sp. CU2551: partial purification and characterization of lutein, canthaxanthin, neochrome and β-carotene. *World Journal of Microbiology and Biotechnology*, **38**, 129, https://doi.org/10.1007/s11274-022-03320-6

Maswanna T. and Maneeruttanarungroj C. (2022). Identification of major carotenoids from green alga Tetraspora sp. CU2551: partial purification and characterization of lutein, canthaxanthin, neochrome, and β-carotene. *World Journal of Microbiology and Biotechnology*, **38**(8), 129.

Mehta P., Singh D., Saxena R., Rani R., Gupta R. P., Puri S. K. and Mathur A. S. (2018). High-value coproducts from algae –an innovational way to deal with advance algal industry. In: Waste to Wealth. Energy, Environment, and Sustainability, R. Singhania, R. Agarwal, R. Kumar and R. Sukumaran (eds.), Springer, Singapore. https://doi.org/10.1007/978-981-10-7431-8_15

Mendes R. L., Nobre B. P., Cardoso M. T., Pereira A. P. and Palavra A. F. (2003). Supercritical carbon dioxide extraction of compounds with pharmaceutical importance from microalgae. *Inorganica Chimica Acta*, **356**, 328–334.

Milovanovic I., Misan A., Saric B., Kos J., Mandic A., Simeunovic J. and Kovac D. (2012). Evaluation of protein and lipid content and determination of fatty acid profile in selected species of cyanobacteria. In *Proceedings of the 6th Central European Congress on Food*, CEFood2012, Novi Sad, Serbia, pp. 23–26.

Mimouni V., Ulmann L., Pasquet V., Mathieu M., Picot L., Bougaran G., Cadoret J. P., Morant-Manceau A. and Schoefs B. (2012). The potential of microalgae for the production of bioactive molecules of pharmaceutical interest. *Current Pharmaceutical Biotechnology*, **13**, 2733–2750, https://doi.org/10.2174/138920112804724828

Mironiuk M. and Chojnacka K. (2018). The environmental benefits arising from the use of algae biomass in industry. In: Algae Biomass: Characteristics and Applications, K. Chojnacka, P. Wieczorek, G. Schroeder and I. Michalak (eds.), Developments in Applied Phycology, Vol. 8. Springer, Cham. https://doi.org/10.1007/978-3-319-74703-3_2.

Molino A., Mehariya S., Karatza D., Chianese S., Iovine A., Casella P. and Musmarra D. (2019). Bench-scale cultivation of microalgae *Scenedesmus almeriensis* for CO_2 capture and lutein production. *Energies*, **12**, 2806, https://doi.org/10.3390/en12142806

Mona S., Malyan S. K., Saini N., Deepak B., Pugazhendhi A. and Kumar S. S. (2021). Towards sustainable agriculture with carbon sequestration and greenhouse gas mitigation using algal biochar. *Chemosphere*, **275**, 129856, https://doi.org/10.1016/j.chemosphere.2021.129856

Moorhead K., Capelli B. and Gerald C. R. (2011). Spirulina Nature's Superfood, published by Cyanotech Corporation, 73-4460 Queen Kaahumanu Hwy# 102. Kailua-Kona, HI, 96740.

Morocho-Jácome A. L., Ruscinc N., Martinez R. M., de Carvalho J. C. M., Santos de Almeida T., Rosado C., Costa J. G., Velasco M. V. R. and Baby A. R. (2020). (Bio)Technological aspects of microalgae pigments for cosmetics. *Applied Microbiology and Biotechnology*, **104**, 9513–9522, https://doi.org/10.1007/s00253-020-10936-x

Mourelle M. L., Gómez C. P. and Legido J. L. (2017). The potential use of marine microalgae and cyanobacteria in cosmetics and thalassotherapy. *Cosmetics*, **4**(4), 46, https://doi.org/10.3390/cosmetics4040046

Mutanda T., Naidoo D., Bwapwa J. K. and Anandraj A. (2020). Biotechnological applications of microalgal oleaginous compounds: current trends on microalgal bioprocessing of products. *Frontiers in Energy Research*, **8**, 598803, https://doi.org/10.3389/fenrg.2020.598803

Nakamura Y., Takahashi J. I., Sakurai A., Inaba Y., Suzuki E., Nihei S., Fujiwara S., Tsuzuki M., Miyashita H., Ikemoto H., Kawachi M., Sekiguchi H. and Kurano N. (2005). Some cyanobacteria synthesize semi-amylopectin type α-polyglucans instead of glycogen. *Plant and Cell Physiology*, **46**(3), 539–545.

Nasrabadi M. R. and Razavi S. H. (2010). Enhancement of canthaxanthin production from dietzianatronolimnaea HS-1 in a fed-batch process using trace elements and statistical methods. *Brazilian Journal of Chemical Engineering*, **27**, 517–529, https://doi.org/10.1590/S0104-66322010000400003

Nowruzi B., Sarvari G. and Blanco S. (2020). The cosmetic application of cyanobacterial secondary metabolites. *Algal Research*, **49**, 101959, https://doi.org/10.1016/j.algal.2020.101959

Pallela R. (2014). Nutraceutical and pharmacological implications of marine carbohydrates. *Advances in Food and Nutrition Research*, **73**, 183–195.

Patel A. K., Vadrale A. P., Tseng Y. S., Chen C. W., Dong C. D. and Singhania R. R. (2022). Bioprospecting of marine microalgae from Kaohsiung Seacoast for lutein and lipid production. *Bioresource Technology*, **351**, 126928, https://doi.org/10.1016/j.biortech.2022.126928

Peng J., Yuan J. P., Wu C. F. and Wang J. H. (2011). Fucoxanthin, a marine carotenoid presents in brown seaweeds and diatoms: metabolism and bioactivities relevant to human health. *Marine Drugs*, **9**, 1806–1828, https://doi.org/10.3390/md9101806

Pérez-López P., González-García S., Jeffryes C., Agathos S. N., McHugh E., Walsh D., Murray P., Moane S., Feijoo G. and Moreira M. T. (2014). Life cycle assessment of the production of the red antioxidant carotenoid astaxanthin by microalgae: from lab to pilot scale. *Journal of Cleaner Production*, **64**, 332–344, https://doi.org/10.1016/j.jclepro.2013.07.011

Perozeni F., Cazzaniga S., Baier T., Zanoni F., Zoccatelli G., Lauersen K. J. and Ballottari M. (2020). Turning a green alga red: engineering astaxanthin biosynthesis by intragenic pseudogene revival in *Chlamydomonas reinhardtii*. *Plant Biotechnology Journal*, **18**, 2053–2067, https://doi.org/10.1111/pbi.13364

Plaza B. M., Gómez-Serrano C., Acién-Fernández F. G. and Jimenez-Becker S. (2018). Effect of microalgae hydrolysate foliar application (*Arthrospira platensis* and *Scenedesmus* sp.) on *Petunia* × *hybrida* growth. *Journal of Applied Phycology*, **30**, 2359–2365, https://doi.org/10.1007/s10811-018-1427-0

Plaza M., Herrero M., Cifuentes A. and Ibanez E. (2009). Innovative natural functional ingredients from microalgae. *Journal of Agricultural and Food Chemistry*, **57**(16), 7159–7170.

Priyadarshani I. and Rath B. (2012). Commercial and industrial applications of micro algae-A review. *Journal of Algal Biomass Utilization*, **3**(4), 89–100.

Pruvost J., Le Gouic B., Lepine O., Legrand J. and Le Borgne F. (2016). Microalgae culture in building-integrated photobioreactors: biomass production modelling and energetic analysis. *Chemical Engineering Journal*, **284**, 850–861, https://doi.org/10.1016/j.cej.2015.08.118

Qu L., Ren L. J. and Huang H. (2013). Scale-up of docosahexaenoic acid production in fed-batch fermentation by Schizochytrium sp. based on volumetric oxygen-transfer coefficient. *Biochemical Engineering Journal*, **77**, 82–87.

Rahman K. M. (2020). Food and high value products from microalgae: market opportunities and challenges. Microalgae Biotechnology for Food, Health and High Value Products, M. Alam, J. L. Xu and Z. Wang (eds.), Springer, Singapore, pp. 3–27, https://doi.org/10.1007/978-981-15-0169-2_1

Ramos-Vega A., Rosales-Mendoza S., Bañuelos-Hernández B. and Angulo C. (2018). Prospects on the use of *Schizochytrium* sp. to develop oral vaccines. *Frontiers in Microbiology*, **9**, 2506, https://doi.org/10.3389/fmicb.2018.02506

Rani L., Thapa K., Kanojia N., Sharma N., Singh S., Grewal A. S., Srivastav A. L. and Kaushal J. (2020). An extensive review on the consequences of chemical pesticides on human health and environment. *Journal of Cleaner Production*, **283**, 124657, https://doi.org/10.1016/j.jclepro.2020.124657

Rastogi R. P., Madamwar D., Nakamoto H. and Incharoensakdi A. (2020). Resilience and self-regulation processes of microalgae under UV radiation stress. *Journal of Photochemistry and Photobiology C: Photochemistry Reviews*, **43**, 100322, https://doi.org/10.1016/j.jphotochemrev.2019.100322

Rebelo B. A., Farrona S., Ventura M. R. and Abranches R. (2020). Canthaxanthin, a red-hot carotenoid: applications, synthesis and biosynthetic evolution. *Plants*, **9**, 1039, https://doi.org/10.3390/plants9081039

Renuka N., Guldhe A., Prasanna R., Singh P. and Bux F. (2018). Microalgae as multifunctional options in modern agriculture: current trends, prospects and challenges. *Biotechnology Advances*, **36**, 1255–1273, https://doi.org/10.1016/j.biotechadv.2018.04.004

Ritika B. and Utpal D. (2014). Biofertilizer, a way towards organic agriculture: a review. *African Journal of Microbiology Research*, **8**(24), 2332–2343, https://doi.org/10.5897/AJMR2013.6374

Ritu J. R., Ambati R. R., Ravishankar G. A., Shahjahan M. and Khan S. (2023). Utilization of astaxanthin from microalgae and carotenoid rich algal biomass as a feed supplement in aquaculture and poultry industry: an overview. *Journal of Applied Phycology*, **35**, 145–171, https://doi.org/10.1007/s10811-022-02817-9

Rodas-Zuluaga L. I., Castillo-Zacarías C., Núñez-Goitia G., Martínez-Prado M. A., Rodríguez-Rodríguez J., López-Pacheco I. Y. and Parra-Saldívar R. (2021). Implementation of kLa-based strategy for scaling up porphyridium purpureum (red marine microalga) to produce high-value phycoerythrin, fatty acids and proteins. *Marine Drugs*, **19**, 290, https://doi.org/10.3390/md19060290

Rodolfi L., Chini Zittelli G., Bassi N., Padovani G., Biondi N., Bonini G. and Tredici M. R. (2009). Microalgae for oil: Strain selection, induction of lipid synthesis and outdoor mass cultivation in a low-cost photobioreactor. *Biotechnology and Bioengineering*, **102**(1), 100–112.

Ronga D., Biazzi E., Parati K., Carminati D., Carminati E. and Tava A. (2019). Microalgal biostimulants and biofertilisers in crop productions. *Agronomy*, **9**, 192, https://doi.org/10.3390/AGRONOMY9040192

Rosic N. N. (2021). Recent advances in the discovery of novel marine natural products and mycosporine-like amino acid UV-absorbing compounds. *Applied Microbiology and Biotechnology*, **105**, 7053–7067, https://doi.org/10.1007/s00253-021-11467-9

Russell C., Rodriguez C. and Yaseen M. (2022). High-value biochemical products and applications of freshwater eukaryotic microalgae. *Science of the Total Environment*, **809**, 151111, https://doi.org/10.1016/j.scitotenv.2021.151111

Saha S. K., Ermis H. and Murray P. (2020). Marine microalgae for potential lutein production. *Applied Sciences*, **10**, 6457, https://doi.org/10.3390/app10186457

Sansone C. and Brunet C. (2019). Promises and challenges of microalgal antioxidant production. *Antioxidants*, **8**, 199, https://doi.org/10.3390/antiox8070199

Santhosh S., Dhandapani R. and Hemalatha N. (2016). Bioactive compounds from Microalgae and its different applications-a review. *Adv Appl Sci Res*, **7**(4), 153–158.

Santiesteban-Romero B., Martínez-Ruiz M., Sosa-Hernández J. E., Parra-Saldívar R. and Iqbal H. M. (2022). Microalgae photo-protectants and related bio-carriers loaded with bioactive entities for skin applications – an insight of microalgae biotechnology. *Marine Drugs*, **20**, 487, https://doi.org/10.3390/md20080487

Sathasivam R. and Ki J. S. (2018). A review of the biological activities of microalgal carotenoids and their potential use in healthcare and cosmetic industries. *Marine Drugs*, **16**, 26, https://doi.org/10.3390/md16010026

Saxena P., Kumar M. and Harish. (2020). Autotrophic cultivation of microalgae for the production of polyunsaturated fatty acid. *Nutraceutical Fatty Acids from Oleaginous Microalgae: A Human Health Perspective*, 165–185.

Sen S. and Mallick N. (2022). Scytonemin: unravelling major progress and prospects. *Algal Research*, **64**, 102678, https://doi.org/10.1016/j.algal.2022.102678

Shah M. M. R., Liang Y., Cheng J. J. and Daroch M. (2016). Astaxanthin-producing green microalga *Haematococcus pluvialis*: from single cell to high value commercial products. *Frontiers in Plant Science*, **7**, 531.

Shahriari A. G., Afsharifar A. and Habibi-Pirkoohi M. (2019). Expression of hemagglutinin-neuraminidase (HN) and fusion (F) epitopes of Newcastle Disease Virus (NDV) in *Chlamydomonas reinhardtii*. *Plant Omics*, **12**, 63–69, https://doi.org/10.21475/poj.12.01.19.pt1864

Shamriz S. and Ofoghi H. (2019). Expression of recombinant PfCelTOS antigen in the chloroplast of *Chlamydomonas reinhardtii* and its potential use in detection of malaria. *Molecular Biotechnology*, **61**, 102–110, https://doi.org/10.1007/s12033-018-0140-1

Shao B., Wang Z., Liu X., Yu J., Lan J., Wang J., Ma L. and Chen Z. (2013). Breeding of a *Chlorella* strain with high yield of polysaccharide and its effect on growth and immunoregulation of *Litopenaeus vannamei*. *Journal of Nuclear Agricultural Sciences*, **27**, 168–172.

Sheibani Madrahi G. and Naeimpoor F. (2023). Effect of synchronized abiotic stressors and UV on scytonemin synthesis by marine cyanobacterium *Leptolynbya mycodia* in sequential two-stage cultivation. *Journal of Applied Phycology*, **35**, 43–56, https://doi.org/10.1007/s10811-022-02864-2

Sheih I. C., Wu T. K. and Fang T. J. (2009). Antioxidant properties of a new antioxidative peptide from algae protein waste hydrolysate in different oxidation systems. *Bioresource Technology*, **100**(13), 3419–3425.

Singh S. and Datta P. (2007). Outdoor evaluation of herbicide resistant strains of *Anabaena variabilis* as biofertilizer for rice plants. *Plant and Soil*, **296**, 95–102, https://doi.org/10.1007/s11104-007-9293-6

Singh D. P., Khattar J. S., Rajput A., Chaudhary R. and Singh R. (2019). High production of carotenoids by the green microalga *Asterarcys quadricellulare* PUMCC 5.1. 1 under optimized culture conditions. *PLoS ONE*, **14**, e0221930.

Singh N., Batghare A. H., Choudhury B. J., Goyal A. and Moholkar V. S. (2020). Microalgae based biorefinery: assessment of wild fresh water microalgal isolate for simultaneous biodiesel and β-carotene production. *Bioresource Technology Reports*, **11**, 100440, https://doi.org/10.1016/j.biteb.2020.100440

Song T., Martensson L., Eriksson T., Zheng W. and Rasmussen U. (2005). Biodiversity and seasonal variation of the cyanobacterial assemblage in a rice paddy field in Fujian, China. *The Federation of European Materials Societies Microbiology Ecology*, **54**, 131–140, https://doi.org/10.1016/j.femsec.2005.03.008

Sosa-Hernández J. E., Rodas-Zuluaga L. I., Castillo-Zacarías C., Rostro-Alanís M., de la Cruz R., Carrillo-Nieves D. and Parra-Saldívar R. (2019). Light intensity and nitrogen concentration impact on the biomass and phycoerythrin production by *Porphyridium purpureum*. *Marine Drugs*, **17**, 460, https://doi.org/10.3390/md17080460

Stirk W. A. and van Staden J. (2020). Potential of phytohormones as a strategy to improve microalgae productivity for biotechnological applications. *Biotechnology Advances*, **44**, 107612.

Stolz P. and Obermayer B. (2005). Manufacturing microalgae for skin care. *Cosmetics and Toiletries*, **120**(3), 99–106.

Talero E., García-Mauriño S., Ávila-Román J., Rodríguez-Luna A., Alcaide A. and Motilva V. (2015). Bioactive compounds isolated from microalgae in chronic inflammation and cancer. *Marine Drugs*, **13**, 6152–6209, https://doi.org/10.3390/md13106152

Thomas N. V. and Kim S. K. (2013). Beneficial effects of marine algal compounds in cosmeceuticals. *Marine Drugs*, **11**, 146–164, https://doi.org/10.3390/md11010146

Tokuşoglu Ö. and Üunal M. K. (2003). Biomass nutrient profiles of three microalgae: Spirulina platensis, Chlorella vulgaris, and Isochrisis galbana. *Journal of Food Science*, **68**(4), 1144–1148.

Tossavainen M., Lahti K., Edelmann M., Eskola R., Lampi A. M., Piironen V. and Romantschuk M. (2019). Integrated utilization of microalgae cultured in aquaculture wastewater: wastewater treatment and production of valuable fatty acids and tocopherols. *Journal of Applied Phycology*, **31**, 1753–1763, https://doi.org/10.1007/s10811-018-1689-6

Tran Q. G., Cho K., Kim U., Yun J. H., Cho D. H., Heo J. and Kim H. S. (2019). Enhancement of β-carotene production by regulating the autophagy-carotenoid biosynthesis seesaw in *Chlamydomonas reinhardtii*. *Bioresource Technology*, **292**, 121937, https://doi.org/10.1016/j.biortech.2019.121937

Transparency Market Research (2018). Beta-carotene market (source – fruits and vegetables, algae and fungi, synthetic; end use – food, aquaculture feed, poultry and pet feed, dietary supplements, pharmaceuticals, cosmetics) – global industry analysis, size, share, growth, trends and forecast 2019–2027.

UNEP (2009). Buildings and climate change: summary for decision-makers. United Nations Environment Programme.

Vanier G., Hempel F., Chan P., Rodamer M., Vaudry D., Maier U. G., Lerouge P. and Bardor M. (2015). Biochemical characterization of human anti-hepatitis B monoclonal antibody produced in the microalgae *Phaeodactylum tricornutum*. *PLoS ONE*, **10**, e0139282, https://doi.org/10.1371/journal.pone.0139282

Vasilopoulou MA, Ioannou E., Roussis V. and Chondrogianni N. (2021). Modulation of the ubiquitin-proteasome system by marine natural products. *Redox Biology*, **41**, 101897, https://doi.org/10.1016/j.redox.2021.101897

Victor P. and Camarena-Bernard C. (2023). Lutein, violaxanthin and zeaxanthin spectrophotometric quantification: a machine learning approach. *Journal of Applied Phycology*, **35**, 73–84, https://doi.org/10.1007/s10811-022-02855-3

Vieira V. V. (2016). The role of the value-chain for the development of high-value products from microalgae. Algal Biomass Summit 2017, European Algal Biomass Association, UABA.

Wan X., Ai C., Chen Y., Gao X., Zhong R., Liu B., Chen X. and Zhao C. (2020). Physicochemical characterization of a polysaccharide from green microalga *Chlorella pyrenoidosa* and its hypolipidemic activity via gut microbiota regulation in rats. *Journal of Agricultural and Food Chemistry*, **68**, 1186–1197, https://doi.org/10.1021/acs.jafc.9b06282

Wang H. M. D., Chen C. C., Huynh P. and Chang J. S. (2015). Exploring the potential of using algae in cosmetics. *Bioresource Technology*, **184**, 355–362, https://doi.org/10.1016/j.biortech.2014.12.001

Wang H. M. D., Li X. C., Lee D. J. and Chang J. S. (2017). Potential biomedical applications of marine algae. *Bioresource Technology*, **244**, 1407–1415, https://doi.org/10.1016/j.biortech.2017.05.198

Wijffels R. H. and Barbosa M. J. (2010). An outlook on microalgal biofuels. *Science*, **329**, 796–799, https://doi.org/10.1126/science.1189003

Wu H., Niu H., Shao A., Wu C., Dixon B. J., Zhang J. and Wang Y. (2015). Astaxanthin as a potential neuroprotective agent for neurological diseases. *Marine Drugs*, **13**, 5750–5766, https://doi.org/10.3390/md13095750

Xia D., Qiu W., Wang X. and Liu J. (2021). Recent advancements and future perspectives of microalgae-derived pharmaceuticals. *Marine Drugs*, **19**, 703, https://doi.org/10.3390/md19120703

Xie Y., Li J., Ma R., Ho S. H., Shi X., Liu L. and Chen J. (2019). Bioprocess operation strategies with mixotrophy/photoinduction to enhance lutein production of microalga *Chlorella sorokiniana* FZU60. *Bioresource Technology*, **290**, 121798, https://doi.org/10.1016/j.biortech.2019.121798

Yadav S. and Sharma S. (2019). Pesticides: problems and remedial measures. In: Environmental Contaminants and Natural Products: A Human Health Perspective, S. Sharma, M. Kumar, S. Kaur and A. K. Nagpal (eds.), Bentham Science Publishers, Sharjah, pp. 94–115.

Yang Z., Wang H., Liu N., Zhao K., Sheng Y., Pang H., Shao K., Zhang M., Li S. and He N. (2022). Algal polysaccharides and derivatives as potential therapeutics for obesity and related metabolic diseases. *Food and Function*, **13**, 11387–11409, https://doi.org/10.1039/D2FO02185D

Zaytseva A., Chekanov K., Zaytsev P., Bakhareva D., Gorelova O., Kochkin D. and Lobakova E. (2021). Sunscreen effect exerted by secondary carotenoids and mycosporine-like amino acids in the aeroterrestrial chlorophyte *Coelastrella rubescens* under high light and UV-A irradiation. *Plants*, **10**, 2601, https://doi.org/10.3390/plants10122601

Zhang Y., Ye Y., Bai F. and Liu J. (2021). The oleaginous astaxanthin-producing alga *Chromochloris zofingiensis*: potential from production to an emerging model for studying lipid metabolism and carotenogenesis. *Biotechnology for Biofuels*, **14**, 119, https://doi.org/10.1186/s13068-021-01969-z

Zhuang D., He N., Khoo K.S., Ng E.P., Chew K.W. and Ling T.C. (2022). Application progress of bioactive compounds in microalgae on pharmaceutical and cosmetics. *Chemosphere* **291**, 132932.

doi: 10.2166/9781789063547_0207

Chapter 11

Production of biopolymers from microalgae and cyanobacteria

M. Sakarika[1,2], E. Sventzouri[3], K. Pispas[3], S. S. Ali[4,5] and M. Kornaros[3]*

[1]*Center for Microbial Ecology and Technology (CMET), Faculty of Bioscience Engineering, Ghent University, Coupure Links 653, 9000 Ghent, Belgium*
[2]*Center for Advanced Process Technology for Urban Resource Recovery (CAPTURE), Frieda Saeysstraat 1, 9052 Ghent, Belgium*
[3]*Laboratory of Biochemical Engineering & Environmental Technology (LBEET), Department of Chemical Engineering, University of Patras, Karatheodori Str. 1, University Campus, 26504 Patras, Greece*
[4]*Biofuels Institute, School of the Environment and Safety Engineering, Jiangsu University, Zhenjiang 212013, China*
[5]*Botany Department, Faculty of Science, Tanta University, Tanta 31527, Egypt*
**Corresponding author: Myrsini.Sakarika@UGent.be*

ABSTRACT

Over the past few decades, plastic-derived pollution has been recognized as a major environmental issue because the use of conventional plastics results in vast amounts of waste as well as in fossil-fuel depletion. Biodegradable and biobased polymers are a promising alternative to conventional plastics. In this context, polyhydroxyalkanoates (PHAs) are bioplastics with similar mechanical and thermal properties to petroleum-based plastics which can be used in a wide range of applications. Several studies have reported the accumulation of PHAs in the biomass of microalgae and cyanobacteria. Under optimal conditions for PHA accumulation, that is, nutrient limitation, and optimal light intensity, PHA content can significantly increase, achieving 85% of dry biomass weight. Downstream recovery of PHAs is also a critical step that affects the properties and the yield of PHAs. Bioplastic production from microalgae and cyanobacteria on a commercial scale is still limited due to its high cost, with the cultivation medium accounting for up to 50% of the total production cost. The use of wastewater as a growth medium can improve the economic feasibility and sustainability of PHA production from microalgae and cyanobacteria and contribute to a more circular economy.

Keywords: biodegradable bioplastics, bioplastic recovery, biorefinery, cyanobacteria, downstream processing, microalgae, PHA blends, polyhydroxyalkanoates, sustainability, upstream processing, wastewater.

11.1 INTRODUCTION

Plastic has made our lives more convenient. This increased convenience has led to an increase in demand, which, in turn, caused an exponential increase in the production of plastics since the beginning of their industrial production, resulting in ~8 billion tons of plastic generated from 1950 onward (European Environmental Agency, 2020). The annual production of plastics steadily increases at a yearly average of 4%, from 279 million tons in 2011 to 391 million tons in 2021 (Statista, 2023). Plastics are polymeric substances, the properties of which depend on the structure of individual monomers and range from flexible to stiff, from permeable to impermeable, from hydrophilic to hydrophobic. Conventional plastics are derived from fossil-based chemicals, and are a cheap solution for strong and durable materials. Commonly used plastics include polyethylene (PE), polyethylene terephthalate (PET), polypropylene (PP), polyvinyl chloride, polystyrene (PS), and polyamides (nylons), and are constituents of a wide variety of products, including medical equipment, agricultural tools, electronic devices, and packaging (Leal Filho *et al.*, 2019).

The unblemished optimism regarding plastics changed around the 1970s. Due to their short usable life and their non-biodegradable nature, the accumulation of plastic in the environment became hard to miss, thereby damaging their reputation (Carpenter & Smith, 1972). From then on, the view on plastics has drastically changed. It is now well-known that due to their long lifetime, for instance up to 800 years as the average reported lifetime for PET (Ward & Reddy, 2020), plastics accumulate in the environment if not properly handled. In more tangible terms, estimations show that the primary plastic waste generation amounted to ~7,500 million tons in 2020, whereas over 200,000 tons of plastic waste enter the Mediterranean Sea every year (European Environmental Agency, 2020), with an economic cost ranging between $3,300 and $33,000 per ton marine plastics per year (Beaumont *et al.*, 2019). When improperly disposed of in the environment, plastics break into small insoluble pieces, referred to as microplastics (with diameters between 1 μm and 5 mm), the size of which makes them difficult to track, trace, and remove (Tirkey & Upadhyay, 2021). Microplastics have thereby entered the food chain (especially via marine animals) and are even found in women's placentas (Ragusa *et al.*, 2021). In addition to this repulsive fact, plastic production consumes ~6% of the global crude oil supply and is responsible for the generation of 2% of the global carbon dioxide (CO_2) emissions (Rosenboom *et al.*, 2022). Therefore, apart from the global health, it is crucial to pursue alternative solutions that also tackle the environmental impact. All these facts call for drastic changes regarding the generation, use, and disposal of plastics.

Biobased bioplastics, polymers derived from biological sources, can be a more sustainable alternative to conventional plastics (European Bioplastics, 2018). They can be divided into two categories, namely (1) biodegradable, for example, polylactic acid (PLA) derived from lactic acid, polyhydroxyalkanoates (PHAs), cellulose, and starch-based bioplastics, and (2) non-biodegradable, such as organic PE and PET (Rosenboom *et al.*, 2022). Advantages of bioplastics over conventional plastics include improved circularity due to the use of renewable resources, lower environmental footprint, biodegradation, and improved properties, which depend on the specific bioplastic type (Rosenboom *et al.*, 2022). Life-cycle assessments show that the substitution of conventional plastics with bioplastics, even from first-generation biofuels, requires 86% less non-renewable energy (Singh *et al.*, 2022). The production of fully biobased bioplastics is currently estimated at ~2 million tons per year (Chen, 2019), and they are expected to play a key role in future circular economy (Cheng & Gross, 2020). In this context, the biomass of microorganisms is increasingly gaining interest as a raw material for biobased products such as bioplastics. Microalgae and cyanobacteria are two microbial groups that have gained a significant share of the attention for this application, due to their potential bioplastic production from recovered resources such as nutrients and organics from wastewater, or CO_2 from off-gasses as well as their high content in targeted biopolymer precursors (Mastropetros *et al.*, 2022).

11.2 STRUCTURE AND PROPERTIES OF BIODEGRADABLE BIOPLASTICS

Biodegradable bioplastics include a range of materials derived from biological processes such as agriculture-derived polysaccharides (e.g., starch- and cellulose-based bioplastics) (Abe *et al.*, 2021), microbial fermentation products (e.g., lactic acid for PLA), and intracellular microbial components (e.g., PHAs), while the feasibility of converting the whole microbial biomass into bioplastic composites has recently been shown as well (Singha *et al.*, 2021). Starch- and cellulose-based bioplastics are interesting due to their abundance, affordability, durability, strength, and biodegradability (Abe *et al.*, 2021; Nanda *et al.*, 2022). Even though cellulose-based biopolymers are water-sensitive and lack interfacial adhesion and thermal stability, research shows that pretreatment can overcome these challenges and increase the popularity of these polymers (Polman *et al.*, 2021). Applications of cellulose-based polymers include packaging films, frames for eyeglasses, and food packaging (Nanda *et al.*, 2022). Starch-based polymers are considered to be promising to produce edible films and have similar mechanical properties and transparency to conventional polymers (Shahabi-Ghahfarrokhi *et al.*, 2019). Similar to cellulose-based polymers, starch-based polymers are also sensitive to moisture, do not have optimal mechanical properties and thermal stability, are gas permeable, and have odor issues (Nanda *et al.*, 2022; Toh *et al.*, 2008). However, combination with other polymers, essential oils, fibers, or plasticizers improves their properties (Syafiq *et al.*, 2020), and enables applications in food-packaging and pharmaceutical fields. Lactic acid monomers are further polymerized to yield PLA, a non-toxic, biocompatible polymer with mechanical properties similar to PET and PS (Karamanlioglu *et al.*, 2017). Owing to its stiffness, mechanical strength, flexibility, thermal stability, lower temperature heat sealing ability, aroma, and flavor resistance, PLA finds applications, among others in food packaging, agriculture, transportation, furniture, electronic appliances, and fabrics (Jamshidian *et al.*, 2010). Finally, the versatility and durability of PHAs has placed them in the spotlight, and their market is increasing, with projections showing an increase from 81 million USD in 2022 to 167 million USD in 2027 (Markets and Markets, 2022).

Microalgae and cyanobacteria produce various types of PHAs, including polyhydroxybutyrate (PHB), poly-3-hydroxybutyrate (P(3HB)), and co-polymers such as poly(3-hydroxybutyrate-*co*-3-hydroxyvalerate) (PHBV) (Mastropetros *et al.*, 2022). These PHAs have properties comparable to conventional plastics such as PP and PE and find applications in the food and bulk-packaging sectors. Furthermore, due to their high biocompatibility and complete biodegradability, they can have high-value applications in the biomedical sector (Costa *et al.*, 2019; Koller, 2018; Paulraj *et al.*, 2018). PHB, the most prevalent PHAs, has a higher melting point and a comparable tensile strength compared to PP and PS (Khanna & Srivastava, 2005). Nevertheless, the low flexibility (i.e., elongation at break), and high brittleness and crystallinity limit the potential applications, excluding their conversion to durable materials (Muneer *et al.*, 2020). Especially regarding crystallinity, levels above 50% yield brittle polymers and are therefore undesirable (Laycock *et al.*, 2013), with microalgal and cyanobacterial PHAs approaching this range (Table 11.1). Additionally, the temperature at which PHB undergoes thermal degradation is very close to its melting point, which causes failures in many applications (Aydemir & Gardner, 2020). Therefore, medium-chain PHA (6–14 carbon atoms) or co-polymers are preferred because they present improved properties (Table 11.1). These properties are correlated with the molecular weight and structure of the monomers (Bugnicourt *et al.*, 2014), the composition of which is determined by the genetic potential of the microorganisms to produce them. Nevertheless, common chemical modification methods have been shown to improve the properties of these microalgal and cyanobacterial PHAs and are recommended to improve the properties and increase the number of applications.

Microalgae and cyanobacteria that are able to produce PHAs have been recently reviewed and summarized by Mastropetros *et al.* (2022), and species with the highest content (up to 78%) belong to the genera *Arthrospira* sp., *Synechocystis* sp., *Synechococcus* sp., *Nostoc* sp., and *Anabaena* sp. Importantly, microalgal and cyanobacterial PHAs can be produced on side-streams, further increasing their sustainability. Despite their good prospects, currently there are only a limited number of studies that show the feasibility of this concept and test the properties of microalgae- and cyanobacterial-derived PHAs, which will be discussed in the following sections.

11.3 EMPLOYING MICROALGAE AND CYANOBACTERIA FOR BIOPLASTIC PRODUCTION

Among the different types of bioplastics currently considered as more sustainable alternatives to conventional plastics, biodegradable and biobased PHAs are considered to be a promising solution

Table 11.1 Average physical properties of conventional, fossil-based polymers, and biopolymers that are produced by microalgae and cyanobacteria.

Polymer	Crystallinity (%)	Elongation at Break (%)	Tensile Strength (MPa)	Melting Point (°C)	Glass Transition Temperature (°C)	References
PP	60	400	38	176	−10	Balaji *et al.* (2013); Hazer and Steinbüchel (2007); Verlinden *et al.* (2007)
HDPE	70	12	—	129	—	Costa *et al.* (2019)
PHB	57	6.2	31	173	1.6	Balaji *et al.* (2013); Verlinden *et al.* (2007); Koller and Rodríguez-Contreras (2015); Garcia-Garcia *et al.* (2016); Simonazzi *et al.* (2021); Bhati and Mallick (2012)
PHBV	53	70	23	153	−2.9	Balaji *et al.* (2013); Hazer and Steinbüchel (2007); Verlinden *et al.* (2007); Bhati and Mallick (2012); Samantaray and Mallick (2014)
PHB/PCL (75/25)	58	11	21	169	—	Garcia-Garcia *et al.* (2017)
PHB/PCL (25/75)	—	125	11	154	—	Przybysz *et al.* (2018)

Source: Adapted from Mastropetros *et al.* (2022).

PP, polypropylene; HDPE, high-density polyethylene; PHB, polyhydroxybutyrate; PHBV, poly(3-hydroxybutyrate-*co*-3-hydroxyvalerate); PCL: polycaprolactone. —, not reported.

with similar thermal and mechanical properties to petroleum-based plastics (Bhatia *et al.*, 2021; Medeiros Garcia Alcântara *et al.*, 2020). The first reported microbial production of PHAs dates back to 1926, from the bacterium *Bacillus megaterium* (Możejko-Ciesielska & Kiewisz, 2016). Even though bacteria have been reported to accumulate up to 90% of cell dry mass in PHAs (Obruča *et al.*, 2022), the high demand for organic carbon results in increased costs that pose a challenge in its widespread application. Microalgae and cyanobacteria can be promising alternative ways to produce PHAs. As photosynthetic microorganisms, they can utilize solar energy and CO_2 for their biomass growth while they have low-nutrient requirements (Costa *et al.*, 2019).

11.3.1 Cultivation conditions

Around 100 strains of microalgae and cyanobacteria have been reported to produce PHAs during their growth. Microalgae and cyanobacteria naturally accumulate these biopolymers as a source of carbon and energy. However, the production of PHAs is a complex metabolic process. The biomass productivity as well as the percentage and the type of PHAs that are produced depend on various parameters such as the selected carbon source and the availability of nutrients and light (Bagatella *et al.*, 2022; Cassuriaga *et al.*, 2018).

11.3.1.1 Photoautotrophic, heterotrophic, or mixotrophic operational mode

In response to shifting environmental conditions, microalgae and cyanobacteria employ different metabolic pathways. In cyanobacteria and microalgae, there are three delineated growth mechanisms. During photoautotrophic metabolism, the cells use CO_2 as a source of carbon and light as a source of energy. Under heterotrophic growth, microalgae and cyanobacteria meet their carbon and energy needs by consuming organic substances. Mixotrophic conditions combine both photoautotrophic and heterotrophic metabolic functions: energy and carbon needs can be covered by light or organic substances and organic or inorganic carbon sources, respectively. The significance of the different metabolic pathways in the cultivation of microalgae lies in their impact on the substrate that is being utilized, the amount of biomass produced, the growth rate, and the macromolecular composition of the cells.

Table 11.2 summarizes PHA production from microalgae and cyanobacteria during their growth by employing natural metabolic pathways. Photoautotrophic microalgae and cyanobacteria can accumulate PHAs (Phalanisong *et al.*, 2021). Apart from the production of these valuable compounds, these photosynthetic microorganisms can capture and utilize atmospheric CO_2 which contributes to carbon fixation (Phalanisong *et al.*, 2021). It has been reported that *Chlorella vulgaris*, *Scenedesmus obliquus*, and *Spirulina* sp. can remove CO_2 with efficiencies of 80, 28, and 53%, respectively (de Morais & Costa, 2007; Sadeghizadeh *et al.*, 2017). Few studies have reported the production of PHAs from eukaryotic microalgae in a photoautotrophic environment. More specifically, during their cultivation, *Botryococcus braunii*, *Chlorella pyrenoidosa*, and *Chlorella fusca* accumulated PHB at a concentration of 16, 27, and 5.5%, respectively (Cassuriaga *et al.*, 2018; Das *et al.*, 2018; Kavitha *et al.*, 2016b). Unlike microalgae, substantial amounts of PHAs are found in many cyanobacteria grown photoautotrophically. Several cyanobacterial strains such as *Nostoc*, *Synechocystis*, *Synechococcus*, and *Spirulina* naturally synthesize PHB at a content lower than 10% (Sirohi *et al.*, 2021).

Perceptibly, the addition of organic substances (e.g., acetic acid, xylose, glucose, and sucrose) increases the yields of PHB in cyanobacteria and microalgae (Price *et al.*, 2020). With the addition

Table 11.2 PHA production from microalgae and cyanobacteria during their cultivation under photoautotrophic, heterotrophic, or mixotrophic conditions.

Microbial Species	Mineral Medium	Carbon Source	Condition	Type of PHA	PHA Content (%)	References
B. braunii	CHU-13	—	p	PHB	16	Kavitha *et al.* (2016b)
S. salina	BG-11	—	p	P(3HB)	6.6	Kovalcik *et al.* (2017)
C. pyrenoidosa	Fogg's	—	p	PHB	27	Das *et al.* (2018)
Spirulina sp.	Zarrouk	—	p	PHB	21	Martins *et al.* (2017)
Nostoc ellipsosporum	BG-11	—	p	PHB	19	Martins *et al.* (2017)
C. fusca	BG-11	—	p	PHB	0.5–5.5	Cassuriaga *et al.* (2018)
N. muscorum	BG-11	—	p	PHB	8.5	Sharma and Mallick (2005)
N. muscorum	NO_3-free BG-11	0.11% acetate + 0.08% propionate	m	PHBV	31	Mallick *et al.* (2007)
C. fusca	BG-11	0.002% xylose	m	PHB	17	Cassuriaga *et al.* (2018)
Aulosira fertilissima	BG-11	1% fructose	m	PHB	16	Samantaray and Mallick (2012)
A. fertilissima	BG-11	0.3% acetate	m	PHB	27	Samantaray and Mallick (2012)
Synechocystis sp.	BG-11	0.4% fructose + 0.4% acetate	h	PHB	38	Panda and Mallick (2007)
Chlorogloeopsis fritschii	BG-11	0.06% acetate	m	P(3HB)	15	Zhang and Bryant (2015)

p, photoautotrophic; m, mixotrophic; h, heterotrophic.—, not reported.

of 20 mg/L xylose, the PHB content in *C. fusca* LEB 111 increased to 17% from the 5.3% that was observed in the photoautotrophic culture under the same conditions (Cassuriaga *et al.*, 2018). A significant increase was observed in the PHB content of *Nostoc muscorum* by adding different sources of organic carbon (Sharma & Mallick, 2005).

11.3.1.2 Nutrient availability

Nitrogen is a key component of proteins, nucleic acids, and chlorophyll which are necessary for the structure and function of cells (Zarrinmehr *et al.*, 2020), and therefore is an important macronutrient that affects growth. In microalgae and cyanobacteria cultivation, the availability of nitrogen must be carefully monitored to achieve optimal growth and productivity. Nitrogen can be obtained from various sources including inorganic nitrogen compounds such as nitrate, nitrite, and ammonium, and organic compounds such as urea. Nitrate is the most commonly used source of nitrogen in microalgae (Yaakob *et al.*, 2021).

Phosphorus is another important macronutrient, where nucleic acids, cell membranes, and energy storage molecules such as adenosine triphosphate are among the many cellular structures that depend on it. To promote their growth, microalgae and cyanobacteria can absorb phosphorus in the form of polyphosphate or orthophosphate, with preference for the latter due to easier assimilation (Yaakob *et al.*, 2021).

Table 11.3 presents the PHA content from different microalgae and cyanobacteria strains during their cultivation under nitrogen and phosphorus deficiency. Nitrogen and phosphorus limitation affect both biomass growth and productivity. When these nutrients are limited, the cells redirect the excess carbon toward the biosynthesis of storage compounds such as PHAs, which can be used as an energy and carbon sources under adverse conditions (Costa *et al.*, 2019). Several studies have shown an increase in PHA content in many species under nitrogen and phosphorus starvation, regardless of the cultivation mode (photoautotrophic, heterotrophic, or mixotrophic) (Dang *et al.*, 2022; Troschl *et al.*, 2017; Yashavanth *et al.*, 2021). Kaewbai-Ngam *et al.* (2016) tested 137 cyanobacterial strains for their ability to accumulate PHB. Under nitrogen limitation conditions, PHB yield increased more than 50% of

Table 11.3 PHA production from microalgae and cyanobacteria under nutrient limitation.

Species	Culture Conditions	Nutrient Limitation	Type of PHA	PHA Content (%)	References
Synechocystis sp.	Photoautotrophic	N-deficiency, P-deficiency	PHB	16	Kamravamanesh *et al.* (2017)
N. muscorum	0.28% acetate, 0.38% glucose, 0.30% valerate	N-deficiency	PHBV	78	Bhati and Mallick (2015)
Scenedesmus sp.	Glucose	P-deficiency	PHB	30	García *et al.* (2021)
Synechococcus sp.	Photoautotrophic	P-deficiency	PHB	55	Nishioka *et al.* (2001)
N. muscorum	Photoautotrophic	N-deficiency, P-deficiency	PHB	23	Panda *et al.* (2005)
N. muscorum	0.20% acetate	N-deficiency, P-deficiency	PHB	35	Sharma and Mallick (2005)
A. fertilissima	0.50% acetate	P-deficiency	PHB	77	Samantaray and Mallick (2012)
Spirulina platensis	0.50% sodium acetate	N-deficiency	P(3HB)	10	Toh *et al.* (2008)
A. fertilissima	0.26% citrate, 0.28% acetate	P-deficiency	PHB	85	Samantaray and Mallick (2012)

the screened cyanobacterial strains. *Synechococcus* sp., a thermophilic cyanobacterium, accumulated 55% PHB when it was grown photoautotrophically and under phosphate limitation (Nishioka *et al.*, 2001). The PHB content in *N. muscorum* grown photoautotrophically and heterotrophically (0.2% acetate) achieved 22.7 and 35% under nitrogen and phosphate deficiency, respectively (Panda *et al.*, 2005; Sharma & Mallick, 2005).

11.3.1.3 Light

The growth of microalgae and cyanobacteria is in most cases significantly affected by light. Under photoautotrophic and mixotrophic conditions, light is essential for photosynthesis, which produces the required energy for cell growth. Under low-light intensities, the growth of microalgae and cyanobacteria can be limited due to limited photosynthetic activity. Excessively high-light intensities can also have negative effects such as photoinhibition and cell damage. Consequently, the intensity and availability of light affect biomass production as well as the accumulation of valuable compounds such as PHAs. However, the optimal light intensity and periodicity vary based on several factors, including the selected strain and the turbidity of the cultivation medium.

Several studies have investigated the impact of light intensity and alternation of light-to-dark cycles on the productivity of PHAs by microalgae and cyanobacteria (Costa *et al.*, 2019; Price *et al.*, 2020). In the study of Ansari and Fatma (2016), *N. muscorum* was cultivated at a light intensity of 25 μmol/m^2/s under three different photoperiods. At 0.4% glucose and in 14:10, 12:12, and 10:14 h light/dark periods, its PHB content was 18, 21, and 24%, respectively. In another study, the PHB content in *C. fusca* increased from 5.3 to 17.4% when light intensity decreased from 58 to 28 μmol/m^2/s under a 6:18 h light/dark period, whereas at the same light intensities, the PHB content was 5.5 and 2.7% under a 12:12 h light/dark period (Cassuriaga *et al.*, 2018). Optimizing the light intensity and photoperiod based on the specific strain being cultivated and culture conditions can be an effective strategy to enhance PHA production.

11.3.1.4 Wastewater as a feedstock for microalgae and cyanobacteria cultivation

Wastewater instead of a potential environmental hazard can be seen as a potential source of nitrogen and phosphorus for the growth of microalgae and cyanobacteria and be upgraded to valuable products (Sakarika *et al.*, 2022). The cultivation of these photosynthetic microorganisms in wastewater is a promising alternative to wastewater treatment as high nutrient removal and high biomass productivity can be achieved (Rizwan *et al.*, 2018). Cultivation of *S. obliquus* in soybean wastewater removed 72% of chemical oxygen demand, 95% total nitrogen, and 54% total phosphorus (Shen *et al.*, 2020). Similarly, *C. vulgaris* cultivated in meat wastewater removed 89% of chemical oxygen demand, 52% of total nitrogen, and 70% of total phosphorus (Hu *et al.*, 2019).

The cost of PHA production from microalgae and cyanobacteria is high compared to the conventional plastics industry. High feedstock and water requirements account for more than 50% of the production cost (Medeiros Garcia Alcântara *et al.*, 2020). To enable cost-effectiveness and feasibility on a larger scale, scientific interest has focused on the utilization of wastewater as a substrate for the cultivation of microalgae and cyanobacteria. Apart from the reduction of the upstream cost of the process and the bioremediation of wastewater, using wastewater as feedstock does not compete with raw materials such as sugars, which can also be used for PHA production (Medeiros Garcia Alcântara *et al.*, 2020). Studies have demonstrated that during the cultivation of microalgae and cyanobacteria in wastewater it is feasible to produce substantial amounts of PHAs with similar properties to conventional plastics. A PHB yield of 247 mg/L was reported by *B. braunii* grown in sewage wastewater at a concentration of 60% (Kavitha *et al.*, 2016a). In another study, *Synechocystis* sp. grown on shrimp wastewater accumulated PHB at a concentration of 33% while the removal efficiency of phosphate was 97% (Krasaesueb *et al.*, 2019).

Table 11.4 summarizes the production of PHAs from microalgae and cyanobacteria cultivated in different types of wastewater. Among the different wastewater types, anaerobic digestion effluents are

Table 11.4 PHA production from microalgae and cyanobacteria cultivated in wastewater.

Species	Type of Wastewater	Temperature (°C)	pH	Type of PHA	PHA Content (%)	References
Synechocystis sp.	Shrimp wastewater	27–30	7.0–9.0	PHB	34	Krasaesueb *et al.* (2019)
N. muscorum	Poultry litter	25 ± 2	7.0	PHB	23	Bhati and Mallick (2016)
N. muscorum	Poultry litter + 10% CO_2	25 ± 2	7.0–8.0	PHBV	65	Bhati and Mallick (2016)
S. salina	Digestate supernatant	25 ± 1	—	PHB	6.3	Meixner *et al.* (2016)
Synechocystis sp.	30% palm oil mill effluent + BG-11 medium	28	8.2	PHB	15	Nur *et al.* (2023)
Synechococcus sp.	30% palm oil mill effluent + BG-11 medium	28	8.2	PHB	15	Nur *et al.* (2023)
B. braunii	50% palm oil mill effluent + glycerol + Fe-EDTA*	30	7.5	PHB	33	Nur (2022)

*EDTA, ethylenediaminetetraacetic acid; —, not reported.

generated in high volumes urging the need to implement a more sustainable disposal way than the current use as fertilizer. Digestates can be upgraded to higher value products when used as a substrate for the cultivation of microalgae and cyanobacteria and the production of value-added compounds as they contain high organic matter and are rich in nutrients such as ammonium-nitrogen and phosphorus (Kaur *et al.*, 2020; Koutra *et al.*, 2018). Only a few studies have investigated the production of PHAs from microalgae and cyanobacteria in digestates. For instance, *Synechocystis salina* was cultivated in diluted digestate and accumulated PHB at a concentration of 6.3% (Meixner *et al.*, 2016).

Overall, the cultivation of microalgae and cyanobacteria in wastewater seems to be an environmentally friendly and promising alternative for sustainable wastewater treatment and production of PHAs. However, fluctuations in the composition of the produced wastewater and the presence of potentially hazardous components can affect the entire process and even inhibit the growth of microalgae and cyanobacteria. Further studies to address these challenges are necessary before the implementation of the process at a larger scale (Mastropetros *et al.*, 2022; Medeiros Garcia Alcântara *et al.*, 2020).

11.3.2 Advantages of PHA production from microalgae and cyanobacteria compared to bacteria

PHAs are naturally produced by various microorganisms, with bacteria in the genera *Pseudomonas*, *Ralstonia*, *Bacillus*, and *Aeromonas* accumulating PHAs at high content. However, PHA production using bacteria is expensive and not feasible for large-scale applications due to the prohibitive cost of the organic carbon sources and oxygen requirements (Możejko-Ciesielska & Kiewisz, 2016; Samantaray & Mallick, 2015).

On the contrary, microalgae and cyanobacteria seem to be promising microorganisms for PHA production, utilizing atmospheric CO_2 and generating energy through photosynthesis. These photosynthetic microorganisms do not need exogenous organic sources for their biomass growth reducing the overall cost of the process by up to 50% (Medeiros Garcia Alcântara *et al.*, 2020; Phalanisong *et al.*, 2021). Microalgae and cyanobacteria can utilize CO_2 that is present in flue gases for PHA production, providing a sustainable solution to greenhouse gas emissions and making the process economically feasible. For instance, *S. salina* and *Synechococcus elongatus* directly utilizing industrial flue gases accumulated PHB at a content of 6.6 and 11%, respectively (Roh *et al.*, 2021; Troschl *et al.*, 2017).

Moreover, microalgae and cyanobacteria can produce more than one bioproduct. For instance, they can accumulate high amounts of lipids, proteins, polyunsaturated fatty acids, and pigments that can be utilized as raw materials for producing bioenergy and other valuable products used in a variety of sectors including food, cosmetics, nutraceutical, and pharmaceutical industries (Kumar *et al.*, 2020). The implementation of a biorefinery concept is a complex procedure, where both upstream and downstream processing can significantly affect the entire process. The type and yield of the produced compounds strongly depend on the selected strains and the cultivation conditions while developing effective methods for extracting and purifying the various compounds from the microbial biomass is challenging and requires a considerable amount of energy (Siddiki *et al.*, 2022). There are only a few experimental data available for the simultaneous production of PHAs and other valuable compounds in a biorefinery concept. *Arthrospira platensis* cultivated in palm oil mill effluent was investigated for the co-production of PHB and C-phycocyanin. Results showed that the productivities of PHB and C-phycocyanin using 50% palm oil mill effluent were 7 and 16 mg/L/day, respectively (Nur, 2022). Another study demonstrated the possibility of cultivating *Synechocystis* sp. in secondary effluent to produce PHB and lipids (Senatore *et al.*, 2023). The ability of microalgae and cyanobacteria to utilize flue gases and wastewater to produce PHAs as well as other valuable compounds in a biorefinery concept can render the microalgae cultivation technology at a large scale economically feasible and environmentally friendly.

11.3.3 PHA blends

PHAs present several environmental benefits as bio-based and biodegradable polymers. However, their industrial application is still limited due to their high production cost. Additionally, for PHA production to become competitive with the petroleum-based plastic industry, the produced biopolymer must have similar properties to conventional plastics. To overcome these limitations, blending PHAs with raw materials and other biodegradable polymers has emerged as a promising and simple approach. The type and properties of the produced polymer blends depend on the choice of the starting constituents and their blending ratio. PHA blending aims to improve the mechanical properties, such as increased tensile strength, elongation at break, and impact resistance, that can be used in a wide range of applications, enhance the biodegradability of the material, reduce the cost, and improve the overall performance (Kumar *et al.*, 2021).

11.3.3.1 PHA blends with raw materials

Starch is considered a highly promising natural polymer because it is biodegradable and widely available in large quantities. PHA/starch blends have improved mechanical properties compared to pure PHAs. In the study of Asl *et al.* (2021), an electrospinning method was used to blend PHB with different concentrations of starch (5–15 wt%). By adding starch at a concentration of up to 10%, the tensile strength of the PHB/starch scaffolds increased from 3 to 16 MPa. The presence of starch also enhanced the thermal stability and degradation rate. The results of this study suggest that electrospun scaffolds produced from PHB/starch could be used in bone tissue engineering applications. In another study, the blend of PHB with modified corn starch was investigated. When the starch concentration increased, an increase in glass transition temperature from 2 to 37°C was observed (Lai *et al.*, 2015). The mechanical and thermal properties of PHA/starch blends can be significantly improved with the addition of cross-linking agents such as citric and adipic acids (Sun *et al.*, 2018).

Lignin is a complex organic polymer and the second most abundant renewable natural polymer on the Earth. PHA blends with lignin offer a promising approach to the development of new materials with improved mechanical properties and biodegradability compared to either material alone. Therefore, by combining lignin with PHAs, it is possible to create new materials with a range of desirable properties that can be used in various applications (Kumar *et al.*, 2021). Lugoloobi *et al.* (2020) reported that

PHB/lignin blends showed higher glass transition temperatures, improved ultraviolet resistance and tensile performance, and higher melt viscosity making them suitable for packaging applications.

Cellulose derivatives are becoming increasingly popular as components that can be blended with PHAs due to their compatibility and their ability to accelerate the degradation of PHAs. Cellulose, acetate, butyrate, ethyl cellulose, and cellulose propionate are cellulose derivatives that are commonly used as drug carriers, blood coagulants, and coatings for pharmaceutical tablets (Sharma *et al.*, 2021). Cellulose-based microfibers (MFs) can be used to enhance the properties of PHA films. According to Mármol *et al.* (2020), the addition of MFs made the PHA film 23% more durable, as both the tensile strength and Young's modulus increased. Overall, PHA blends with raw materials result in new biopolymers with improved properties when compared to either of the individual components.

11.3.3.2 PHA blends with biodegradable polymers

PLA is a biodegradable polymer derived from renewable resources. Blending PHAs with PLA is the most studied approach as it can result in material with improved mechanical properties and biodegradability, with the specific properties depending on the PHA to PLA ratio. The PHA/PLA blend has been used in three-dimensional printing, where the printed materials exhibited favorable mechanical properties and thermal stability (Ausejo *et al.*, 2018). In another study, the PHA/PLA blend was demonstrated to have the capacity to absorb oil from water, which is similar to that of currently utilized absorbents (Iordanskii *et al.*, 2019).

Polycaprolactone (PCL) is a synthetic biodegradable polymer with a low melting point. Blending PCL with other biopolymers, such as PHAs, can improve biodegradability and decrease the production cost. The degradation rate of the blend as well as the mechanical properties can be controlled by adjusting the ratio of PCL to PHA. For instance, higher maximum stress was exhibited in blends rich in PHB, whereas blends rich in PCL led to greater strain at break. The PCL/PHA blend can be utilized in various biomedical applications, especially in tissue engineering (Kumar *et al.*, 2021; Li *et al.*, 2016).

Blending PHAs with poly(butylene adipate-*co*-terephthalate) (PBAT) is another approach to developing biodegradable polymer blends with improved properties and increasing the field of their application (Tian & Wang, 2020). Similar to PCL, PBAT is a synthetic biodegradable polymer. The blend can be processed using common techniques such as injection molding and extrusion. During injection molding of PHBV with PBAT, as the PBAT content increased the toughness and strain at break increased, while the specific modulus and strength decreased (Javadi *et al.*, 2010). In another study, the addition of PBAT increased the shear storage modules of the PHB/PBAT blends and decreased the tensile storage modulus (Larsson *et al.*, 2016). Overall, blending PHAs with other biodegradable polymers is a promising alternative to reduce production costs and develop new materials with improved properties and biodegradability, making them attractive for a wide range of applications.

11.4 DOWNSTREAM PROCESSING OF BIOPLASTIC RECOVERY FROM MICROALGAE AND CYANOBACTERIA

During the past few years, increased research efforts aim at developing more efficient harvesting, pretreatment, and extraction techniques, with the hope of lowering the cost of microalgal bioplastics production. These efforts include exploring and evaluating various methods and technologies that can be used to optimize the downstream process. By reducing the costs associated with these production stages, the development of sustainable bioplastics derived from microalgae and cyanobacteria can become economically viable and contribute to a more sustainable future. In addition, this can also lead to the development of new and more efficient approaches to produce other valuable products from these microorganisms toward a biorefinery concept.

11.4.1 Harvesting

The relatively low final biomass concentrations in microalgal and cyanobacterial cultures (with values from 0.5 g/L in open-pond systems to 5 g/L in photobioreactors), resulting from light restriction due to shading from cell growth, lead to the urge for separation of the biomass from a large water volume (Pahl *et al.*, 2013; Vandamme *et al.*, 2013). To achieve the desired solid–liquid separation during harvesting, various mechanical-, chemical-, biological-, or electrical-based techniques can be used through one or more steps (Mata *et al.*, 2010; Morais Junior *et al.*, 2020). The selection of an appropriate harvesting method depends on several factors, such as the microalgal cell morphology (e.g., filamentous, spherical, or elongated), the biomass concentration in the culture medium, the specific gravity, and size (typically microalgae will be in the range of 0.5–200 μm) of cells (Caroppo & Pagliara, 2022; Gerardo *et al.*, 2015; Roy & Mohanty, 2019). Additionally, the surface charge of microalgae and cyanobacteria, which is estimated by their zeta potential, plays a key role in downstream processing by preventing cells from clumping together and leading to a stable cell suspension (Krishnan *et al.*, 2022). This potential can fluctuate significantly, depending on factors such as cell age and culture conditions (e.g., salinity and pH), and ranges from −5 to −80 mV (Greenwell *et al.*, 2010; Yang *et al.*, 2022; Zhang *et al.*, 2013). Considering the above, harvesting biomass is one of the main obstacles in downstream processing as it requires large amounts of energy and it has been stated that the cost of collecting and drying the biomass from wet cultures is ~20–30% of the total operational cost of biomass production (Molina Grima *et al.*, 2003; Price *et al.*, 2022).

Highly efficient and minimally damaging methods for separating the biomass from the culture medium are essential during harvesting. There is a plethora of available methods for harvesting microalgae and cyanobacteria (Vasistha *et al.*, 2021), where the appropriate method depends on the characteristics of the microorganism, the properties of the culture medium, and the intended application of the harvested biomass. Furthermore, it is a widespread practice to combine two or more methods to achieve a higher separation efficiency while reducing the costs involved (Barros *et al.*, 2015). Next, we discuss various separation techniques for microalgal and cyanobacterial biomass harvesting.

11.4.1.1 Centrifugation

Centrifugation methods use force to separate particles based on the different densities between the particles. This allows microbial cells, that are denser than the culture medium, to settle (Pahl *et al.*, 2013). Centrifugation is one of the most common harvesting methods on lab scale and can be applied to most microalgae and cyanobacteria. Compared to gravity sedimentation, centrifugal force accelerates sedimentation, leading to a higher biomass recovery efficiency. Additionally, centrifugation eliminates the need for chemicals (e.g., flocculants), which could decrease the quality of the biomass. However, the high energy requirements (up to 8 kWh/m^3) limit its large-scale application to high-value products (Barros *et al.*, 2015; Laamanen *et al.*, 2016; Pahl *et al.*, 2013).

11.4.1.2 Filtration

Membrane filtration is a commonly used technique for biomass separation and can be considered a viable harvesting option. During this process, the liquid fraction of the culture is allowed to pass through a porous membrane, usually by applying pressure or a vacuum to the system, while the cells are retained. The ability of solute or solid to pass through a particular porous membrane is dependent on its dimensions, electrical charge, and morphology. Additionally, factors such as the viscosity and mixing rate of the suspension can impact this process (Mathimani & Mallick, 2018). Due to the relatively low-energy requirements (0.2–0.88 kWh/m^3) and cost, combined with the ease of scalability, this method is highly advantageous (Pahl *et al.*, 2013). Also, similar to centrifugation, no chemicals are needed, thereby avoiding the qualitative degradation of the recovered biomass. However, the accumulation of microalgal deposits on the filter, leading to fouling (or clogging) of the membrane is the primary limitation of these methods, and it raises their operational costs. Membrane fouling is primarily caused by extracellular polymeric substances (EPSs), which are organic compounds secreted by microalgae during their growth or released upon cell lysis (Singh & Patidar, 2018).

11.4.1.3 Flocculation and coagulation

Flocculation/coagulation is an economical method for harvesting microalgae and cyanobacterial biomass due to large culture volumes and the need for a universal process that can be applied to various species. Flocculation/coagulation involves the use of inorganic (e.g., $Al_2(SO_4)_3$, $FeCl_3$, and $Fe_2(SO_4)_3$) or organic (e.g., poly(diallyldimethylammonium chloride), PDADMAC) salts, which work by neutralizing the negative charges of cells, resulting in clustering of particles, allowing the suspension to concentrate up to 100 times (Mubarak *et al.*, 2019; Singh & Patidar, 2018; Vandamme *et al.*, 2013). Combining this technique with gravity sedimentation reduces the energy demand of the overall operation, leading to an economically viable harvesting process (Barros *et al.*, 2015). However, a major disadvantage of using aluminum or iron salts as flocculants is that any remaining chemicals can be a potential environmental and health hazard. Also, the use of organic flocculants appears to negatively affect the levels of unsaturated fatty acids in the recovered biomass (Laamanen *et al.*, 2016). In recent years, several studies have been conducted on bioflocculation, in which microalgae cluster together with various microorganisms, including bacteria, fungi, or other microalgae (Kumar *et al.*, 2023). The above procedure can be carried out with the use of bioflocculants, which are usually EPSs produced by several microorganisms (Moreira *et al.*, 2022).

11.4.1.4 Gravity sedimentation

One of the simplest methods for liquid–solid separation is gravity sedimentation. Although this form requires low operating and designing costs, the fluctuating densities and consequently the low sedimentation rates (0.1–2.6 cm/h) of most microalgae, make the process relatively time-consuming, with the risk of degrading the collected biomass (Barros *et al.*, 2015; Greenwell *et al.*, 2010). Therefore, in most cases, gravity sedimentation takes place after a flocculation/coagulation step (Chatsungnoen & Chisti, 2016). Finally, the high self-sedimentation property of some species, such as cyanobacteria *Chlorogloea fritschii*, *Phormidium* sp., and microalga *Golenkinia* sp., eliminates the need for additional energy and reduces the cost and time required to harvest biomass (Hotos *et al.*, 2023; Monshupanee *et al.*, 2016; Nie *et al.*, 2018).

11.4.1.5 Flotation

Flotation is another separation technique based on air or gas bubbles that adhere to the surface of the particles, achieving their transport to the surface and facilitating the separation (Pahl *et al.*, 2013). Furthermore, some cyanobacteria float on their own, as they possess intracellular gas vesicles (aerotopes) (Duval *et al.*, 2021). Flotation is often combined with flocculation/coagulation techniques for optimal harvesting results. Flotation cells are typically supplied with air via dispersed air, dissolved air, or electrolytic mechanisms. Currently, the most common flotation methods are dissolved air flotation (with bubble diameters less than 100 μm), dispersed air flotation or foam flotation (bubble diameters between 100 and 1,000 μm), electrolytic flotation, and dispersed ozone flotation (Barros *et al.*, 2015). The efficiency of smaller sized gas bubbles increased, compared to larger bubbles, as they possess a larger surface area per unit volume. The larger the surface area, the greater is the chance of collision between air bubbles and particles (Pahl *et al.*, 2013). Qi *et al.* (2022) achieved a harvesting efficiency of 96% for the microalga *Tribonema* sp. using flotation, with a significantly lower amount of energy (0.19 kWh/kg biomass) compared to other harvesting methods.

11.4.2 Drying

The extraction techniques for most bioplastics (especially PHAs) from microalgae and cyanobacteria presupposes the drying of the biomass, as the residual water can have a significant effect on their efficiency. Thus, a reliable drying method such as freeze drying, convective drying, spray drying, or solar drying is necessary (Levett *et al.*, 2016). It is estimated that biomass drying can account for up to 20% of the overall cost of producing PHAs from cyanobacteria, posing a barrier to upscaling commercial production (Costa *et al.*, 2019). Solar drying is an inexpensive dehydration technique but requires extended drying periods, due to the low temperature, and a large land area. In addition, the slow dehydration rate can promote bacterial growth and consequently degradation of the microalgal biomass (Chen *et al.*, 2015). However, in closed solar systems an increase in the drying rate can be achieved, leading to drying of the biomass in 3–5 h at a temperature of 60°C (Prakash *et al.*, 1997). Lyophilization (freeze drying) and spray drying are techniques commonly used to remove water from microalgal biomass. Unlike convective drying, these methods preserve all cellular components without damaging the cell wall (Chen *et al.*, 2015). Spray drying is generally considered more advantageous than lyophilization due to its faster drying speed, ability for continuous operation, and lower cost, but there is a greater possibility of oxidation of carotenoids (Zhang *et al.*, 2022). Nevertheless, until now there is no evidence on how each drying method affects the structure and physicochemical characteristics of the recovered PHAs from microalgal/cyanobacterial biomass.

11.4.3 Extraction

Recent research efforts have focused on developing extraction techniques that reduce the overall cost of producing bioplastics. The commonly used extraction methods are based on organic solvents, usually halogenated. Chloroform and dichloromethane are commonly used solvents as they dissolve bioplastics but no other biological products (Levett *et al.*, 2016). After biomass dehydration, the disruption of the cell membrane takes place, which can be achieved using organic solvents or physical stress, so that the solvent can come into contact with the PHA granules, which are trapped intracellularly (Mastropetros *et al.*, 2022). Additionally, a pretreatment step could be applied prior to extraction to enhance the recovery, usually using sodium hypochlorite (Kosseva & Rusbandi, 2018). Following the extraction of bioplastics from the dry biomass, a suitable solvent such as methanol is used for the recovery, and partial purification of the product, a method known as liquid antisolvent precipitation. Although organic solvents are effective in creating a product with minimal reduction in the molecular weight of the polymer, they are costly and are an environmental hazard (Kosseva & Rusbandi, 2018). Therefore, new environmental-friendly, sustainable, and profitable technologies are needed to scale up and commercialize bioplastic production.

Biomass hydrolysis could be a potential method for recovering biopolymers. By using an acid or base solution, the cells are hydrolyzed, leaving the bioplastic granules undissolved (López-Abelairas *et al.*, 2015). However, some chemical compounds used for this process seem to have a negative effect on the molecular weight and characteristics of the recovered bioplastics (Mastropetros *et al.*, 2022). To prevent such issues, the use of enzymes (e.g., trypsin, bromelain and lysozyme) has been proposed, because they can denature the cell wall during biomass treatment without degrading PHAs. Enzymatic methods for PHA extraction typically involve a heat pretreatment and enzymatic hydrolysis (Kapritchkoff *et al.*, 2006).

Supercritical fluids, such as supercritical CO_2, have been suggested as substitutes for organic solvents for extracting and purifying PHAs. More specifically, supercritical CO_2 can extract up to 90% of the PHA content at purity levels ranging from 86 to 99% and can be used as a secondary step to remove oily biomass residues and refine the bioplastics (Kosseva & Rusbandi, 2018; Mastropetros *et al.*, 2022). However, the high operational costs associated with supercritical fluid extraction and purification processes have impeded their widespread implementation. Nevertheless, the non-hazardous, non-combustible, and low-reactivity nature of supercritical fluids makes them an attractive alternative to organic solvent extraction methods (Mastropetros *et al.*, 2022).

There are various biodegradable, eco-friendly, and recyclable solvents that can be used for the extraction and purification of PHAs, including alcohols, acetone, ketones, and ethylene carbonate. Dimethyl carbonate is another green solvent that shows good performance and does not cause degradation of PHAs such as halogenated solvents. Ethylene carbonate is also used to recover a higher quantity of PHAs without causing degradation (Kurian & Das, 2021). A recent study compared various solvents and found that dimethyl carbonate is a more environmentally friendly and less hazardous choice for PHA extraction from biomass (Koller, 2020). In addition, ionic liquids are being increasingly favored as a solvent for extraction, and they have the potential to replace traditional organic solvents, as they behave similarly because of their electrically charged ions (Mastropetros *et al.*, 2022). It has been noted that the use of ionic liquids as solvents for extraction offers the benefit of being able to recover the ionic liquid, thereby increasing the viability of the process (Dubey *et al.*, 2018).

11.5 CHALLENGES AND FUTURE PERSPECTIVES

One of the main bottlenecks in the widespread adoption of PHAs from microalgae and cyanobacteria is their accumulation at low percentages on a dry weight basis. Therefore, strategies to enhance the productivity of PHAs are necessary. Process optimization by controlling the cultivation conditions, such as light intensity, pH, and temperature can improve microalgal growth and PHA accumulation. Supplementation of organic carbon sources (e.g., simple sugars) and nutrient limitation (e.g., nitrogen or/and phosphorus starvation) have also been reported to increase PHA accumulation by microalgae and cyanobacteria (Costa *et al.*, 2019). Strain improvement via genetic engineering could be another option to enhance the production of PHAs. For instance, genetic modification of *Synechocystis* sp. enhanced PHB production up to 35% in dry cell weight (Sirohi *et al.*, 2021). However, there are concerns related to the safety and ethical implications of using genetically modified microorganisms (Chia *et al.*, 2020; Sirohi *et al.*, 2021).

The industrial application of PHA production is still limited due to its high cost. Despite the technological advances, PHA costs 5€/kg compared to the production cost of synthetic plastics, which ranges from 0.8€ to 1.5€/kg. One way to reduce the production cost is to reduce the cultivation cost. The feedstock used for the cultivation of microalgae and cyanobacteria represents more than half of the production cost. The utilization of wastewater streams as raw materials seems to be a promising alternative as they are widely available and enriched in organic carbon and nutrients that microalgae and cyanobacteria need for their growth and the production of valuable compounds. This approach will not only diminish the cost of PHA production but also contribute to the bioremediation of wastewaters. However, several issues need to be addressed. The feedstock composition strongly affects the type and yield of PHA produced. The combination of different streams of wastewater or their dilution with water can assure its constant characteristics and decrease the turbidity caused by suspended particles. Furthermore, cultivation in wastewater can affect the end-life of the produced PHA as it may contain impurities that could potentially compromise the biocompatibility of the resulting plastics (Khatami *et al.*, 2021; Medeiros Garcia Alcântara *et al.*, 2020). As discussed in Section 11.3.1.1, microalgae and cyanobacteria can photoautotrophically accumulate PHAs using CO_2 as the sole carbon source. The capture of flue gases, which are rich in CO_2, for PHA synthesis can reduce the production costs while promoting CO_2 mitigation and the reduction of greenhouse gases with several environmental benefits (Sirohi *et al.*, 2021).

The downstream processing is also a critical and costly step in the production of PHAs from microalgae and cyanobacteria. The properties, purity, and yield of the produced PHAs, apart from the potential for specific microalgae to produce them, also depend on the extraction methods used. The most common strategy is the extraction of PHAs with organic solvents such as chloroform and acetone. However, this method creates waste and need extra costs. Therefore, it is necessary to investigate alternative extraction methods, such as enzymatic ones, and the use of different solvents that are recyclable to establish downstream processes that are both cost-effective and environmentally sustainable without affecting the efficiency of the process (Kurian & Das, 2021).

The implementation of a biorefinery concept is a sustainable and economically viable method for PHA production. Cultivation of microalgae and cyanobacteria using carbon flue gases and wastewater for the production of PHAs and other value-added compounds has several environmental and economic benefits. The downstream processing in this approach is still challenging as the separation of the various compounds is difficult and a high amount of energy is required. In conclusion, PHA production from microalgae has the potential to be a more sustainable and environmentally friendly alternative to petroleum-based plastics. However, several challenges need to be addressed to enable cost-effective and scalable PHA production technologies.

11.6 CONCLUSION

To enable a more sustainable future, the transition toward the utilization of biodegradable bioplastic materials derived from renewable sources is necessary. Microalgae and cyanobacteria are promising candidates for PHA production which can have similar thermal and mechanical properties to conventional plastics. Considering that the downstream processing significantly affects the yield and the properties of the produced PHAs, further research is required to optimize the extraction methods as well as to decrease the dependence on organic solvents. However, the industrial application of bioplastics is still limited due to their high cost. Ongoing research has focused on enhancing the PHA productivity and reducing the cost of the process. PHA production is possible during the cultivation of microalgae and cyanobacteria in various types of wastewaters and side-streams, which could increase the sustainability of the process. Valorization of wastewater and CO_2 from flue gases in the cultivation of microalgae and cyanobacteria to produce PHAs and other valuable co-products such as biofuels and pigments can be the key to the application of bioplastics on an industrial scale.

REFERENCES

Abe M. M., Martins J. R., Sanvezzo P. B., Macedo J. V., Branciforti M. C., Halley P., Botaro V. R. and Brienzo M. (2021). Advantages and disadvantages of bioplastics production from starch and lignocellulosic components. *Polymers*, **13**(15), 2484, https://doi.org/10.3390/polym13152484

Ansari S. and Fatma T. (2016). Cyanobacterial polyhydroxybutyrate (PHB): screening, optimization and characterization. *PLoS ONE*, **11**(6), e0158168, https://doi.org/10.1371/journal.pone.0158168

Asl M. A., Karbasi S., Beigi-Boroujeni S., Benisi S. Z. and Saeed M. (2021). Evaluation of the effects of starch on polyhydroxybutyrate electrospun scaffolds for bone tissue engineering applications. *International Journal of Biological Macromolecules*, **191**, 500–513, https://doi.org/10.1016/j.ijbiomac.2021.09.078

Ausejo J. G., Rydz J., Musioł M., Sikorska W., Sobota M., Włodarczyk J., Adamus G., Janeczek H., Kwiecień I. and Hercog A. (2018). A comparative study of three-dimensional printing directions: the degradation and toxicological profile of a PLA/PHA blend. *Polymer Degradation and Stability*, **152**, 191–207, https://doi.org/10.1016/j.polymdegradstab.2018.04.024

Aydemir D. and Gardner D. J. (2020). Biopolymer blends of polyhydroxybutyrate and polylactic acid reinforced with cellulose nanofibrils. *Carbohydrate Polymers*, **250**, 116867, https://doi.org/10.1016/j.carbpol.2020.116867

Bagatella S., Ciapponi R., Ficara E., Frison N. and Turri S. (2022). Production and characterization of polyhydroxyalkanoates from wastewater via mixed microbial cultures and microalgae. *Sustainability*, **14**(6), 3704, https://doi.org/10.3390/su14063704

Balaji S., Gopi K. and Muthuvelan B. (2013). A review on production of poly β hydroxybutyrates from cyanobacteria for the production of bio plastics. *Algal Research*, **2**(3), 278–285, https://doi.org/10.1016/j.algal.2013.03.002

Barros A. I., Gonçalves A. L., Simões M. and Pires J. C. M. (2015) Harvesting techniques applied to microalgae: a review. *Renewable and Sustainable Energy Reviews*, **41**, 1489–1500, https://doi.org/10.1016/j.rser.2014.09.037

Beaumont N. J., Aanesen M., Austen M. C., Börger T., Clark J. R., Cole M., Hooper T., Lindeque P. K., Pascoe C. and Wyles K. J. (2019). Global ecological, social and economic impacts of marine plastic. *Marine Pollution Bulletin*, **142**, 189–195, https://doi.org/10.1016/j.marpolbul.2019.03.022

Bhati R. and Mallick N. (2012). Production and characterization of poly(3-hydroxybutyrate-*co*-3-hydroxyvalerate) co-polymer by a N_2-fixing cyanobacterium, *Nostoc muscorum* Agardh. *Journal of Chemical Technology and Biotechnology*, **87**(4), 505–512, https://doi.org/10.1002/jctb.2737

Bhati R. and Mallick N. (2015). Poly(3-hydroxybutyrate-*co*-3-hydroxyvalerate) copolymer production by the diazotrophic cyanobacterium *Nostoc muscorum* Agardh: process optimization and polymer characterization. *Algal Research*, **7**, 78–85, https://doi.org/10.1016/j.algal.2014.12.003

Bhati R. and Mallick N. (2016). Carbon dioxide and poultry waste utilization for production of polyhydroxyalkanoate biopolymers by *Nostoc muscorum* Agardh: a sustainable approach. *Journal of Applied Phycology*, **28**(1), 161–168, https://doi.org/10.1007/s10811-015-0573-x

Bhatia S. K., Otari S. V., Jeon J. M., Gurav R., Choi Y. K., Bhatia R. K., Pugazhendhi A., Kumar V., Rajesh Banu J., Yoon J. J., Choi K. Y. and Yang Y. H. (2021). Biowaste-to-bioplastic (polyhydroxyalkanoates): conversion technologies, strategies, challenges, and perspective. *Bioresource Technology*, **326**, 124733, https://doi.org/10.1016/j.biortech.2021.124733

Bugnicourt E., Cinelli P., Lazzeri A. and Alvarez V. (2014). Polyhydroxyalkanoate (PHA): review of synthesis, characteristics, processing and potential applications in packaging. *eXPRESS Polymer Letters*, **8**(11), 791–808, https://doi.org/10.3144/expresspolymlett.2014.82

Caroppo C. and Pagliara P. (2022). Microalgae: a promising future. *Microorganisms*, **10**(8), 1448, https://doi.org/10.3390/microorganisms10081488

Carpenter E. J. and Smith K. L. (1972). Plastics on the Sargasso Sea surface. *Science*, **175**(4027), 1240–1241, https://doi.org/10.1126/science.175.4027.1240

Cassuriaga A. P. A., Freitas B. C. B., Morais M. G. and Costa J. A. V. (2018). Innovative polyhydroxybutyrate production by *Chlorella fusca* grown with pentoses. *Bioresource Technology*, **265**, 456–463, https://doi.org/10.1016/j.biortech.2018.06.026

Chatsungnoen T. and Chisti Y. (2016). Harvesting microalgae by flocculation–sedimentation. *Algal Research*, **13**, 271–283, https://doi.org/10.1016/j.algal.2015.12.009

Chen J. (2019). Global Markets and Technologies for Bioplastics. BCC Publishing, PLS050E 1–231.

Chen C. L., Chang J. S. and Lee D. J. (2015). Dewatering and drying methods for microalgae. *Drying Technology*, **33**(4), 443–454, https://doi.org/10.1080/07373937.2014.997881

Cheng H. N. and Gross R. A. (2020). Sustainability and green polymer chemistry – an overview. In: Sustainability & Green Polymer Chemistry Volume 1: Green Products and Processes, ACS Symposium Series; American Chemical Society: Washington, DC. https://doi.org/10.1021/bk-2020-1372

Chia W. Y., Tang D. Y. Y., Khoo K. S., Lup A. N. K. and Chew K. W. (2020). Nature's fight against plastic pollution: algae for plastic biodegradation and bioplastics production. *Environmental Science and Ecotechnology*, **4**, 100065, https://doi.org/10.1016/j.ese.2020.100065

Costa S. S., Miranda A. L., de Morais M. G., Costa J. A. V. and Druzian J. I. (2019). Microalgae as source of polyhydroxyalkanoates (PHAs) – a review. *International Journal of Biological Macromolecules*, **131**, 536–547, https://doi.org/10.1016/j.ijbiomac.2019.03.099

Dang B. T., Bui X. T., Tran D. P. H., Hao Ngo H., Nghiem L. D., Hoang T. K. D., Nguyen P. T., Nguyen H. H., Vo T. K. Q., Lin C., Yi Andrew Lin K. and Varjani S. (2022). Current application of algae derivatives for bioplastic production: a review. *Bioresource Technology*, **347**, 126698, https://doi.org/10.1016/j.biortech.2022.126698

Das S. K., Sathish A. and Stanley J. (2018). Production of biofuel and bioplastic from *Chlorella pyrenoidosa*. *Materials Today: Proceedings*, **5**, 16774–16781, https://doi.org/10.1016/j.matpr.2018.06.020

de Morais M. G. and Costa J. A. V. (2007). Biofixation of carbon dioxide by *Spirulina* sp. and *Scenedesmus obliquus* cultivated in a three-stage serial tubular photobioreactor. *Journal of Biotechnology*, **129**(3), 439–445, https://doi.org/10.1016/j.jbiotec.2007.01.009

Dubey S., Bharmoria P., Gehlot P. S., Agrawal V., Kumar A. and Mishra S. (2018). 1-Ethyl-3-methylimidazolium diethylphosphate based extraction of bioplastic 'polyhydroxyalkanoates' from bacteria: green and sustainable approach. *ACS Sustainable Chemistry & Engineering*, **6**(1), 766–773, https://doi.org/10.1021/acssuschemeng.7b03096

Duval C., Hamlaoui S., Piquet B., Toutirais G., Yéprémian C., Reinhardt A., Duperron S., Marie B., Demay J. and Bernard C. (2021) Diversity of cyanobacteria from thermal muds (Balaruc-Les-Bains, France) with the description of *Pseudochroococcus coutei* gen. nov., sp. nov. *FEMS Microbes*, **2**, xtab006, https://doi.org/10.1093/femsmc/xtab006

European Bioplastics. (2018). New market data: The positive trend for the bioplastics industry remains stable [online]. Available from: https://www.european-bioplastics.org/new-market-data-the-positive-trend-for-the-bioplastics-industry-remains-stable/ [Accessed April 5, 2023]

European Environmental Agency. (2020). Plastics, the Circular Economy and Europe's Environment – A Priority for Action. Publications Office of the European Union, Luxembourg.

García G., Sosa-Hernández J. E., Rodas-Zuluaga L. I., Castillo-Zacarías C., Iqbal H. and Parra-Saldívar R. (2021). Accumulation of PHA in the microalgae *Scenedesmus sp.* under nutrient-deficient conditions. *Polymers*, **13**(1), 1–23, https://doi.org/10.3390/polym13010131

Garcia-Garcia D., Ferri J. M., Boronat T., Lopez-Martinez J. and Balart R. (2016). Processing and characterization of binary poly(hydroxybutyrate) (PHB) and poly(caprolactone) (PCL) blends with improved impact properties. *Polymer Bulletin*, **73**(12), 3333–3350, https://doi.org/10.1007/s00289-016-1659-6

Garcia-Garcia D., Rayón E., Carbonell-Verdu A., López-Martínez J. and Balart R. (2017). Improvement of the compatibility between poly(3-hydroxybutyrate) and poly(ε-caprolactone) by reactive extrusion with dicumyl peroxide. *European Polymer Journal*, **86**, 41–57, https://doi.org/10.1016/j.eurpolymj.2016.11.018

Gerardo M. L., Van Den Hende S., Vervaeren H., Coward T. and Skill S. C. (2015). Harvesting of microalgae within a biorefinery approach: a review of the developments and case studies from pilot-plants. *Algal Research*, **11**, 248–262, https://doi.org/10.1016/j.algal.2015.06.019

Greenwell H. C., Laurens L. M. L., Shields R. J., Lovitt R. W. and Flynn K. J. (2010). Placing microalgae on the biofuels priority list: a review of the technological challenges. *Journal of the Royal Society Interface*, **7**(46), 703–726, https://doi.org/10.1098/rsif.2009.0322

Hazer B. and Steinbüchel A. (2007). Increased diversification of polyhydroxyalkanoates by modification reactions for industrial and medical applications. *Applied Microbiology and Biotechnology*, **74**(1), 1–12, https://doi.org/10.1007/s00253-006-0732-8

Hotos G., Avramidou D., Mastropetros S. G., Tsigkou K., Kouvara K., Makridis P. and Kornaros M. (2023). Isolation, identification, and chemical composition analysis of nine microalgal and cyanobacterial species isolated in lagoons of Western Greece. *Algal Research*, **69**, 102935, https://doi.org/10.1016/j.algal.2022.102935

Hu X., Meneses Y. E., Stratton J. and Wang B. (2019). Acclimation of consortium of micro-algae help removal of organic pollutants from meat processing wastewater. *Journal of Cleaner Production*, **214**, 95–102, https://doi.org/10.1016/j.jclepro.2018.12.255

Iordanskii A. L., Samoilov N. A., Olkhov A. A., Markin V. S., Rogovina S. Z., Kildeeva N. R. and Berlin A. A. (2019). New fibrillar composites based on biodegradable poly(3-hydroxybutyrate) and polylactide polyesters with high selective absorption of oil from water medium. Doklady Physical Chemistry, **487**(5), 106–108, https://doi.org/10.31857/S0869-56524875528-531

Jamshidian M., Tehrany E. A., Imran M., Jacquot M. and Desobry S. (2010). Poly-lactic acid: production, applications, nanocomposites, and release studies. *Comprehensive Reviews in Food Science and Food Safety*, **9**(5), 552–571, https://doi.org/10.1111/j.1541-4337.2010.00126.x

Javadi A., Kramschuster A. J., Pilla S., Lee J., Gong S. and Turng L. (2010). Processing and characterization of microcellular PHBV/PBAT blends. *Polymer Engineering & Science*, **50**(7), 1440–1448, https://doi.org/10.1002/pen.21661

Kaewbai-Ngam A., Incharoensakdi A. and Monshupanee T. (2016). Increased accumulation of polyhydroxybutyrate in divergent cyanobacteria under nutrient-deprived photoautotrophy: an efficient conversion of solar energy and carbon dioxide to polyhydroxybutyrate by *Calothrix scytonemicola* TISTR 8095. *Bioresource Technology*, **212**, 342–347, https://doi.org/10.1016/j.biortech.2016.04.035

Kamravamanesh D., Pflügl S., Nischkauer W., Limbeck A., Lackner M. and Herwig C. (2017). Photosynthetic poly-β-hydroxybutyrate accumulation in unicellular cyanobacterium *Synechocystis* sp. PCC 6714. *AMB Express*, **7**(1), 143, https://doi.org/10.1186/s13568-017-0443-9

Kapritchkoff F. M., Viotti A. P., Alli R. C. P., Zuccolo M., Pradella J. G. C., Maiorano A. E., Miranda E. A. and Bonomi A. (2006). Enzymatic recovery and purification of polyhydroxybutyrate produced by *Ralstonia eutropha*. *Journal of Biotechnology*, **122**(4), 453–462, https://doi.org/10.1016/j.jbiotec.2005.09.009

Karamanlioglu M., Preziosi R. and Robson G. D. (2017). Abiotic and biotic environmental degradation of the bioplastic polymer poly(lactic acid): a review. *Polymer Degradation and Stability*, **137**, 122–130, https://doi.org/10.1016/j.polymdegradstab.2017.01.009

Kaur G., Wong J. W. C., Kumar R., Patria R. D., Bhardwaj A., Uisan K. and Johnravindar D. (2020). Value addition of anaerobic digestate from biowaste: thinking beyond agriculture. *Current Sustainable/Renewable Energy Reports*, **7**(2), 48–55, https://doi.org/10.1007/s40518-020-00148-2

Kavitha G., Kurinjimalar C., Sivakumar K., Kaarthik M., Aravind R., Palani P. and Rengasamy R. (2016a). Optimization of polyhydroxybutyrate production utilizing waste water as nutrient source by *Botryococcus braunii Kütz* using response surface methodology. *International Journal of Biological Macromolecules*, **93**, 534–542, https://doi.org/10.1016/j.ijbiomac.2016.09.019

Kavitha G., Kurinjimalar C., Sivakumar K., Palani P. and Rengasamy R. (2016b). Biosynthesis, purification and characterization of polyhydroxybutyrate from *Botryococcus braunii kütz*. *International Journal of Biological Macromolecules*, **89**, 700–706, https://doi.org/10.1016/j.ijbiomac.2016.04.086

Khanna S. and Srivastava A. K. (2005). Recent advances in microbial polyhydroxyalkanoates. *Process Biochemistry*, **40**(2), 607–619, https://doi.org/10.1016/j.procbio.2004.01.053

Khatami K., Perez-Zabaleta M., Owusu-Agyeman I. and Cetecioglu Z. (2021). Waste to bioplastics: how close are we to sustainable polyhydroxyalkanoates production? *Waste Management*, **119**, 374–388, https://doi.org/10.1016/j.wasman.2020.10.008

Koller M. (2018). Biodegradable and biocompatible polyhydroxy-alkanoates (PHA): auspicious microbial macromolecules for pharmaceutical and therapeutic applications. *Molecules*, **23**, 362, https://doi.org/10.3390/molecules23020362

Koller M. (2020). Established and advanced approaches for recovery of microbial polyhydroxyalkanoate (PHA) biopolyesters from surrounding microbial biomass. *The EuroBiotech Journal*, **4**(3), 113–126, https://doi.org/10.2478/ebtj-2020-0013

Koller M. and Rodríguez-Contreras A. (2015). Techniques for tracing PHA-producing organisms and for qualitative and quantitative analysis of intra- and extracellular PHA. *Engineering in Life Sciences*, **15**(6), 558–581, https://doi.org/10.1002/elsc.201400228

Kosseva M. R. and Rusbandi E. (2018). Trends in the biomanufacture of polyhydroxyalkanoates with focus on downstream processing. *International Journal of Biological Macromolecules*, **107**, 762–778, https://doi.org/10.1016/j.ijbiomac.2017.09.054

Koutra E., Grammatikopoulos G. and Kornaros M. (2018). Selection of microalgae intended for valorization of digestate from agro-waste mixtures. *Waste Management*, **73**, 123–129, https://doi.org/10.1016/j.wasman.2017.12.030

Kovalcik A., Meixner K., Mihalic M., Zeilinger W., Fritz I., Fuchs W., Kucharczyk P., Stelzer F. and Drosg B. (2017). Characterization of polyhydroxyalkanoates produced by *Synechocystis salina* from digestate supernatant. *International Journal of Biological Macromolecules*, **102**, 497–504, https://doi.org/10.1016/j.ijbiomac.2017.04.054

Krasaesueb N., Incharoensakdi A. and Khetkorn W. (2019). Utilization of shrimp wastewater for poly-β-hydroxybutyrate production by *Synechocystis* sp. PCC 6803 strain ΔSphU cultivated in photobioreactor. *Biotechnology Reports*, **23**, e00345, https://doi.org/10.1016/j.btre.2019.e00345

Krishnan A., Devasya R., Hu Y. and Bassi A. (2022). Fundamental investigation of bio-surfactants-assisted harvesting strategy for microalgae. *Biomass and Bioenergy*, **158**, 106364, https://doi.org/10.1016/j.biombioe.2022.106364

Kumar M., Sun Y., Rathour R., Pandey A., Thakur I. S. and Tsang D. C. W. (2020). Algae as potential feedstock for the production of biofuels and value-added products: opportunities and challenges. *Science of the Total Environment*, **716**, 137116, https://doi.org/10.1016/j.scitotenv.2020.137116

Kumar V., Sehgal R. and Gupta R. (2021). Blends and composites of polyhydroxyalkanoates (PHAs) and their applications. *European Polymer Journal*, **161**, 110824, https://doi.org/10.1016/j.eurpolymj.2021.110824

Kumar N., Banerjee C., Negi S. and Shukla P. (2023). Microalgae harvesting techniques: updates and recent technological interventions. *Critical Reviews in Biotechnology*, **43**(3), 342–368, https://doi.org/10.1080/07388551.2022.2031089

Kurian N. S. and Das B. (2021). Comparative analysis of various extraction processes based on economy, eco-friendly, purity and recovery of polyhydroxyalkanoate: a review. *International Journal of Biological Macromolecules*, **183**, 1881–1890, https://doi.org/10.1016/j.ijbiomac.2021.06.007

Laamanen C. A., Ross G. M. and Scott J. A. (2016). Flotation harvesting of microalgae. *Renewable and Sustainable Energy Reviews*, **58**, 75–86, https://doi.org/10.1016/j.rser.2015.12.293

Lai S., Sun W. and Don T. (2015). Preparation and characterization of biodegradable polymer blends from poly(3-hydroxybutyrate)/poly(vinyl acetate)-modified corn starch. *Polymer Engineering & Science*, **55**(6), 1321–1329, https://doi.org/10.1002/pen.24071

Larsson M., Markbo O. and Jannasch P. (2016). Melt processability and thermomechanical properties of blends based on polyhydroxyalkanoates and poly(butylene adipate-*co*-terephthalate). *RSC Advances*, **6**(50), 44354–44363, https://doi.org/10.1039/C6RA06282B

Laycock B., Halley P., Pratt S., Werker A. and Lant P. (2013). The chemomechanical properties of microbial polyhydroxyalkanoates. *Progress in Polymer Science*, **38**(3–4), 536–583, https://doi.org/10.1016/j.progpolymsci.2012.06.003

Leal Filho W., Saari U., Fedoruk M., Iital A., Moora H., Klöga M. and Voronova V. (2019). An overview of the problems posed by plastic products and the role of extended producer responsibility in Europe. *Journal of Cleaner Production*, **214**, 550–558, https://doi.org/10.1016/j.jclepro.2018.12.256

Levett I., Birkett G., Davies N., Bell A., Langford A., Laycock B., Lant P. and Pratt S. (2016). Techno-economic assessment of poly-3-hydroxybutyrate (PHB) production from methane – the case for thermophilic bioprocessing. *Journal of Environmental Chemical Engineering*, **4**(4, Part A), 3724–3733, https://doi.org/10.1016/j.jece.2016.07.033

Li Z., Yang J. and Loh X. J. (2016). Polyhydroxyalkanoates: opening doors for a sustainable future. *NPG Asia Materials*, **8**(4), e265–e265, https://doi.org/10.1038/am.2016.48

López-Abelairas M., García-Torreiro M., Lú-Chau T., Lema J. M. and Steinbüchel A. (2015). Comparison of several methods for the separation of poly(3-hydroxybutyrate) from *Cupriavidus necator* H16 cultures. *Biochemical Engineering Journal*, **93**, 250–259, https://doi.org/10.1016/j.bej.2014.10.018

Lugoloobi I., Li X., Zhang Y., Mao Z., Wang B., Sui X. and Feng X. (2020). Fabrication of lignin/poly(3-hydroxybutyrate) nanocomposites with enhanced properties via a Pickering emulsion approach. *International Journal of Biological Macromolecules*, **165**, 3078–3087, https://doi.org/10.1016/j.ijbiomac.2020.10.156

Mallick N., Gupta S., Panda B. and Sen R. (2007). Process optimization for poly(3-hydroxybutyrate-*co*-3-hydroxyvalerate) co-polymer production by *Nostoc muscorum*. *Biochemical Engineering Journal*, **37**(2), 125–130, https://doi.org/10.1016/j.bej.2007.04.002

Markets and Markets. (2022). Polyhydroxyalkanoate (PHA) Market Share, Size, 2022–2027. MarketsandMarkets Research Private Ltd. CH 1610.

Mármol G., Gauss C. and Fangueiro R. (2020). Potential of cellulose microfibers for PHA and PLA biopolymers reinforcement. *Molecules*, **25**(20), 4653, https://doi.org/10.3390/molecules25204653

Martins R. G., Gonçalves I. S., De Morais M. G. and Vieira Costa J. A. (2017). New technologies from the bioworld: selection of biopolymer-producing microalgae. *Polimeros*, **27**(4), 285–289, https://doi.org/10.1590/0104-1428.2375

Mastropetros S. G., Pispas K., Zagklis D., Ali S. S. and Kornaros M. (2022). Biopolymers production from microalgae and cyanobacteria cultivated in wastewater: *recent advances*. *Biotechnology Advances*, **60**, 107999, https://doi.org/10.1016/j.biotechadv.2022.107999

Mata T. M., Martins A. A. and Caetano N. S. (2010). Microalgae for biodiesel production and other applications: a review. *Renewable and Sustainable Energy Reviews*, **14**(1), 217–232, https://doi.org/10.1016/j.rser.2009.07.020

Mathimani T. and Mallick N. (2018). A comprehensive review on harvesting of microalgae for biodiesel – key challenges and future directions. *Renewable and Sustainable Energy Reviews*, **91**, 1103–1120, https://doi.org/10.1016/j.rser.2018.04.083

Medeiros Garcia Alcântara J., Distante F., Storti G., Moscatelli D., Morbidelli M. and Sponchioni M. (2020). Current trends in the production of biodegradable bioplastics: the case of polyhydroxyalkanoates. *Biotechnology Advances*, **42**, 107582, https://doi.org/10.1016/j.biotechadv.2020.107582

Meixner K., Fritz I., Daffert C., Markl K., Fuchs W. and Drosg B. (2016). Processing recommendations for using low-solids digestate as nutrient solution for poly-β-hydroxybutyrate production with *Synechocystis salina*. *Journal of Biotechnology*, **240**, 61–67, https://doi.org/10.1016/j.jbiotec.2016.10.023

Molina Grima E., Belarbi E.-H., Acién Fernández F. G., Robles Medina A. and Chisti Y. (2003). Recovery of microalgal biomass and metabolites: process options and economics. *Biotechnology Advances*, **20**(7), 491–515, https://doi.org/10.1016/S0734-9750(02)00050-2

Monshupanee T., Nimdach P. and Incharoensakdi A. (2016). Two-stage (photoautotrophy and heterotrophy) cultivation enables efficient production of bioplastic poly-3-hydroxybutyrate in auto-sedimenting cyanobacterium. *Scientific Reports*, **6**(1), 37121, https://doi.org/10.1038/srep37121

Morais Junior W. G., Gorgich M., Corrêa P. S., Martins A. A., Mata T. M. and Caetano N. S. (2020). Microalgae for biotechnological applications: cultivation, harvesting and biomass processing. *Aquaculture*, **528**, 735562, https://doi.org/10.1016/j.aquaculture.2020.735562

Moreira J. B., Kuntzler S. G., Bezerra P. Q. M., Cassuriaga A. P. A., Zaparoli M., da Silva J. L. V., Costa J. A. V. and de Morais M. G. (2022). Recent advances of microalgae exopolysaccharides for application as bioflocculants. *Polysaccharides*, **3**(1), 264–276, https://doi.org/10.3390/polysaccharides3010015

Możejko-Ciesielska J. and Kiewisz R. (2016). Bacterial polyhydroxyalkanoates: still fabulous? *Microbiological Research*, **192**(2016), 271–282, https://doi.org/10.1016/j.micres.2016.07.010

Mubarak M., Shaija A. and Suchithra T. V. (2019). Flocculation: an effective way to harvest microalgae for biodiesel production. *Journal of Environmental Chemical Engineering*, **7**(4), 103221, https://doi.org/10.1016/j.jece.2019.103221

Muneer F., Rasul I., Azeem F., Siddique M. H., Zubair M. and Nadeem H. (2020). Microbial polyhydroxyalkanoates (PHAs): efficient replacement of synthetic polymers. *Journal of Polymers and the Environment*, **28**(9), 2301–2323, https://doi.org/10.1007/s10924-020-01772-1

Nanda S., Patra B. R., Patel R., Bakos J. and Dalai A. K. (2022). Innovations in applications and prospects of bioplastics and biopolymers: a review. *Environmental Chemistry Letters*, **20**(1), 379–395, https://doi.org/10.1007/s10311-021-01334-4

Nie C., Pei H., Jiang L., Cheng J. and Han F. (2018). Growth of large-cell and easily-sedimentation microalgae *Golenkinia* SDEC-16 for biofuel production and campus sewage treatment. *Renewable Energy*, **122**, 517–525, https://doi.org/10.1016/j.renene.2018.02.005

Nishioka M., Nakai K., Miyake M., Asada Y. and Taya M. (2001). Production of poly-β-hydroxybutyrate by thermophilic cyanobacterium, *Synechococcus* sp. MA19, under phosphate-limited conditions. *Biotechnology Letters*, **23**, 1095–1p099, https://doi.org/10.1023/A:1010551614648

Nur M. M. A. (2022). Co-production of polyhydroxybutyrate and C-phycocyanin from *Arthrospira platensis* growing on palm oil mill effluent by employing UV-C irradiation. *Journal of Applied Phycology*, **34**(3), 1389–1396, https://doi.org/10.1007/s10811-022-02738-7

Nur M. M. A., Achmad Z., Jaya D., Setyoningrum T. M., Widayati T. W., Kholisoh S. D. and Djarot I. N. (2023). Screening and optimization of cyanobacteria cultivated on palm oil mill effluent (POME) to produce polyhydroxybutyrate. Journal of Applied Phycology, **35**(3), 1213–1221. https://doi.org/10.1007/s10811-023-02954-9

Obruča S., Dvořák P., Sedláček P., Koller M., Sedlář K. and Pernicová I. and Šafránek, D. (2022). Polyhydroxyalkanoates synthesis by halophiles and thermophiles: towards sustainable production of microbial bioplastics. *Biotechnology Advances*, **58**, 107906, https://doi.org/10.1016/j.biotechadv.2022.107906

Pahl S. L., Lee A. K., Kalaitzidis T., Ashman P. J., Sathe S. and Lewis D. M. (2013). Harvesting, thickening and dewatering microalgae biomass. In: Algae for Biofuels and Energy, M. Borowitzka and N. Moheimani N. (eds), Springer Netherlands, Dordrecht, pp. 165–185.

Panda B. and Mallick N. (2007). Enhanced poly-β-hydroxybutyrate accumulation in a unicellular cyanobacterium, *Synechocystis* sp. PCC 6803. *Letters in Applied Microbiology*, **44**(2), 194–198, https://doi.org/10.1111/j.1472-765X.2006.02048.x

Panda B., Sharma L. and Mallick N. (2005) Poly-β-hydroxybutyrate accumulation in *Nostoc muscorum* and *Spirulina platensis* under phosphate limitation. *Journal of Plant Physiology*, **162**(12), 1376–1379, https://doi.org/10.1016/j.jplph.2005.05.002

Paulraj P., Vnootheni N., Chandramohan M. and Thevarkattil M. J. P. (2018). Exploration of global trend on biomedical application of polyhydroxyalkanoate (PHA): a patent survey. *Recent Patents on Biotechnology*, **12**(3), 186–199, https://doi.org/10.2174/1872208312666180131114125

Phalanisong P., Plangklang P. and Reungsang A. (2021). Photoautotrophic and mixotrophic cultivation of polyhydroxyalkanoate-accumulating microalgae consortia selected under nitrogen and phosphate limitation. *Molecules*, **26**(24), 7613, https://doi.org/10.3390/molecules26247613

Polman E. M. N., Gruter G. J. M., Parsons J. R. and Tietema A. (2021). Comparison of the aerobic biodegradation of biopolymers and the corresponding bioplastics: a review. *Science of the Total Environment*, **753**, 141953, https://doi.org/10.1016/j.scitotenv.2020.141953

Prakash J., Pushparaj B., Carlozzi P., Torzillo G., Montaini E. and Materassi R. (1997). Microalgal biomass drying by a simple solar device. *International Journal of Solar Energy*, **18**(4), 303–311, https://doi.org/10.1080/01425919708914325

Price S., Kuzhiumparambil U., Pernice M. and Ralph P. J. (2020). Cyanobacterial polyhydroxybutyrate for sustainable bioplastic production: critical review and perspectives. *Journal of Environmental Chemical Engineering*, **8**(4), 104007, https://doi.org/10.1016/j.jece.2020.104007

Price S., Kuzhiumparambil U., Pernice M. and Ralph P. (2022). Techno-economic analysis of cyanobacterial PHB bioplastic production. *Journal of Environmental Chemical Engineering*, **10**(3), 107502, https://doi.org/10.1016/j.jece.2022.107502

Przybysz M., Marć M., Klein M., Saeb M. R. and Formela K. (2018). Structural, mechanical and thermal behavior assessments of PCL/PHB blends reactively compatibilized with organic peroxides. *Polymer Testing*, **67**, 513–521, https://doi.org/10.1016/j.polymertesting.2018.03.014

Qi S., Chen J., Hu Y., Hu Z., Zhan X. and Stengel D. B. (2022). Low energy harvesting of hydrophobic microalgae (*Tribonema* sp.) by electro-flotation without coagulation. *Science of the Total Environment*, **838**, 155866, https://doi.org/10.1016/j.scitotenv.2022.155866

Ragusa A., Svelato A., Santacroce C., Catalano P., Notarstefano V., Carnevali O., Papa F., Rongioletti M. C. A., Baiocco F., Draghi S., D'Amore E., Rinaldo D., Matta M. and Giorgini E. (2021). Plasticenta: first evidence of microplastics in human placenta. *Environment International*, **146**, 106274, https://doi.org/10.1016/j.envint.2020.106274

Rizwan M., Mujtaba G., Memon S. A., Lee K. and Rashid N. (2018). Exploring the potential of microalgae for new biotechnology applications and beyond: a review. *Renewable and Sustainable Energy Reviews*, **92**, 394–404, https://doi.org/10.1016/j.rser.2018.04.034

Roh H., Lee J. S., Choi H. Il, Sung Y. J., Choi S. Y., Woo H. M. and Sim S. J. (2021). Improved CO_2-derived polyhydroxybutyrate (PHB) production by engineering fast-growing cyanobacterium *Synechococcus elongatus* UTEX 2973 for potential utilization of flue gas. *Bioresource Technology*, **327**, 124789, https://doi.org/10.1016/j.biortech.2021.124789

Rosenboom J. G., Langer R. and Traverso G. (2022). Bioplastics for a circular economy. *Nature Reviews Materials*, **7**(2), 117–137, https://doi.org/10.1038/s41578-021-00407-8

Roy M. and Mohanty K. (2019). A comprehensive review on microalgal harvesting strategies: current status and future prospects. *Algal Research*, **44**, 101683, https://doi.org/10.1016/j.algal.2019.101683

Sadeghizadeh A., Farhad dad F., Moghaddasi L. and Rahimi R. (2017). CO_2 capture from air by *Chlorella vulgaris* microalgae in an airlift photobioreactor. *Bioresource Technology*, **243**, 441–447, https://doi.org/10.1016/j.biortech.2017.06.147

Sakarika M., Koutra E., Mastropetros S. G., Giannakopoulos S. and Kornaros M. (2022). Wastewater treatment coupled to algal biomass production. In: An Integration of Phycoremediation Processes in Wastewater Treatment, M. Shah, S. Rodriguez-Couto, C. B. De La Cruz and J. Biswas (eds.), Elsevier Inc., Amsterdam, Netherlands, pp. 203–204, https://doi.org/10.1016/B978-0-12-823499-0.00020-1

Samantaray S. and Mallick N. (2012). Production and characterization of poly-β-hydroxybutyrate (PHB) polymer from *Aulosira fertilissima*. *Journal of Applied Phycology*, **24**(4), 803–814, https://doi.org/10.1007/s10811-011-9699-7

Samantaray S. and Mallick N. (2014). Production of poly(3-hydroxybutyrate-*co*-3-hydroxyvalerate) co-polymer by the diazotrophic cyanobacterium *Aulosira fertilissima* CCC 444. *Journal of Applied Phycology*, **26**(1), 237–245, https://doi.org/10.1007/s10811-013-0073-9

Samantaray S. and Mallick N. (2015). Impact of various stress conditions on poly-β-hydroxybutyrate (PHB) accumulation in *Aulosira fertilissima* CCC 444. *Current Biotechnology*, **4**, 366–372, https://doi.org/10.2174/2211550104666150806000642

Senatore V., Rueda E., Bellver M., Díez-Montero R., Ferrer I., Zarra T., Naddeo V. and García J. (2023). Production of phycobiliproteins, bioplastics and lipids by the cyanobacteria *Synechocystis* sp. treating secondary effluent in a biorefinery approach. *Science of the Total Environment*, **857**, 159343, https://doi.org/10.1016/j.scitotenv.2022.159343

Shahabi-Ghahfarrokhi I., Goudarzi V. and Babaei-Ghazvini A. (2019). Production of starch based biopolymer by green photochemical reaction at different UV region as a food packaging material: physicochemical characterization. *International Journal of Biological Macromolecules*, **122**, 201–209, https://doi.org/10.1016/j.ijbiomac.2018.10.154

Sharma L. and Mallick N. (2005). Accumulation of poly-β-hydroxybutyrate in *Nostoc muscorum*: regulation by pH, light-dark cycles, N and P status and carbon sources. *Bioresource Technology*, **96**(11), 1304–1310, https://doi.org/10.1016/j.biortech.2004.10.009

Sharma V., Sehgal R. and Gupta R. (2021). Polyhydroxyalkanoate (PHA): properties and modifications. *Polymer*, **212**, 123161, https://doi.org/10.1016/j.polymer.2020.123161

Shen X. F., Gao L. J., Zhou S. B., Huang J. L., Wu C. Z., Qin Q. W. and Zeng R. J. (2020). High fatty acid productivity from *Scenedesmus obliquus* in heterotrophic cultivation with glucose and soybean processing wastewater via nitrogen and phosphorus regulation. *Science of the Total Environment*, **708**, 134596, https://doi.org/10.1016/j.scitotenv.2019.134596

Siddiki S. Y. A., Mofijur M., Kumar P. S., Ahmed S. F., Inayat A., Kusumo F., Badruddin I. A., Khan T. M. Y., Nghiem L. D., Ong H. C. and Mahlia T. M. I. (2022). Microalgae biomass as a sustainable source for biofuel, biochemical and biobased value-added products: an integrated biorefinery concept. *Fuel*, **307**, 121782, https://doi.org/10.1016/j.fuel.2021.121782

Simonazzi M., Pezzolesi L., Galletti P., Gualandi C., Pistocchi R., De Marco N., Paganelli Z. and Samorì C. (2021). Production of polyhydroxybutyrate by the cyanobacterium cf. *Anabaena* sp. *International Journal of Biological Macromolecules*, **191**, 92–99, https://doi.org/10.1016/j.ijbiomac.2021.09.054

Singh G. and Patidar S. K. (2018). Microalgae harvesting techniques: a review. *Journal of Environmental Management*, **217**, 499–508, https://doi.org/10.1016/j.jenvman.2018.04.010

Singh N., Ogunseitan O. A., Wong M. H. and Tang Y. (2022). Sustainable materials alternative to petrochemical plastics pollution: a review analysis. *Sustainable Horizons*, **2**, 100016, https://doi.org/10.1016/j.horiz.2022.100016

Singha S., Mahmutovic M., Zamalloa C., Stragier L., Verstraete W., Svagan A. J., Das O. and Hedenqvist M. S. (2021). Novel bioplastic from single cell protein as a potential packaging material. *ACS Sustainable Chemistry and Engineering*, **9**(18), 6337–6346, https://doi.org/10.1021/acssuschemeng.1c00355

Sirohi R., Lee J. S., Yu B. S., Roh H. and Sim S. J. (2021). Sustainable production of polyhydroxybutyrate from autotrophs using CO_2 as feedstock: challenges and opportunities. *Bioresource Technology*, **341**, 125751, https://doi.org/10.1016/j.biortech.2021.125751

Statista. (2023). Annual production of plastics worldwide from 1950 to 2021 [online]. Available from: https://www.statista.com/statistics/282732/global-production-of-plastics-since-1950/ [Accessed April 1, 2023]

Sun S., Liu P., Ji N., Hou H. and Dong H. (2018). Effects of various cross-linking agents on the physicochemical properties of starch/PHA composite films produced by extrusion blowing. *Food Hydrocolloids*, **77**, 964–975, https://doi.org/10.1016/j.foodhyd.2017.11.046

Syafiq R., Sapuan S. M., Zuhri M. Y. M., Ilyas R. A., Nazrin A., Sherwani S. F. K. and Khalina A. (2020). Antimicrobial activities of starch-based biopolymers and biocomposites incorporated with plant essential oils: a review. *Polymers*, **12**, 2403, https://doi.org/10.3390/polym12102403

Tian Y. and Wang J. H. (2020). Polyhydroxyalkanoates for biodegradable mulch films applications. In: Sustainability & Green Polymer Chemistry Volume 2: Biocatalysis and Biobased Polymers, H.N. Cheng and R. A. Gross (eds.), ACS Publications, Washington, pp. 145–160, https://doi.org/10.1021/bk-2020-1373.ch008

Tirkey A. and Upadhyay L. S. B. (2021). Microplastics: an overview on separation, identification and characterization of microplastics. *Marine Pollution Bulletin*, **170**, 112604, https://doi.org/10.1016/j.marpolbul.2021.112604

Toh P. S. Y., Jau M.-H., Yew S.-P., Abed R. M. M. and Sudesh K. (2008). Comparison of polyhydroxyalkonates biosynthesis, mobilization and the effects of cellular morphology in *Spirulina platensis* and *Synechocystis* sp. UNIWG. *Tropical Life Sciences Research*, **19**(2), 21–38. https://ejournal.usm.my/tlsr/article/view/tlsr_vol19-no-2-2008_3

Troschl C., Meixner K. and Drosg B. (2017). Cyanobacterial PHA production – review of recent advances and a summary of three years' working experience running a pilot plant. *Bioengineering*, **4**(4), 26, https://doi.org/10.3390/bioengineering4020026

Vandamme D., Foubert I. and Muylaert K. (2013). Flocculation as a low-cost method for harvesting microalgae for bulk biomass production. *Trends in Biotechnology*, **31**(4), 233–239, https://doi.org/10.1016/j.tibtech.2012.12.005

Vasistha S., Khanra A., Clifford M. and Rai M. P. (2021). Current advances in microalgae harvesting and lipid extraction processes for improved biodiesel production: a review. *Renewable and Sustainable Energy Reviews*, **137**, 110498, https://doi.org/10.1016/j.rser.2020.110498

Verlinden R. A. J., Hill D. J., Kenward M. A., Williams C. D. and Radecka I. (2007). Bacterial synthesis of biodegradable polyhydroxyalkanoates. *Journal of Applied Microbiology*, **102**(6), 1437–1449, https://doi.org/10.1111/j.1365-2672.2007.03335.x

Ward C. P. and Reddy C. M. (2020). We need better data about the environmental persistence of plastic goods. *Proceedings of the National Academy of Sciences of the United States of America*, **117**(26), 14618–14621, https://doi.org/10.1073/pnas.2008009117

Yaakob M. A., Mohamed R. M. S. R., Al-Gheethi A., Aswathnarayana Gokare R. and Ambati R. R. (2021). Influence of nitrogen and phosphorus on microalgal growth, biomass, lipid, and fatty acid production: an overview. *Cells*, **10**(2), 393, https://doi.org/10.3390/cells10020393

Yang Z.-Y., Gao F., Liu J.-Z., Yang J.-S., Liu M., Ge Y.-M., Chen D.-Z. and Chen J.-M. (2022). Improving sedimentation and lipid production of microalgae in the photobioreactor using saline wastewater. *Bioresource Technology*, **347**, 126392, https://doi.org/10.1016/j.biortech.2021.126392

Yashavanth P. R., Das M. and Maiti S. K. (2021). Recent progress and challenges in cyanobacterial autotrophic production of polyhydroxybutyrate (PHB), a bioplastic. *Journal of Environmental Chemical Engineering*, **9**(4), 105379, https://doi.org/10.1016/j.jece.2021.105379

Zarrinmehr M. J., Farhadian O., Heyrati F. P., Keramat J., Koutra E., Kornaros M. and Daneshvar E. (2020). Effect of nitrogen concentration on the growth rate and biochemical composition of the microalga, *Isochrysis galbana*. *Egyptian Journal of Aquatic Research*, **46**(2), 153–158, https://doi.org/10.1016/j.ejar.2019.11.003

Zhang S. and Bryant D. A. (2015). Biochemical validation of the glyoxylate cycle in the cyanobacterium *Chlorogloeopsis fritschii* strain PCC 9212. *Journal of Biological Chemistry*, **290**(22), 14019–14030, https://doi.org/10.1074/jbc.M115.648170

Zhang X., Jiang Z., Chen L., Chou A., Yan H., Zuo Y. Y. and Zhang X. (2013). Influence of cell properties on rheological characterization of microalgae suspensions. *Bioresource Technology*, **139**, 209–213, https://doi.org/10.1016/j.biortech.2013.03.195

Zhang H., Gong T., Li J., Pan B., Hu Q., Duan M. and Zhang X. (2022). Study on the effect of spray drying process on the quality of microalgal biomass: a comprehensive biocomposition analysis of spray-dried *S. acuminatus* biomass. *BioEnergy Research*, **15**(1), 320–333, https://doi.org/10.1007/s12155-021-10343-8

doi: 10.2166/9781789063547_0229

Chapter 12

Processes and biorefinery approach for enhanced algal bioproduct recovery in the form of lipid and UV protectant

Rashmi Chandra and Masoomeh Hooshmand*

Department of Civil and Environmental Engineering, University of Waterloo, 200 University Avenue West, Waterloo, Ontario, N2L 3G1, Canada
**Corresponding author: rashmichandrabhu@gmail.com; rashmi.chandra@uwaterloo.ca*

ABSTRACT

This chapter discusses various methodologies for lipid extraction, including solvent extraction, enzymatic treatment, ultrasonic aided extraction, and supercritical carbon dioxide extraction, underscoring the need for further research and optimization for large-scale applications. The chapter further explores the potential symbiotic relationship between algal fuel production and waste treatment. This strategy effectively utilizes microalgae's natural ability to thrive in adverse conditions and sequester CO_2 and other pollutants. This approach can simultaneously reduce the environmental footprint while generating valuable biomass for biodiesel production. Another noteworthy point the chapter brings forward is the ability of microalgae to produce valuable compounds under environmental stress, particularly UV radiation. The UV-absorbing compounds such as mycosporine-like amino acids (MAAs) and scytonemin, present substantial potential for use in the cosmetic and pharmaceutical sectors due to their potent UV absorption and photoprotective properties.

12.1 INTRODUCTION

The idea of a 'biorefinery' has arisen as a set of integrated processes for turning microalgal biomass into fuel and other high-value products (Cherubini, 2010; Thomassen *et al.*, 2018). A more sustainable and cost-effective strategy that just concentrates on fuel production is made possible by the diverse and complementary outputs (Salama *et al.*, 2018). Based on current capital costs per unit of fuel production, the generation of biofuel from microalgae is not economically viable. As a result, producing high-value co-products is necessary to increase a microalgae biorefinery's profitability.

Microalgae are microbial factories that can produce a variety of substances besides lipids for biodiesel, having a lipid (7–23%), carbohydrate (5–23%), and protein (6–52%) composition (Chandra *et al.*, 2014). Microalgae can be an excellent source of raw materials for commercially significant value-added products utilized in the food, nutraceutical, cosmetic, and pharmaceutical industries (Haznedaroglu *et al.*, 2016). An integrated biorefinery can maximize product outputs from a single biological source, capturing the value of numerous components (Oh *et al.*, 2018).

The concept of a biorefinery was inspired by petroleum refineries, which provide fuels, oils, and other materials used in the chemical industry (Roux *et al.*, 2017). A biorefinery uses a series of interconnected processes to utilize all the components of the raw materials without causing any loss or harm to the finished goods. In an algae-based biorefinery, there are major hurdles to the sustainable extraction of these chemicals when taking green chemistry principles into account (Yellapu *et al.*, 2018). Maximizing microalgae biomass utilization requires a lot of energy, while utilizing the least amount of energy is still the key goal (Bakonyi *et al.*, 2018). For instance, the Department of Energy's (DOE) primary goal, as stated in the outlook provided in the U.S. multi-year program plan, is cost reduction to produce algal biofuel (Barry *et al.*, 2016).

In this chapter, the most recent research on how to effectively use algal biomass in a sustainable way by using biochemical processes and a bio-refinery technique is discussed. In addition, the framework enables an algal bio-refinery to effectively create value-added products like oil and UV protectant. This chapter also provides a thorough overview of current advancements in the processing of algal biomass utilizing various sustainable methods in an integrated biorefinery.

12.2 FERMENTATION

12.2.1 Selective fermentation

Despite its benefits, microalgal biofuel has not been commercially successful in part because of technical and financial difficulties with algae harvesting and lipid extraction. Pretreatment techniques including pulsed electric fields (PEFs), ultrasound, and acid/alkaline hydrolysis can be effective but are typically too energy-intensive and therefore expensive (Lai *et al.*, 2014; Laurens *et al.*, 2015; Sheng *et al.*, 2011a; Zbinden *et al.*, 2013). To lessen risks to the environment and workers, the present 'gold standards' for lipid extraction, Folch (1:1 chloroform:methanol) and Bligh & Dyer (1:1:0.5 chloroform:methanol:water), must be replaced by non-toxic 'green' solvents. Hexane and isopropanol mixed 1:1 (v/v) is an illustration of a non-toxic solvent (Lai *et al.*, 2014, 2016a, 2016b).

A revolutionary biological strategy for simplifying and improving the economics of lipid extraction is called selective fermentation (SF) (Lai *et al.*, 2016a, 2016b). SF takes advantage of the fact that, under anaerobic conditions, lipids typically biodegrade more slowly than do carbohydrates and proteins. Because they grow slowly, lipid-fermenting bacteria (Christ *et al.*, 2000) can be removed from a reactor with a short solids retention time (SRT) by their washout (Lai *et al.*, 2016a, 2016b). As a result, SF permits the fermentation of carbohydrates and protein in microalgae cells while leaving lipids unaltered. Yet this results in a condition that is much easier to extract because the 'protection' provided by the carbohydrates and proteins has been removed (Lai *et al.*, 2016a, 2016b). Protein fermentation may be a bottleneck in SF since it proceeds more slowly than carbohydrate fermentation (Lu *et al.*, 2012).

The process of biohydrogenation, which transforms long-chain fatty acids (LCFAs) into saturated forms like C18:0, C16:0, and C14:0, is another advantage of SF. Because they have a higher energy content, a higher-octane number for improved combustion efficiency, and a stronger resistance to oxidation, saturated fatty acids are advantageous for the production of transportation fuel (Knothe, 2011). There are two main ways that biohydrogenation can take place. One method is the direct conversion of unsaturated bonds to saturated bonds. In this process, H_2 serves as the electron donor and the LCFA molecule's carbon content remains constant (Lai *et al.*, 2016a, 2016b). An example of direct biohydrogenation is the reduction of C18:1 to C18:0.

Low H_2 concentrations can thermodynamically limit direct biohydrogenation, whereas high H_2 concentrations can accelerate direct biohydrogenation (Cavaleiro *et al.*, 2016; Lai *et al.*, 2016a, 2016b). Strains of the genera *Butyrivibrio* and *Pseudobutyrivibrio* (both in the order Clostridiales) can directly biohydrogenate C18:2 n-6 and C18:3 n-3 to C18:0 (John Wallace *et al.*, 2006; Van De Vossenberg & Joblin, 2003). The family Porphyromonodaceae (Order Bacteroides) and Ruminococcaceae (Order Clostridiales) are involved in direct biohydrogenation in ruminants, according to in vivo research (Castro-Carrera *et al.*, 2014; Huws *et al.*, 2011).

The second method involves the beta-oxidation process, which converts an unsaturated LCFA into a saturated LCFA with the loss of two C atoms as acetate (Cavaleiro *et al.*, 2016). An example of the second route is transformation of C18:1 to C16:0.

As H_2 is produced during beta-oxidation, this pathway does not require an external source of H_2. In theory, processes that utilize H_2, acetate, or both might give this method of biohydrogenation a thermodynamic boost (Cavaleiro *et al.*, 2016).

Lipid conservation is valuable and varies with the biohydrogenation route. β-Oxidation reduces saturated LCFA chain length by two C atoms per step, and the loss is more substantial if multiple steps of beta-oxidation occur. An example is the transformation from C16:0 to C14:0, which produces 2 moles of H_2 and 1 mole of acetate per 1 mole of C14:0 produced.

12.2.2 Electrofermentation

Anode respiring bacteria (ARB) establish a biofilm on the anode in microbial electrolysis cells (MECs), oxidize short-chain fatty acids (SCFAs), and then extracellularly transmit the extracted electrons to the anode (EET) (Reguera *et al.*, 2005; Torres *et al.*, 2009b, 2009a; Yang *et al.*, 2012). Through the external circuit, electrons move to the cathode, where they are absorbed by water molecules to create H_2, which emerges from the cathode as a gas. The MEC is a potential technique for accelerating protein biodegradation in substrates made of complex organic molecules in case of microalgae (Lu *et al.*, 2012).

Due to the need for pretreatment and the use of harmful solvents, extracting lipids from microalgae has been shown to be both commercially and environmentally unfeasible. By selectively biodegrading proteins and carbohydrates while preserving lipids, SF aids in the resolution of these issues. Electro-selective fermentation (ESF) enhances the fermentation performance through anode respiration in a microbial electrolysis cell (MEC) (Liu *et al.*, 2019). ESF was assessed and compared to SF using biomass from *Scenedesmus acutus*. Even though anode respiration only accounted for 1% of the total electrons supplied, ESF boosted protein breakdown three times more than SF did. Although ESF increased the total lipid loss, it tripled the effectiveness of lipid wet extraction with a non-toxic solvent.

The long-chain fatty acid (LCFA) profile changed from C18:1 to C16:0 and C14:0 as a result of lipid loss caused by beta-oxidation associated with biohydrogenation. Anode-respiring bacteria (ARB) on the ESF anode and protein-degrading bacteria and biohydrogenaters in the ESF suspension were highlighted by microbial community analysis. Overall, ESF enhanced the quality of biofuel and lipid extractability.

12.2.3 Coupling SF and electrofermentation

A combination ESF is created with the aid of the MEC and SF. It aids in enhancing protein and carbohydrate conversions while protecting lipids for extraction. A strong biofilm of ARB oxidizes SCFAs quickly in the ESF (Ki *et al.*, 2015; Torres *et al.*, 2007), leading to a low concentration of SCFAs in the anode liquid. By reducing a thermodynamic barrier, a lower concentration of SCFAs should encourage upstream fermentation reactions (Fukuzaki *et al.*, 1990; Jones *et al.*, 2015; Pratt *et al.*, 2012). Unfortunately, this method could potentially speed up beta-oxidation as well, which would lead to a loss of all LCFAs.

H_2 is also an ARB substrate, either directly or indirectly through its homo-acetogenic conversion to acetate (Parameswaran *et al.*, 2009). A well-known strategy for overcoming thermodynamic obstacles to fermentation is scavenging H_2 (Cavaleiro *et al.*, 2016; Parameswaran *et al.*, 2010). ESF may therefore hasten the fermentation of proteins and carbohydrates. The loss of protein could cause the cell membrane to rupture and release intracellular lipids for easy extraction by interfering with the hydrogen bonding between membrane proteins and lipids (Cooney *et al.*, 2009; Sheng *et al.*, 2011b).

12.3 BIODIESEL EXTRACTION FROM MICROALGAE

12.3.1 Pretreatment

Several techniques can be used to algae biomass in order to extract intracellular substances. There are various conversion techniques, but the mechanical-based techniques are among the most significant (Cherubini *et al.*, 2009). The biomass is concentrated once microalgae cultures reach the stationary growth phase, and the desired products can be recovered using either dry or wet biomass. To break down the cellular walls and encourage the release of microalgae components that are not released outside the cell, the initial biomass can be dewatered by centrifugation, which is then followed by a cell disruption technique. The approaches employed typically involve a disturbance, break, or breakdown (Dong *et al.*, 2016). The biomass is then thermally dried to obtain a dried form after the dewatering process, which typically results in a paste-like biomass with a dry weight above 85% (Xu *et al.*, 2011).

12.3.2 Extraction

12.3.2.1 Principle of solvent extraction

One of the primary methods for recovering valuable compounds from microalgae is organic solvent extraction. Based on the polarity of the target chemicals, solvents should be selected. Because TAGs, the primary lipid target for the manufacture of biodiesel, are non-polar molecules, a non-polar solvent is an appropriate option for extraction. The majority of solvent-based extraction methods used to extract lipids from microalgae are based on conventional procedures for extracting plant oils, including organic solvent extraction, the Folch method, and the Soxhlet method. Organic solvents penetrate the cell wall of the microalgae, where they promote swelling and cell rupture, releasing the contents of the cell for further separation steps (Grima *et al.*, 2003). When selecting a solvent to extract lipids from microalgae, the primary factors to take into account are polarity or extractability, lipid solubility, water miscibility (ability to operate in two-phase systems), and low toxicity (Bensalem *et al.*, 2018).

12.3.2.2 Solvent extraction methods

12.3.2.2.1 Folch method

The Folch method, which is the foundation of many solvent extraction techniques currently in use, uses a 2:1 chloroform–methanol mixture to extract intracellular lipids. A cell homogenate is first stirred to equilibrate with 25% volume of saline solution. The lipids are allowed to settle on the top layer of this mixture until biphasic separation (Ranjith Kumar *et al.*, 2015). This procedure requires the breaking of microalgae cell walls as a preliminary step. It was initially intended for animal cells and tissues (Grima *et al.*, 2003).

12.3.2.2.2 Soxhlet extraction

In the Soxhlet extraction (SE) procedure, components of a solid sample that are only partially soluble are transported to a liquid phase (solvent) using a Soxhlet extractor. This method uses hexane and other non-polar solvents to produce neutral lipids. The extraction process involves inserting the solid sample into the Soxhlet apparatus's main chamber in a filter paper thimble. The solvent is then heated to reflux and forced into the main chamber, where the less soluble chemicals are recovered. Due to the recovery of complex lipids and pigments, a greater extraction yield from microalgae can be attained when the extraction solvent polarity increases (Baumgardt *et al.*, 2016). This is a crucial factor to take into account because whole lipid extracts using polar solvents are complicated and contain other metabolites besides lipids. The characteristics of a Soxhlet extraction are the solvent of choice, sample particle size, and extraction time (Sharif *et al.*, 2014). SE is typically done on a small scale in the lab and requires a lot of solvent and a long extraction period.

12.3.2.3 Bligh and Dyer method

The Bligh and Dyer method involves partitioning and extracting lipids simultaneously, with protein precipitation occurring at the interface of two liquid phases. This extraction method is comparable to the Folch method, but with a different solvent combination composition and ratio. A cell homogenate's lipids are first extracted with a 1:2 solution of chloroform and methanol, and the chloroform phase – which is rich in lipids – is then recovered. Lipids from microalgae are removed and quantified using gravimetry. Both pilot and large-scale operations use this approach (Ranjith Kumar *et al.*, 2015). Instead of using water, this approach can be improved by adding 1 M NaCl to prevent the binding of acidic lipids to denatured lipids. The addition of 0.2 M phosphoric acid and HCl has resulted in shorter separation times. By adding 0.5% acetic acid (v/v), acidic phospholipid recovery has been improved (Ranjith Kumar *et al.*, 2015).

12.3.3 Mechanical methods

Solid shear, cavitation and collapse, PEFs, chemical hydrolysis, enzymatic digestion, subcritical water extraction, high-pressure homogenization, and bead milling are a few techniques used to destroy cells and thus release their content.

12.3.3.1 Milling

Bead milling is the process of breaking down the walls of microalgae cells by agitating and grinding the cells over a surface made of glass beads (Ghasemi Naghdi *et al.*, 2016). A disruption needs beads that are between 0.3 and 0.5 mm in size. Typically, zirconia–silica or zirconium oxide can be used to create the beads. The temperature, biomass concentration, flow rate, agitator movement type, and speed all affect how effectively the process works.

Milling can be carried out using agitated beads or shaken vessels. In the shaking vessel method, a vibrating platform is used to shake the culture vessel, which causes the microalgae cells to migrate and crash into one another. When Ryckebosch *et al.* (2012) used this technique, they were able to recover 40% of the lipids from a culture of *Phaeodactylum tricronutum*, which was the highest lipid recovery achieved. On the other hand, Zheng *et al.* (2012) used a bead milling vessel to extract 11% of lipids from a culture of *Chlorella vulgaris*. According to Lee *et al.* (2010) agitated beads use a method in which the beads and the culture are stirred around by a rotating agitator inside the culture vessel while also being heated to aid in the disruption process. Using cultures of *Botryococcus* sp., *Chlorella vulgaris*, and *Scenedesmus* sp., the authors employed this methodology and obtained an oil yield between 7.9 and 8.1 g/L.

12.3.3.2 Pressing

One of the traditional techniques for obtaining value-added goods from a variety of sources is the use of presses (Kumar *et al.*, 2020a, 2020b). The mechanical crushing of materials with a very low moisture content is the foundation of this technique. Dried biomass is first put under intense mechanical pressure to shatter and crush the cells, and then it is squeezed to extract the oil. Variations in the pressure force, algae strain, and press and piston arrangement can all increase the extraction efficiency (Kumar *et al.*, 2020a, 2020b). With the gel-press approach, algae are first rinsed before employing diluted alkali to extract the carbohydrates. Centrifugation is used to separate the residues, then they are filtered through porous silica, and finally concentrated using evaporation. The recovered material is extruded into a cold potassium chloride solution using spinnerets, and the threads that have gelled are then compressed to remove water (Amin, 2009).

High pressures are used by shear-based machines like the French press and Hughes press to push a biomass solution through a tiny aperture. The average oil recovery is between 70% and 75% (Kumar *et al.*, 2020a, 2020b). Mechanical crushing is occasionally employed in addition to chemical procedures for better oil recovery. The primary limitations of this technology are that it requires expensive maintenance and is less effective than other mechanical extraction methods (Ranjith Kumar *et al.*, 2015).

12.3.3.3 Freeze–thaw method

Since the loss of volatile lipids owing to evaporation is reduced to a minimum with the freeze–thaw process, lipid extraction from microalgae biomass is favored. By freezing the wet biomass at a temperature of −80°C, the intracellular water crystallizes in this process. The samples are then thawed, causing the ice crystals to expand and lyse the frozen cells. To maximize yield efficiency, this process is typically used in conjunction with another technique, such as ultrasonication, microwave-assisted extraction (MAE), or bead milling (Esquivel-Hernández *et al.*, 2017; Parfati *et al.*, 2018). Cycles of freezing and thawing must be carefully controlled, though. Unfrozen samples showed a 10% decrease in reproducibility after the first cycle and a further 7% decrease after the second, according to a study of the metabolic profile of marine microalgae after freeze-thawing under standard freeze-storage temperatures (–20°C and –78°C) for 1 and 2 cycles of 7 days each (Chr. Eilertsen *et al.*, 2014).

12.3.4 Enzymatic methods

A mixture of enzymes is used in enzymatic extraction procedures to dissolve the algal cell wall, expel lipid bodies from the cell, and separate the lipid fraction from the lipid/protein matrix. An alternative to mechanical cell destruction is enzymatic lysis. Due to the presence of polysaccharides like cellulose and hemicellulose in algal cell walls and lipids, packed in a sac surrounded by phospholipids, in algal cell walls, the lytic enzymes must be specific for the microalgae species, with cellulase and lipase being the most prevalent (Parfati *et al.*, 2018).

Microalgae lipids can be extracted using the cell disruption method known as aqueous enzymatic aided extraction (AEAE). High selectivity, gentle reaction conditions (neutral pH, incubation from 25°C to 37°C), and the lack of labor-intensive drying processes are noteworthy characteristics (Sierra *et al.*, 2017). The best extraction parameters were determined to be 37°C, pH 5.0, 1.3% cellulase, liquid/solid ratio 15 mL/g, and 5 h. An improved approach for enzymatic lysis combined with thermal treatment for extracting lipids from *N. oceanica*. Up to 28.8% of lipids were produced under these circumstances (Amin, 2009).

Biomass collection, enzyme conditioning and addition, stirring incubation to break down algal cell walls, solvent addition (if necessary), centrifugation, and lipid fraction recovery are the primary steps in the enzymatic extraction of lipids from microalgae (Lee *et al.*, 2010). Moreover, after the removal of lipids, the carbohydrate biomass can be saccharified via enzymatic digestion to produce bioethanol (Parfati *et al.*, 2018).

12.3.5 Physical extraction methods

12.3.5.1 Supercritical fluid extraction

By exerting pressure and temperature above a compound's or mixture's critical point, supercritical fluid extraction (SFE) makes use of a supercritical fluid's solvating capability. Solvent, temperature, pressure, solvent flow rate, extraction time, sample size, usage of a modifier, and particle size are some of the adjustable parameters to take into account for SFE (Sharif *et al.*, 2014).

To avoid using hazardous solvents, supercritical fluid extraction with carbon dioxide (SFE-CO_2) has been used as an alternative green extraction technique (Hernández *et al.*, 2014). SFE-CO_2 has several benefits, including being generally recognized as safe (GRAS) by the Food and Drug Administration (FDA), having a low critical point of CO_2 at near room temperature and relatively low pressure (30.9°C and 73.9 bar), and being ecologically benign (Reverchon & de Marco, 2006). Moreover, CO_2 is converted to gas after depressurization, which allows it to be removed from the sample without leaving any traces of solvent behind. This allows it to be recycled for additional extraction cycles, which has both financial and environmental advantages. Supercritical CO_2, which is especially helpful for the extraction of biodiesel, is very selective for non-polar lipids like triglycerides and does not solubilize phospholipids (Hernández *et al.*, 2014). Hydrocarbons (hexane, pentane, and butane), nitrous oxide, sulfur hexafluoride, and fluorinated hydrocarbons are some of the additional solvents employed in SFE (Reverchon & de Marco, 2006).

12.3.5.2 Microwave-assisted extraction

MAE depends on the interaction of a dielectric polar substance (such as water) and a rapidly oscillating electric field created by microwaves (Esquivel-Hernández *et al.*, 2017; Moretto *et al.*,

2022). This electric field generates heat as a result of the friction created by the movement of the molecules within and between it. The cell begins to produce water vapor as a result of the heat, which finally ruptures the cell and leads to increased intracellular component leakage and release, driven by the electroporation action (Ghasemi Naghdi *et al.*, 2016). As a result, MAE is recognized as a quick, easy, safe, efficient, and affordable approach for the extraction of lipids that does not necessitate sample dewatering beforehand (Bensalem *et al.*, 2018). Moreover, microalgae processed with microwaves have numerous microfissures in the cell wall, which increases the amount of bio-oil recovered (Šoštarič *et al.*, 2012).

In addition to oil extraction, microwaves can be used to transesterify oils into biodiesel, which is a desirable alternative due to its quick reaction time (15–20 min), low operating costs, and effective extraction of algal oils. The substantial maintenance costs associated with using this technology on a commercial scale are a significant downside (Kumar *et al.*, 2015). The primary factors to be considered for MAE are extraction time, temperature, the process mixture's dielectric characteristics, the solid/liquid ratio, and the kind and concentration of the solvent (Ghasemi Naghdi *et al.*, 2016).

12.3.5.3 Ultrasound-assisted extraction

Using cavitation, ultrasonic-assisted extraction (UAE) can recover oils from microalgae cells (Harun *et al.*, 2010). Little vacuum bubbles with a high intensity are produced in the liquid during the low-pressure cycle. A high-pressure cycle occurs when the bubbles violently collapse once they reach a particular size. Locally, extremely high pressures and fast-moving liquid jets are created during the implosion, and the ensuing shear stresses cause the mechanical breakdown of the cell structure. The extraction of lipids from algae is supported by this outcome (Wei *et al.*, 2008). Solvent diffusion into the cell structure is supported by the high-pressure cycles of the ultrasonic waves. Lipids are more easily transferred from the cell into the solvent when using ultrasound because it mechanically breaches the cell membrane through cavitation shear pressures (Cravotto *et al.*, 2008).

By extending the exposure period and combining polar and non-polar solvents, lipid recovery can be improved. UAE also supports mass transfer and solvent penetration inside the cell to release the contents of the cells into the solvent. UAE may be carried out at low temperatures, which is ideal when dealing with the extraction of compounds that are thermally sensitive (Ghasemi Naghdi *et al.*, 2016).

12.3.5.4 Pressurized liquid extraction

Wet algal biomass is used in the wet lipid extraction procedure along with a corresponding amount of solvent (Al-Jabri *et al.*, 2022; Sathish & Sims, 2012). Although it differs depending on the biomass type, this technique is similar to the wet solvent extraction procedure (see Section 3.2). Biomass is transformed into liquid biocrude through the process of hydrothermal liquefaction in hot, compressed water (Biller *et al.*, 2012; Zhang *et al.*, 2022). Because the water must remain in the subcritical area to prevent the latent heat of vaporization, processing temperatures vary from 200°C to 350°C with pressures of around 15–20 MPa (Biller *et al.*, 2012). Complex molecules are disassembled and repolymerized to oily substances under these circumstances. This process eliminates the need to dry the feedstock, making it suitable for converting high-moisture biomass like microalgae.

12.3.5.5 Osmotic pressure

A quick shift in the solute concentration around a cell results in a rapid alteration in the transport of water across the cell membrane, which is known as osmotic shock or osmotic stress (Fajardo *et al.*, 2007). The microalgae's cellular contents are released as a result of this shock. The technique works better with strains grown in maritime conditions (e.g. *Nannochloropsis* sp.). To release cellular components for biochemical examination, osmotic stress is also generated (Alami *et al.*, 2021; Shen *et al.*, 2010). *Halorubrum* sp. isolated from saltern ponds can likewise be treated using this technique. Increased lipid productivities and different lipid compositions were observed (Lopalco *et al.*, 2004).

12.3.5.6 PEF technologies

A technique called pulsed electric field (PEF) processing uses brief bursts of a powerful electric field to process cells (Chittapun *et al.*, 2020). Between two electrodes, algae biomass is positioned, and a PEF is applied (Käferböck *et al.*, 2020). The holes in cell membranes are made larger and release their contents when exposed to an electric field (Wang *et al.*, 2023).

12.4 BIOREFINERY

12.4.1 Cyanobacterial biorefineries

The most basic type of photosynthetic microorganisms, cyanobacteria, have a significant potential for the generation of bioenergy as well as high-value food and pharmaceutical items (Thajuddin & Subramanian, 2005). Since it remains a difficult task, extensive research is being done to turn cyanobacterial lipid into a significant industrial process (Patnaik & Mallick, 2015). In recent years, the process of scaling up and commercializing cyanobacterial or microalgal products has begun, but cautious and methodical development is required to make it a sustainable industrial process.

At an industrial scale, cyanobacteria have enormous promise for recovering value-added products and biofuels. They are rich in lipids, susceptible to metabolic engineering, and contain value-added components like antioxidants (Esquivel-Hernández *et al.*, 2017), phycobiliproteins (Chandra *et al.*, 2017), UV protectants (Rastogi & Incharoensakdi, 2014), and vitamins (Esquivel-Hernández *et al.*, 2017). This makes it a potential feed stock for biorefinery (Sheng *et al.*, 2011b; Vermaas, 1996). Despite the fact that cyanobacteria are ideally suited for biorefining due to the diverse composition of their biomass, recovering co-products from cyanobacteria remains a difficult problem. Therefore, it is necessary to investigate moderate and sequential extraction techniques that maintain the usefulness of different cell compounds such as UV protectant, protein, vitamins, lipid, and carbohydrates.

12.4.2 Cyanobacterial biorefinery products

12.4.2.1 Biodiesel

A possible renewable feedstock for the manufacture of value-added goods and ethanol appears to be cyanobacteria. A comprehensive assessment of multiple product recovery is required to guarantee the economic and sustainability of biofuel production. Chandra *et al.* (2019, 2020) provided strong evidence in favor of producing mycosporine-like amino acids (MAAs) and high-quality biodiesel in succession. A procedure in which *Lyngbya* biomass was sequentially collected from all experimental variations after UV irradiation and treated with 100% HPLC-grade methanol for 12 h at 4°C before being centrifuged at 4000 rpm for 15 min at 4°C. The supernatant was gathered, dried at 38°C, and combined with the pellet for lipid extraction. It was determined that the dried methanolic residue is MAAs. To remove the photosynthetic pigment, this MAA was washed with chloroform and water. Together with the residue from the previous stage, the aqueous phase used to collect MAAs and the chloroform phase were both treated for lipid recovery. According to this method, recovering UV protectant after UV exposure with biodiesel is a more sustainable solution for high fuel productivity. The manufacture of algae biodiesel gains value as a result of this procedure. The problem of heterotrophic bacterial contamination is lessened by UV exposure, and the lipids content and saturation index of biodiesel are increased. This results in a biorefinery that is both affordable and sustainable (Chandra *et al.*, 2020).

12.4.2.2 UV protectant

Due to its great market demand and value, UV protectant could be a significant product of an algal biorefinery. For instance, it is predicted that by 2024, the global demand for this kind of goods will increase to $13.2 billion from more than $7.6 billion in 2012 (Oilgae, 2014). By doing so, it is possible to enhance the value of the process, the number of products with added value, and the environmental impact all at once. Because of their special makeup, microalgae can contain a variety of pigments, such as UV filters and carotenoids, astaxanthin, lutein, zeaxantin, phycocyanin, and phycoerythrin.

Oil accumulation in microalgae is known to be significantly affected by exposing cells to environmental stress (ultraviolet radiation) (Arakaki *et al.*, 2017), nutrient depletion (Pancha *et al.*, 2014), salinity (BenMoussa-Dahmen *et al.*, 2016), oxidative stress (Yilancioglu *et al.*, 2014), temperature or pH changes. In order to reduce the stress condition, these parameters also have an impact on the other cell components and trigger the production of new molecules. Ultraviolet radiations (UVR) are a key method in this regard to generate UV inhibitors like mycosporin-like amino acids (MAAs) and investigate their impact on lipid productivity (Chandra *et al.*, 2019). By manufacturing MAAs and scytonemin, cyanobacteria are known to defend themselves against photochemical harm (Chandra *et al.*, 2020). They reside in the sheath of cyanobacteria and are lipophilic. Because the growth medium is not changed and heterotrophic contamination may be minimized, UVR provides various benefits for oil production.

A range of defense mechanisms are used by the cyanobacteria and marine algae group to endure and thrive under high UV fluxes. These organisms synthesize UV-absorbing substances like mycosporine-like amino acids (MAAs) and scytonemin as a mitigation mechanism (Chandra *et al.*, 2019, 2020). By scavenging significant amounts of reactive oxygen produced by supersaturated oxygen in deep, light-exposed water, MAAs display substantial antioxidant activity. The 3-dehydroquinate and 4-deoxygadusol precursors of the shikimate pathway are the sources of the main MAA, mycosporine-glycine (Chandra *et al.*, 2019). Mycosporine-glycine is converted into secondary MAAs via the addition or subtraction of amino acids as well as metabolic processes. However, it has been discovered that the cyanobacterium *A. variabilis* has an MAA biosynthetic gene cluster that transforms sedoheptulose-7-phosphate to shinorine via 4-deoxygadusol and switches the precursor 3-dehydroquinite (Balskus & Walsh, 2010). Tryptophan and tyrosine, two aromatic amino acids that are byproducts of the shikimate pathway, are thought to be the source of scytonemin. Moreover, a cluster of genes for scytonemin production has been found, and UV-A activation of these genes has been demonstrated (Rastogi *et al.*, 2015). MAAs and scytonemin can be used as active ingredients in the cosmetic and pharmaceutical sectors due to their potent UV absorption and photoprotective qualities.

12.5 CONCLUSION

Establishing a sustainable algal biomass-based biorefinery requires multidisciplinary research on biorefinery methodologies. A multi-product, integrated, and sustainable approach is crucial for efficient product recovery and process development. Microalgae biomass serves as a flexible feedstock for biodiesel production, utilizing techniques such as photobioreactors, fermenters, and open raceway ponds. Various lipid extraction techniques have been explored, including solvent extraction, enzymatic treatment, ultrasonic-aided extraction, and supercritical carbon dioxide extraction. Optimization work is needed for large-scale applications, particularly for supercritical carbon dioxide extraction (SFE-CO_2), PEFs, osmotic shock, hydrothermal liquefaction, and wet lipid extraction. Integrating algal fuel production with wastewater and waste treatment enhances economic viability. UV exposure stimulates UV defense synthesis and lipid productivity in *Lyngbya*, with UVB favoring fuel qualities and UVA benefiting food properties. The recovery of lipids and UV protectants from the same feedstock promotes cost-effective and environmentally responsible options in a sustainable biorefinery.

REFERENCES

Alami A. H., Tawalbeh M., Alasad S., Ali M., Alshamsi M. and Aljaghoub H. (2021). Cultivation of *Nannochloropsis* algae for simultaneous biomass applications and carbon dioxide capture. *Energy Sources, Part A: Recovery, Utilization, and Environmental Effects*, 1–12, https://doi.org/10.1080/15567036.2021.1933267

Al-Jabri H., Das P., Khan S., AbdulQuadir M., Thaher M. I., Hoekman K. and Hawari A. H. (2022). A comparison of bio-crude oil production from five marine microalgae – using life cycle analysis. *Energy*, **251**, 123954, https://doi.org/10.1016/j.energy.2022.123954

Amin S. (2009). Review on biofuel oil and gas production processes from microalgae. *Energy Conversion and Management*, **50**, 1834–1840, https://doi.org/10.1016/j.enconman.2009.03.001

Arakaki A., Matsumoto T., Tateishi T., Matsumoto M., Nojima D., Tomoko Y. and Tanaka T. (2017). UV-C irradiation accelerates neutral lipid synthesis in the marine oleaginous diatom *Fistulifera solaris*. *Bioresource Technology*, **245**, 1520–1526, https://doi.org/10.1016/j.biortech.2017.05.188

Bakonyi P., Kumar G., Koók L., Tóth G., Rózsenberszki T., Bélafi-Bakó K. and Nemestóthy N. (2018). Microbial electrohydrogenesis linked to dark fermentation as integrated application for enhanced biohydrogen production: a review on process characteristics, experiences and lessons. *Bioresource Technology*, **251**, 381–389, https://doi.org/10.1016/j.biortech.2017.12.064

Balskus E. P. and Walsh C. T. (2010). The genetic and molecular basis for sunscreen biosynthesis in cyanobacteria. *Science*, **329**, 1653–1656, https://doi.org/10.1126/science.1193637

Barry A., Wolfe A., English C., Ruddick C. and Lambert D. (2016). 2016 National Algal Biofuels Technology Review.

Baumgardt F. J. L., Filho A. Z., Brandalize M. V., da Costa D. C., Antoniosi Filho N. R., Abreu P. C. O. V., Corazza M. L. and Ramos L. P. (2016). Lipid content and fatty acid profile of *Nannochloropsis oculata* before and after extraction with conventional solvents and/or compressed fluids. *The Journal of Supercritical Fluids*, **108**, 89–95, https://doi.org/10.1016/j.supflu.2015.11.003

BenMoussa-Dahmen I., Chtourou H., Rezgui F., Sayadi S. and Dhouib A. (2016). Salinity stress increases lipid, secondary metabolites and enzyme activity in *Amphora subtropica* and *Dunaliella* sp. for biodiesel production. *Bioresource Technology*, **218**, 816–825, https://doi.org/10.1016/j.biortech.2016.07.022

Bensalem S., Lopes F., Bodénès P., Pareau D., Français O. and le Pioufle B. (2018). Understanding the mechanisms of lipid extraction from microalga *Chlamydomonas reinhardtii* after electrical field solicitations and mechanical stress within a microfluidic device. *Bioresource Technology*, **257**, 129–136, https://doi.org/10.1016/j.biortech.2018.01.139

Biller P., Ross A. B., Skill S. C., Lea-Langton A., Balasundaram B., Hall C., Riley R. and Llewellyn C. A. (2012). Nutrient recycling of aqueous phase for microalgae cultivation from the hydrothermal liquefaction process. *Algal Research*, **1**, 70–76, https://doi.org/10.1016/j.algal.2012.02.002

Castro-Carrera T., Toral P. G., McEwan N. R., Hervas G., Abecia L., Pinloche E., Girdwood S. E. and B. A. (2014). Rumen bacterial community evaluated by 454 pyrosequencing and terminal restriction fragment length polymorphism analyses in dairy sheep fed marine algae. *Journal of Dairy Science*, **97**, 1661–1669, https://doi.org/10.3168/jds.2013-7243

Cavaleiro A. J., Pereira M. A., Guedes A. P., Stams A. J. M., Alves M. M. and Sousa D. Z. (2016). Conversion of Cn-unsaturated into Cn-2-saturated LCFA can occur uncoupled from methanogenesis in anaerobic bioreactors. *Environmental Science & Technology*, **50**, 3082–3090, https://doi.org/10.1021/acs.est.5b03204

Chandra R., Rohit M. V., Swamy Y. V. and Venkata Mohan S. (2014). Regulatory function of organic carbon supplementation on biodiesel production during growth and nutrient stress phases of mixotrophic microalgae cultivation. *Bioresource Technology*, **165**, 279–287, https://doi.org/10.1016/j.biortech.2014.02.102

Chandra R., Parra R. M. N. and Iqbal H. (2017). Phycobiliproteins: a novel green tool from marine origin blue-green algae and red algae. *Protein & Peptide Letters*, **24**, 118–125, https://doi.org/10.2174/0929866523666160802160222

Chandra R., Das P., Vishal G. and Nagra S. (2019). Factors affecting the induction of UV protectant and lipid productivity in *Lyngbya* for sequential biorefinery product recovery. *Bioresource Technology*, **278**, 303–310, https://doi.org/10.1016/j.biortech.2019.01.084

Chandra R., Pons-Faudoa F. P., Parra Saldívar R. and Rittmann B. E. (2020). Effect of ultra-violet exposure on production of mycosporine-like amino acids and lipids by *Lyngbya purpurem*. *Biomass and Bioenergy*, **134**, 105475, https://doi.org/10.1016/j.biombioe.2020.105475

Cherubini F. (2010). The biorefinery concept: using biomass instead of oil for producing energy and chemicals. *Energy Conversion and Management*, **51**, 1412–1421, https://doi.org/10.1016/j.enconman.2010.01.015

Cherubini F., Jungmeier G., Wellisch M., Willke T., Skiadas I., van Ree R. and de Jong E. (2009). Toward a common classification approach for biorefinery systems. *Biofuels, Bioproducts and Biorefining*, **3**, 534–546, https://doi.org/10.1002/bbb.172

Chittapun S., Jonjaroen V., Khumrangsee K. and Charoenrat T. (2020). C-phycocyanin extraction from two freshwater cyanobacteria by freeze thaw and pulsed electric field techniques to improve extraction efficiency and purity. *Algal Research*, ***46***, 101789, https://doi.org/10.1016/j.algal.2020.101789

Chr. Eilertsen H., Huseby S., Degerlund M., Eriksen G. K., Ingebrigtsen R. A. and Hansen E. (2014). The effect of freeze/thaw cycles on reproducibility of metabolic profiling of marine microalgal extracts using direct infusion high-resolution mass spectrometry (HR-MS). *Molecules*, **19**, 16373–16380, https://doi.org/10.3390/molecules191016373

Christ O., Wilderer P. A., Angerhofer R. and Faulstich M. (2000). Mathematical modeling of the hydrolysis of anaerobic processes. *Water Science and Technology*, **41**, 61–65, https://doi.org/10.2166/wst.2000.0056

Cooney M., Young G. and Nagle N. (2009). Extraction of bio-oils from microalgae. *Separation & Purification Reviews*, **38**, 291–325, https://doi.org/10.1080/15422110903327919

Cravotto G., Boffa L., Mantegna S., Perego P., Avogadro M. and Cintas P. (2008). Improved extraction of vegetable oils under high-intensity ultrasound and/or microwaves. *Ultrasonics Sonochemistry*, **15**, 898–902, https://doi.org/10.1016/j.ultsonch.2007.10.009

Dong T., Knoshaug E. P., Davis R., Laurens L. M. L., van Wychen S., Pienkos P. T. and Nagle N. (2016). Combined algal processing: a novel integrated biorefinery process to produce algal biofuels and bioproducts. *Algal Research*, **19**, 316–323, https://doi.org/10.1016/j.algal.2015.12.021

Esquivel-Hernández D. A., Rodríguez-Rodríguez J., Rostro-Alanis M., Cuéllar-Bermúdez S. P., Mancera-Andrade E. I., Núñez-Echevarría J. E., García-Pérez J. S., Chandra R. and Parra-Saldívar R. (2017). Advancement of green process through microwave-assisted extraction of bioactive metabolites from *Arthrospira Platensis* and bioactivity evaluation. *Bioresource Technology*, **224**, 618–629, https://doi.org/10.1016/j.biortech.2016.10.061

Fajardo A. R., Cerdán L. E., Medina A. R., Fernández F. G. A., Moreno P. A. G. and Grima E. M. (2007). Lipid extraction from the microalga *Phaeodactylum tricornutum*. *European Journal of Lipid Science and Technology*, **109**, 120–126, https://doi.org/10.1002/ejlt.200600216

Fukuzaki S., Nishio N., Shobayashi M. and Nagai S. (1990). Inhibition of the fermentation of propionate to methane by hydrogen, acetate, and propionate. *Applied and Environmental Microbiology*, **56**, 719–723, https://doi.org/10.1128/aem.56.3.719-723.1990

Ghasemi Naghdi F., González González L. M., Chan W. and Schenk P. M. (2016). Progress on lipid extraction from wet algal biomass for biodiesel production. *Microbial Biotechnology*, **9**(6), 718–726, https://doi.org/10.1111/1751-7915.12360

Grima E. M., Fernández F.G.A. and Medina A.R. (2003). Downstream processing of cell mass and products. In: Handbook of Microalgal Culture, A. Richmond (ed.), Wiley Blackwell, pp. 267–309, https://doi.org/10.1002/9780470995280.ch10

Harun R., Danquah M. K. and Forde G. M. (2010). Microalgal biomass as a fermentation feedstock for bioethanol production. *Journal of Chemical Technology & Biotechnology*, **85**, 199–203, https://doi.org/10.1002/jctb.2287

Haznedaroglu B. Z., Rismani-Yazdi H., Allnutt F. C. T., Reeves D. and Peccia J. (2016). Algal biorefinery for high-value platform chemicals. In *Platform Chemical Biorefinery*, Elsevier, pp. 333–360. https://doi.org/10.1016/B978-0-12-802980-0.00018-3

Hernández D., Solana M., Riaño B., García-González M. C. and Bertucco A. (2014). Biofuels from microalgae: lipid extraction and methane production from the residual biomass in a biorefinery approach. *Bioresource Technology*, **170**, 370–378, https://doi.org/10.1016/j.biortech.2014.07.109

Huws S. A., Kim E. J., Lee M. R. F., Scott M. B., Tweed J. K. S., Pinloche E., Wallace R. J. and Scollan N. D. (2011). As yet uncultured bacteria phylogenetically classified as *Prevotella, Lachnospiraceae incertae sedis* and unclassified *Bacteroidales, Clostridiales* and *Ruminococcaceae* may play a predominant role in ruminal biohydrogenation. *Environmental Microbiology*, **13**, 1500–1512, https://doi.org/10.1111/j.1462-2920.2011.02452.x

John Wallace R., Chaudhary L. C., McKain N., McEwan N. R., Richardson A. J., Vercoe P. E., Walker N. D. and Paillard D. (2006). *Clostridium proteoclasticum*: a ruminal bacterium that forms stearic acid from linoleic acid. *FEMS Microbiology Letters*, **265**, 195–201, https://doi.org/10.1111/j.1574-6968.2006.00487.x

Jones R. J., Massanet-Nicolau J., Guwy A., Premier G. C., Dinsdale R. M. and Reilly M. (2015). Removal and recovery of inhibitory volatile fatty acids from mixed acid fermentations by conventional electrodialysis. *Bioresource Technology*, **189**, 279–284, https://doi.org/10.1016/j.biortech.2015.04.001

Käferböck A., Smetana S., de Vos R., Schwarz C., Toepfl S. and Parniakov O. (2020). Sustainable extraction of valuable components from *Spirulina* assisted by pulsed electric fields technology. *Algal Research*, ***48***, 101914, https://doi.org/10.1016/j.algal.2020.101914

Ki D., Parameswaran P., Popat S. C., Rittmann B. E. and Torres C. I. (2015). Effects of pre-fermentation and pulsed-electric-field treatment of primary sludge in microbial electrochemical cells. *Bioresource Technology*, **195**, 83–88, https://doi.org/10.1016/j.biortech.2015.06.128

Knothe G. (2011). A technical evaluation of biodiesel from vegetable oils vs. algae. Will algae-derived biodiesel perform? *Green Chemistry*, **13**, 3048, https://doi.org/10.1039/c0gc00946f

Kumar M., Sun Y., Rathour R., Pandey A., Thakur I. S. and Tsang D. C. W. (2020a). Algae as potential feedstock for the production of biofuels and value-added products: opportunities and challenges. *Science of the Total Environment*, **716**, 137116, https://doi.org/10.1016/j.scitotenv.2020.137116

Kumar R., Ghosh A. K. and Pal P. (2020b). Synergy of biofuel production with waste remediation along with value-added co-products recovery through microalgae cultivation: a review of membrane-integrated green approach. *Science of the Total Environment*, **698**, 134169, https://doi.org/10.1016/j.scitotenv.2019.134169

Kumar R., Hanumantha Rao P. and Arumugam M. (2015). Lipid extraction methods from microalgae: a comprehensive review. *Frontiers in Energy Research*, **2**, 61. https://doi.org/10.3389/fenrg.2014.00061

Lai Y., Parameswaran P., Li A., Baez M. and Rittmann B. E. (2014). Effects of pulsed electric field treatment on enhancing lipid recovery from the microalga *Scenedesmus*. *Bioresource Technology*, **173**, 457–461, https://doi.org/10.1016/j.biortech.2014.09.124

Lai Y. J. S., McCaw A., Ontiveros-Valencia A., Shi Y., Parameswaran P. and Rittmann B. E. (2016a). Multiple synergistic benefits of selective fermentation of *Scenedesmus* biomass for fuel recovery via wet-biomass extraction. *Algal Research*, **17**, 253–260, https://doi.org/10.1016/j.algal.2016.05.001

Lai Y. S., Parameswaran P., Li A., Aguinaga A. and Rittmann B. E. (2016b). Selective fermentation of carbohydrate and protein fractions of *Scenedesmus*, and biohydrogenation of its lipid fraction for enhanced recovery of saturated fatty acids. *Biotechnology and Bioengineering*, **113**, 320–329, https://doi.org/10.1002/bit.25714

Laurens L. M. L., Nagle N., Davis R., Sweeney N., Van Wychen S., Lowell A. and Pienkos P. T. (2015). Acid-catalyzed algal biomass pretreatment for integrated lipid and carbohydrate-based biofuels production. *Green Chemistry*, **17**, 1145–1158, https://doi.org/10.1039/C4GC01612B

Lee J.-Y., Yoo C., Jun S.-Y., Ahn C.-Y. and Oh H.-M. (2010). Comparison of several methods for effective lipid extraction from microalgae. *Bioresource Technology*, **101**, S75–S77, https://doi.org/10.1016/j.biortech.2009.03.058

Liu Y., Lai Y.-J. S., Barbosa T. S., Chandra R., Parameswaran P. and Rittmann B. E. (2019). Electro-selective fermentation enhances lipid extraction and biohydrogenation of *Scenedesmus acutus* biomass. *Algal Research*, **38**, 101397, https://doi.org/10.1016/j.algal.2018.101397

Lopalco P., Lobasso S., Babudri F. and Corcelli A. (2004). Osmotic shock stimulates de novo synthesis of two cardiolipins in an extreme halophilic archaeon. *Journal of Lipid Research*, **45**, 194–201, https://doi.org/10.1194/jlr.M300329-JLR200

Lu L., Xing D., Liu B. and Ren N. (2012). Enhanced hydrogen production from waste activated sludge by cascade utilization of organic matter in microbial electrolysis cells. *Water Research*, **46**, 1015–1026, https://doi.org/10.1016/j.watres.2011.11.073

Moretto J. A., de Souza A. O., Berneira L. M., Brigagão L. G. G., de Pereira C. M. P., Converti A. and Pinto E. (2022). Microwave-assisted extraction of fatty acids from cultured and commercial phytoplankton Species. *Applied Sciences*, **12**, 2407, https://doi.org/10.3390/app12052407

Oh Y. K., Hwang K. R., Kim C., Kim J. R. and Lee J. S. (2018). Recent developments and key barriers to advanced biofuels: a short review. *Bioresource Technology*, **257**, 320–333, https://doi.org/10.1016/j.biortech.2018.02.089

Oilgae (2014). Emerging Algae Product and Business Opportunities. Oilgae 15.

Pancha I., Chokshi K., George B., Ghosh T., Paliwal C., Maurya R. and Mishra S. (2014). Nitrogen stress triggered biochemical and morphological changes in the microalgae *Scenedesmus* sp. CCNM 1077. *Bioresource Technology*, **156**, 146–154, https://doi.org/10.1016/j.biortech.2014.01.025

Parameswaran P., Torres C. I., Lee H.-S., Krajmalnik-Brown R. and Rittmann B. E. (2009). Syntrophic interactions among anode respiring bacteria (ARB) and non-ARB in a biofilm anode: electron balances. *Biotechnology and Bioengineering*, **103**, 513–523, https://doi.org/10.1002/bit.22267

Parameswaran P., Zhang H., Torres C. I., Rittmann B. E. and Krajmalnik-Brown R. (2010). Microbial community structure in a biofilm anode fed with a fermentable substrate: the significance of hydrogen scavengers. *Biotechnology and Bioengineering*, **105**, 69–78, https://doi.org/10.1002/bit.22508

Parfati N., Rani K. C. and Meilany (2018). The effect of coprocessed superdisintegrants ratio (crospovidone-sodium starch glycolate) to the physicochemical characteristics of atenolol orally disintegrating tablets. *Asian Journal of Pharmaceutical and Clinical Research*, **11**, 318–, https://doi.org/10.22159/ajpcr.2018.v11i2.23010

Patnaik R. and Mallick N. (2015). Utilization of *Scenedesmus obliquus* biomass as feedstock for biodiesel and other industrially important co-products: an integrated paradigm for microalgal biorefinery. *Algal Research*, **12**, 328–336, https://doi.org/10.1016/j.algal.2015.09.009

Pratt S., Liew D., Batstone D. J., Werker A. G., Morgan-Sagastume F. and Lant P. A. (2012). Inhibition by fatty acids during fermentation of pre-treated waste activated sludge. *Journal of Biotechnology*, **159**, 38–43, https://doi.org/10.1016/j.jbiotec.2012.02.001

Ranjith Kumar R., Hanumantha Rao P. and Arumugam M. (2015). Lipid extraction methods from microalgae: a comprehensive review. *Frontiers in Energy Research*, **2**, 1–, https://doi.org/10.3389/fenrg.2014.00061

Rastogi R. P. and Incharoensakdi A. (2014). Characterization of UV-screening compounds, mycosporine-like amino acids, and scytonemin in the cyanobacterium *Lyngbya* sp. CU2555. *FEMS Microbiology Ecology*, **87**, 244–256, https://doi.org/10.1111/1574-6941.12220

Rastogi R. P., Sonani R. R. and Madamwar D. (2015). Cyanobacterial *Sunscreen Scytonemin*: role in photoprotection and biomedical research. *Applied Biochemistry and Biotechnology*, **176**, 1551–1563, https://doi.org/10.1007/s12010-015-1676-1

Reguera G., McCarthy K. D., Mehta T., Nicoll J. S., Tuominen M. T. and Lovley D. R. (2005). Extracellular electron transfer via microbial nanowires. *Nature*, **435**, 1098–1101, https://doi.org/10.1038/nature03661

Reverchon E. and de Marco I. (2006). Supercritical fluid extraction and fractionation of natural matter. *The Journal of Supercritical Fluids*, **38**, 146–166, https://doi.org/10.1016/j.supflu.2006.03.020

Roux J.-M., Lamotte H. and Achard J.-L. (2017). An overview of microalgae lipid extraction in a biorefinery framework. *Energy Procedia*, **112**, 680–688, https://doi.org/10.1016/j.egypro.2017.03.1137

Ryckebosch E., Muylaert K. and Foubert I. (2012). Optimization of an analytical procedure for extraction of lipids from microalgae. *Journal of the American Oil Chemists' Society*, **89**, 189–198, https://doi.org/10.1007/s11746-011-1903-z

Salama E. S., Hwang J. H., El-Dalatony M. M., Kurade M. B., Kabra A. N., Abou-Shanab R. A. I., Kim K. H., Yang I. S., Govindwar S. P., Kim S. and Jeon B. H. (2018). Enhancement of microalgal growth and biocomponent-based transformations for improved biofuel recovery: a review. *Bioresource Technology*, **258**, 365–375, https://doi.org/10.1016/j.biortech.2018.02.006

Sathish A. and Sims R. C. (2012). Biodiesel from mixed culture algae via a wet lipid extraction procedure. *Bioresource Technology*, **118**, 643–647, https://doi.org/10.1016/j.biortech.2012.05.118

Sharif K. M., Rahman M. M., Azmir J., Mohamed A., Jahurul M. H. A., Sahena F. and Zaidul I. S. M. (2014). Experimental design of supercritical fluid extraction – a review. *Journal of Food Engineering*, **124**, 105–116, https://doi.org/10.1016/j.jfoodeng.2013.10.003

Shen Y., Yuan W., Pei Z. and Mao E. (2010). Heterotrophic culture of *Chlorella protothecoides* in various nitrogen sources for lipid production. *Applied Biochemistry and Biotechnology*, **160**, 1674–1684, https://doi.org/10.1007/s12010-009-8659-z

Sheng J., Vannela R. and Rittmann B. E. (2011a). Evaluation of cell-disruption effects of pulsed-electric-field treatment of *Synechocystis* PCC 6803. *Environmental Science & Technology*, **45**, 3795–3802, https://doi.org/10.1021/es103339x

Sheng J., Vannela R. and Rittmann B. E. (2011b). Evaluation of methods to extract and quantify lipids from *Synechocystis* PCC 6803. *Bioresource Technology*, **102**, 1697–1703, https://doi.org/10.1016/j.biortech.2010.08.007

Sierra L. S., Dixon C. K. and Wilken L. R. (2017). Enzymatic cell disruption of the microalgae Chlamydomonas reinhardtii for lipid and protein extraction. *Algal Research*, **25**, 149–159. https://doi.org/10.1016/j.algal.2017.04.004

Šoštarič M., Klinar D., Bricelj M., Golob J., Berovič M. and Likozar B. (2012). Growth, lipid extraction and thermal degradation of the microalga *Chlorella vulgaris*. *New Biotechnology*, **29**, 325–331, https://doi.org/10.1016/j.nbt.2011.12.002

Thajuddin N. and Subramanian G. (2005). Cyanobacterial biodiversity and potential applications in biotechnology. *Current Science*, **89**, 47–57.

Thomassen G., Van Dael M. and Van Passel S. (2018). The potential of microalgae biorefineries in Belgium and India: an environmental techno-economic assessment. *Bioresource Technology*, **267**, 271–280, https://doi.org/10.1016/j.biortech.2018.07.037

Torres C. I., Marcus A. K. and Rittmann B. E. (2007). Kinetics of consumption of fermentation products by anode-respiring bacteria. *Applied Microbiology and Biotechnology*, **77**, 689–697, https://doi.org/10.1007/s00253-007-1198-z

Torres C. I., Krajmalnik-Brown R., Parameswaran P., Marcus A. K., Wanger G., Gorby Y. A. and Rittmann B. E. (2009a). Selecting anode-respiring bacteria based on anode potential: phylogenetic, electrochemical, and microscopic characterization. *Environmental Science & Technology*, **43**, 9519–9524, https://doi.org/10.1021/es902165y

Torres C. I., Lee H., Krajmalnik-brown R. and Rittmann B. E. (2009b). A kinetic perspective on extracellular electron transfer by anode-respiring bacteria. *FEMS Microbiology Reviews*, **34**, 3–17, https://doi.org/10.1111/j.1574-6976.2009.00191.x

Van De Vossenberg J. L. C. M. and Joblin K. N. (2003). Biohydrogenation of C18 unsaturated fatty acids to stearic acid by a strain of *Butyrivibrio hungatei* from the bovine rumen. *Letters in Applied Microbiology*, **37**, 424–428, https://doi.org/10.1046/j.1472-765X.2003.01421.x

Vermaas W. (1996). Molecular genetics of the cyanobacterium *Synechocystis* sp. PCC 6803: principles and possible biotechnology applications. *Journal of Applied Phycology*, **8**, 263–273, https://doi.org/10.1007/BF02178569

Wang M., Zhou J., Castagnini J. M., Berrada H. and Barba F. J. (2023). Pulsed electric field (PEF) recovery of biomolecules from *Chlorella*: extract efficiency, nutrient relative value, and algae morphology analysis. *Food Chemistry*, ***404***, 134615, https://doi.org/10.1016/j.foodchem.2022.134615

Wei F., Gao G.-Z., Wang X.-F., Dong X.-Y., Li P.-P., Hua W., Wang X., Wu X.-M. and Chen H. (2008). Quantitative determination of oil content in small quantity of oilseed rape by ultrasound-assisted extraction combined with gas chromatography. *Ultrasonics Sonochemistry*, **15**, 938–942, https://doi.org/10.1016/j.ultsonch.2008.04.003

Xu L., (Wim) Brilman D. W. F., Withag J. A. M., Brem G. and Kersten S. (2011). Assessment of a dry and a wet route for the production of biofuels from microalgae: energy balance analysis. *Bioresource Technology*, **102**, 5113–5122, https://doi.org/10.1016/j.biortech.2011.01.066

Yang Y., Xu M., Guo J. and Sun G. (2012). Bacterial extracellular electron transfer in bioelectrochemical systems. *Process Biochemistry*, **47**, 1707–1714, https://doi.org/10.1016/j.procbio.2012.07.032

Yellapu S. K., Bharti Kaur R., Kumar L. R., Tiwari B., Zhang X. and Tyagi R. D. (2018). Recent developments of downstream processing for microbial lipids and conversion to biodiesel. *Bioresource Technology*, **256**, 515–528, https://doi.org/10.1016/j.biortech.2018.01.129

Yilancioglu K., Cokol M., Pastirmaci I., Erman B. and Cetiner S. (2014). Oxidative stress is a mediator for increased lipid accumulation in a newly isolated *Dunaliella salina* strain. *PLoS One*, **9**, 1–, https://doi.org/10.1371/journal.pone.0091957

Zbinden M. D. A., Sturm B. S. M., Nord R. D., Carey W. J., Moore D., Shinogle H. and Stagg-Williams S. M. (2013). Pulsed electric field (PEF) as an intensification pretreatment for greener solvent lipid extraction from microalgae. *Biotechnology and Bioengineering*, **110**, 1605–1615, https://doi.org/10.1002/bit.24829

Zhang P., Feng L., Su B. and Li X. (2022). Microalgae cultivated in wastewater catalytic hydrothermal liquefaction: effects of process parameter on products and energy balance. *Journal of Cleaner Production*, **341**, 130895, https://doi.org/10.1016/j.jclepro.2022.130895

Zheng G., Li C., Guo L., Ruo W. and Wang S. (2012). Purification of extracted fatty acids from the microalgae *Spirulina*. *Journal of the American Oil Chemists' Society*, **89**, 561–566, https://doi.org/10.1007/s11746-011-1956-z

Index

IWA Publishing's authorised EU representative for General Product Safety Regulations is Diane D'Arras, 15 rue Duret, 75116 Paris, France, e-mail: safety@iwap.co.uk.

Printed and bound by CPI Group (UK) Ltd, Croydon, CR0 4YY
12/05/2026
02108869-0010